Ecology and Field Biology

Ecology and Field Biology

Roger J. Lederer
California State University at Chico

The Benjamin/Cummings Publishing Company, Inc.

Menlo Park, California • Reading, Massachusetts
London • Amsterdam • Don Mills, Ontario • Sydney

Sponsoring Editor: James W. Behnke
Associate Editor: Bonnie Garmus
Production Editors: Fannie Toldi and Wendy Calmenson
Designer: Wendy Calmenson
Cover Designer: Michael Rogondino
Cover Photograph: Larry R. Ditto/Tom Stack and Associates
Illustrator: Karen Daniels
Photo Researcher: Jo Andrews

Library of Congress Cataloging in Publication Data

Lederer, Roger J.
  Ecology and field biology

  (The Benjamin/Cummings series in the life sciences)
  Bibliography: p.
  Includes indexes.
    1. Ecology.   2. Biology—Field work.   I. Title.
II. Series.
QH541.L373 1984   574.5   83-22290
ISBN 0-8053-5718-1

HIJ-HA-89

The Benjamin/Cummings Publishing Company, Inc.
2727 Sand Hill Road
Menlo Park, California 94025

To my children,
Melinda and Joey,
who are still small enough to be in awe
of the simplest of nature's wonders.

The Benjamin/Cummings Series in the Life Sciences

F. J. Ayala, *Population and Evolutionary Genetics: A Primer* (1982)

F. J. Ayala and J. A. Kiger, Jr., *Modern Genetics*, second edition (1984)

F. J. Ayala and J. W. Valentine, *Evolving: The Theory and Processes of Organic Evolution* (1979)

C. L. Case and T. R. Johnson, *Laboratory Experiments in Microbiology* (1984)

R. E. Dickerson and I. Geis, *Hemoglobin* (1983)

L. E. Hood, I. L. Weissman, W. B. Wood, and J. H. Wilson, *Immunology*, second edition (1984)

J. B. Jenkins, *Human Genetics* (1983)

K. D. Johnson, D. L. Rayle, and H. L. Wedberg, *Biology: An Introduction* (1984)

R. J. Lederer, *Ecology and Field Biology* (1984)

A. L. Lehninger, *Bioenergetics: The Molecular Basis of Biological Energy Transformations*, second edition (1971)

S. E. Luria, S. J. Gould, and S. Singer, *A View of Life* (1981)

E. N. Marieb, *Human Anatomy and Physiology Lab Manual: Brief Edition* (1983)

E. N. Marieb, *Human Anatomy and Physiology Lab Manual: Cat and Fetal Pig Versions* (1981)

E. B. Mason, *Human Physiology* (1983)

A. P. Spence, *Basic Human Anatomy* (1982)

A. P. Spence and E. B. Mason, *Human Anatomy and Physiology*, second edition (1983)

G. J. Tortora, B. R. Funke, and C. L. Case, *Microbiology: An Introduction* (1982)

J. D. Watson, N. Hopkins, J. Roberts, and J. Steitz, *Molecular Biology of the Gene*, fourth edition (1985)

W. B. Wood, J. H. Wilson, R. M. Benbow, and L. E. Hood, *Biochemistry: A Problems Approach*, second edition (1981)

Preface

Ecology and Field Biology is designed for a one-term course in ecology and/or natural history for liberal arts, social science, and business students. It assumes no prerequisites. Ecology is a unifying subscience of the natural sciences, pulling together many areas of biology as it addresses physiology, anatomy, genetics, behavior, and evolution, as well as some geology, chemistry, and physics. A course in ecology is therefore perhaps representative of the sciences as a whole, and thus is an ideal course for the nonscience major.

The ecology texts available today are generally too sophisticated for students with little scientific background. In addition, I have never been able to find a single book that explained both the principles and the organisms on which they are based. My goal in writing this book was to bring together many ideas and create a comprehensible whole that will provide a life-long understanding of ecology. The intent is to provide a coherent presentation of ecological principles as they apply to biological systems and their environments.

ORGANIZATION

The first section of the book (Chapters 1–11) presents ecological principles. The second section (Chapters 12–17) focuses on organisms and their ecological adaptations, deemphasizing the taxonomic and anatomical treatment usually found in basic biology texts and emphasizing instead adaptations of organisms to the environment.

Because there are many ways in which a course in ecology and field biology and/or natural history can be organized, this text is arranged for flexibility. The chapters progress logically, but can be read in almost any order without detracting significantly from a student's understanding. For example, Part Two of the text, Natural History, could be read first, or omitted for a short course. Potentially, Part Two could be used in the laboratory portion of a natural history course while the first part of the text is simultaneously being read for the lecture. The book's length should allow the professor to complete the book and course within the confines of a quarter or semester system.

SPECIAL FEATURES

Clear, Nontechnical Writing Style

Concepts, principles, laws, and theories of ecology are presented in straightforward, minimally technical prose. For example, the text covers population growth without reference to the logarithmic form of the population equation or the mosaic model of population regulation. Each concept is presented as simply and completely as possible, and is elucidated by actual or probable occurrences in nature. This practical context makes the principles accessible.

Laboratory and Field Exercises

The best way to understand concepts and clarify issues is to learn by doing. To this end I have provided ideas for laboratory and field activities and exercises for each chapter. You will find these exercises in the margins alongside the concept in the text which they attempt to clarify. Each exercise is contained in a shaded box, a field glasses emblem set above the exercise for easy identification. All of these exercises have been used in my own classroom and I have found them effective as demonstrations of ideas presented in the text.

Human Impact on the Environment

Although this text focuses on natural ecosystems, I have included the human impact on ecosystems. Pesticides, air pollution, sewage treatment, strip mining, and many other activities affect organisms, population, communities, even entire biomes. Discussing any ecosystem without mentioning human incursion through industry and development would fail to give an accurate picture of the ecological situation since few natural systems are pristine.

Art Program

The line art is both clear and instructive and will help simplify difficult concepts to the reader. Equally important are the photographs, which have been chosen to instruct rather than to decorate the text.

In-Text Learning Aids

The pedagogical features in this book can enhance student understanding of the text. *Key words* are boldfaced within the text. Each chapter ends with a short and concise *summary* and *end-chapter questions*. Annotated *suggested readings* can be found at the back of the book for those students interested in pursuing further information. A comprehensive *glossary* facilitates quick reference to terminology. An *index of scientific names* will help students match common names of organisms mentioned in the text to their genus or species.

My personal fascination with nature has deepened in the preparation and writing of this book. I will consider myself a success if the reader captures the extravagant richness of natural adaptations and grasps the overall scheme of how and why ecosystems work.

ACKNOWLEDGEMENTS

There are many people to whom I am indebted for their professional and technical skills, and their opinions, ideas, advice, and information. Drs. Richard Demaree, Wesley Dempsey, Patricia Edelmann, David Kistner, Donald Kowalski, Vernon Oswald, Robert Schlising, and Kingsley Stern of the Department of Biological Sciences, Dr. Carol Burr, Department of English, and Mr. Tim Devine, California State University, Chico; Dr. Patricia Rich, Department of Earth Sciences, Monash University, Victoria, Australia; Ms. Sylvia Wenger; and especially Dr. Stephen Wilson of the Department of Biology, Central Missouri State University.

I am particularly indebted to the following reviewers for their helpful comments.

David C. Culver, *Northwestern University*
Arthur Driscoll, *Westfield State College*
William E. Dunscombe, *Union College*
Gary James, *Orange Coast College*
Barbara Klemenok, *Napa College*
David A. Lovejoy, *Westfield State College*
W. Elinor Lowell, *Tunxis Community College*
Marschall Stevens, *Citrus College*
Stephen Wilson, *Central Missouri State University*

I also would like to thank the several hundred individuals who re-sponded to a questionnaire at the beginning of my writing process.

Many diagrams and photographs were provided by individuals and or-ganizations, and I appreciated their helpfulness and cooperation. Karen Daniels did most of the book's artwork and I am sincerely indebted to her for the time and effort she devoted to the text, and very much appreciate her highly professional skills.

Elaine Erdmann, Joyce Hall, and Jackie Ruskjer typed large portions of several drafts of the manuscript. They are all to be commended for inter-preting my nearly illegible handwriting and their ability to produce a beau-tiful manuscript from a pile of cut and pasted papers.

I would especially like to thank the very professional staff of Benjamin/Cummings. James Behnke got me started and oversaw the whole project; Bonnie Garmus worked with me through two years of writing and review-ing, and Fannie Toldi had the uncanny ability to produce a beautiful book from what I perceived as an incomprehensible mass of words, photos, sketches, and editorial symbols. They are a delightful group of people with a dedication to high standards. And thanks to Wendy Calmenson for step-ping in at the last moment to help see the book to its conclusion.

I would also like to thank Dr. S. Charles Kendeigh, Professor Emeritus at the University of Illinois, whose boundless energy and enthusiasm start-ed me on the road to becoming an ecologist.

Brief Contents

Part One:
Ecology
2

Part Two:
Natural History
256

Detailed Contents

Part Two:

**Natural
History**
256

Part One

Ecology

*T*he aim of science is to discover basic principles that explain observed phenomena. Ecology is not a new science, but only recently has sufficient information been amassed to derive some fundamentals of this specialized, but broadly applicable, science. **Ecology** can be defined as the study of the relationships between living organisms and their environment. It is much more than simply describing the life cycle of a plant or animal; it is an examination of the development, traits, and effects of organisms and the evolution of their life cycles in changing environments.

The science of ecology has existed at least since Aristotle (384–322 B.C.), who surmised that plants obtain their food from the soil, and animals get their food by eating plants and other animals. In 1886 Ernst Haeckel defined the word ecology in the way we use it today. There have been many studies done since Aristotle's time that could be described as ecological, but it has only been within the last century that ecology has come to be recognized as a major branch of the biological sciences.

The early studies in ecology were essentially descriptions of plants and animals, their adaptations, and their life cycles. Today we term these studies "natural history." Ecology in many ways is simply natural history information extended by

quantitative analysis to derive general principles. Most of these general principles have come to light only in the last half-century or so, and in that time ecology has become a recognized science. This recognition is due in part to classic works by Frederick Clements and Victor Shelford, who are sometimes known as the "fathers" of plant ecology and animal ecology, respectively.

Ecology has become such a broad and complex science that it now contains subsciences. Physiological ecology deals with how physiological mechanisms of organisms allow them to survive under certain environmental conditions. Population ecology is concerned with topics such as population growth and decline, reproductive rates and cycles. Community ecology is an examination of the interactions of many species of organisms. Paleoecologists are interested in the ecological relationships of organisms in the geologic past. The list is long, but the purpose of each is similar: How do organisms coexist with each other in the environment?

A living organism must cope with its continually changing surroundings. The **environment** of an organism can be defined as the sum of all external forces that affect an organism. All of these external forces can be categorized as biotic or abiotic. The

3

biotic *factors are living organisms such as trees, squirrels, mushrooms, bacteria, and the bodies of these organisms for a short time after death. The* **abiotic** *components are the nonliving physical factors such as water, rocks, wind, temperature, chemical nutrients, and pH. Since all living organisms are formed from nonliving matter and decompose into nonliving chemicals after death, the distinction is not as clear as it appears, but is it useful nevertheless.*

Over many generations, the survival of a type of organism depends on whether it has been able to change with the altering environment. There are over two million species of living plants and animals. Of all the organisms that have ever lived, however, 95% have become extinct because they were unable to adapt to changing conditions or were displaced by better adapted organisms. A few organisms, such as some sharks, have existed relatively unchanged since their appearance 300 million years ago, at least partly because the ocean environment has changed little in that time. Others, like the dinosaurs, were successful for scores of millions of years before disappearing. Most organisms probably existed relatively unchanged for only a short period of geologic time (several hundred thousand years or less, for example). They either evolved new adaptations or ultimately disappeared. The fossil record of birds indicates that the average bird species had a longevity of one-half million to 3 million years.

Each species of plant and animal is tolerant to some degree of environmental components that impinge upon it and is able to reduce or avoid some of the effects of these environmental forces. One of the reasons for the existence of millions of species of organisms is that there is an infinite variety of environmental situations, each of which can be dealt with in numerous ways. In short, there are lots of ways to make a living in the biological world.

As unimaginative and stoic as it sounds, the sole function of an organism is to reproduce. An organism's anatomy, physiology, and behavior are all adaptations that allow it to survive. To be a successful creature, however, it must also reproduce, since nonreproduction and death produce equal results in a generation's time. Through the mechanics of the evolutionary process, the offspring that have the genetic makeup best suited to their environment survive and reproduce; the less fit produce fewer offspring and may ultimately disappear.

Evolution is a dynamic and continual process. What we observe today are simply the best evolved adaptations to date for the present environment.

*An **ecosystem**, or ecological system, is a complex of living and nonliving environmental components interacting and closely interdependent in any kind of fairly stable situation. Ecosystems may be broadly defined; their size is not a factor. The coniferous forests covering much of Canada can be considered an ecosystem as can a goldfish bowl with aquatic plants and fish. A rotting log with a myriad of creatures living in it is another ecosystem. Human skin, with living and dead epidermal cells, bacteria, viruses, and perhaps fungi, is an ecosystem. Even a human's intestinal tract, sterile at birth, becomes an ecosystem with a billion bacteria per gram of intestinal fluid in a few days.*

As organisms are the result of many years of evolution, so are ecosystems; ecosystems continually change since the organisms that comprise them continually change. Geologic and climatic changes are the greatest physical forces instigating change. It was only about 12,000 years ago that sheets of ice from the last glaciation began to recede from the midwestern United States, which probably looked then like the Alaskan tundra does now. Continual transformations of the earth's surface, climate, and organisms are the rule. This appears to imply instability, but nature is not capricious, and most changes occur in a gradual, orderly fashion with exceptions such as floods, volcanic eruptions, and earthquakes.

Ecological phenomena will be considered by first examining the components of the physical environment, then the biological components and their dynamic processes. The growth, death, movement, and distribution of organisms will be studied in light of the evolutionary process to provide an understanding of why organisms are as they are.

Chapter 1

The Bases of Ecology and the Physical Environment

To understand why organisms are built as they are, act as they do, and live where they live, we must first consider the environment in which they live. The environment is a sculptor in many ways—shaping the form and function of organisms, and continually adding, rejecting, and reorganizing. The physical surroundings of organisms pose a continual challenge that only the best adapted can meet.

Organisms are not just simple slaves to their environment. The external biotic and abiotic factors that impinge upon plants and animals are attenuated by the internal environment of the organism. Oak trees in California's great central valley can withstand high temperatures and dry conditions because they reduce water loss by having small leaves that are light-colored to reflect heat, and long roots to tap water sources deep in the soil. Squirrels survive the low temperatures of a Russian winter by hibernating. Some birds and bats migrate to more hospitable areas. Carp are tolerant of a wide range of water temperatures. Lungfish of the southern continents can burrow into the mud when their ponds dry up and survive until rainwater refills their habitat. Thus, the impact of environmental factors on living creatures depends upon the physical, physiological, and behavioral adaptations of these organisms. Let's look at this more closely.

In 1840 a German scientist by the name of Liebig, who was studying the effects of single environmental factors on plants, found that the growth of a plant depends on the availability of necessary

Figure 1.1 Liebig's Law of the Minimum. All organisms need minimal amounts of nutrients to survive, grow, and reproduce. Liebig derived the law of the minimum by providing plants with what he discovered to be minimal amounts of nutrients; he later expanded the law to include minimal amounts of heat, light, water, and other requisites. Note in the graph that some amount of necessary nutrient must be present for growth to occur at all. Notice also that a constant increase in that nutrient does not continue to induce growth; more than a sufficient amount has no effect and may even be deleterious. This graph assumes that all other necessary nutrients are present in adequate amounts.

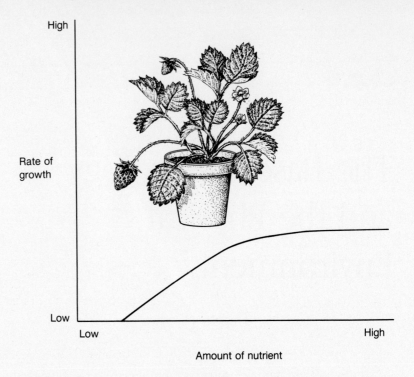

elements, such as phosphorus and nitrogen; without a minimum amount of each element a plant cannot grow. This concept became known as the "law of the minimum" (Figure 1.1). In 1905 it was demonstrated that even if a sufficient quantity of a necessary factor is abundantly available, growth may not occur if another necessary factor is not present in sufficient abundance. In 1921 this idea was expanded into the "law of tolerance": Not only do organisms need a minimum quantity of certain elements, but there is also a maximum amount they are able to tolerate. Physical factors such as light, heat, and moisture were also included as necessary elements. Most organisms survive best within a zone of tolerance below or above which survival rates decrease (Figure 1.2a, b). Environmental factors also interact with each other. On a hot, calm day a fox breathes faster to lose body heat and needs more water than on a cooler or windier day. Plants lose less water from their leaves on humid than on dry days, and less water on calm than windy days that increase evaporative rates.

Acclimation, a nonhereditary adaptation to the environment, can occur over the lifetime of an organism as well as over a short term. Consider a sun-tolerant tree. It is essential to the tree that its foliage be exposed to sunlight. If it grows in the center of a dense forest, it becomes straight, tall, and symmetrical as it grows toward

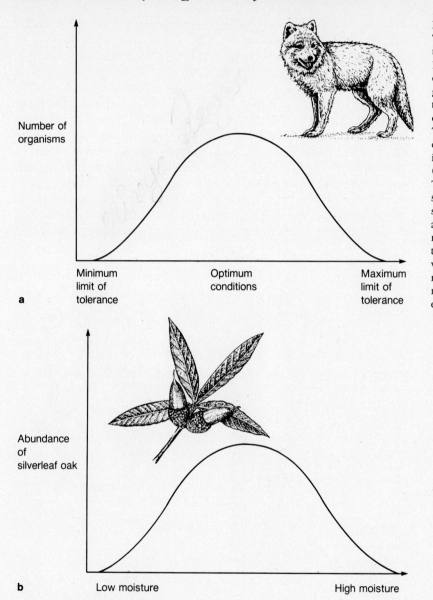

Number of
organisms

Minimum
limit of
tolerance

Optimum
conditions

Maximum
limit of
tolerance

a

Abundance
of
silverleaf oak

b   Low moisture   High moisture

Figure 1.2 The Law of Tolerance. (a)Application of the Law of Tolerance to Distribution. The abundance of animals and plants is greatest where their tolerance to environmental conditions is also greatest. The graph represents the distribution and number of individuals of any species. (b) Example of the Law of Tolerance. Abundance of silver leaf oak in the southwestern United States and northern Mexico along a moisture gradient from dry to wet. The trees do best with an intermediate moisture regime and thus are most abundant in that environment.

the sunlight above. The lower branches, shaded by the higher ones and other trees, have fewer leaves; as the tree grows taller by adding branches to the top, some lower branches die. On the forest edge, where one side of the tree is exposed to full sunlight, that side is well developed, while the shaded side of the tree toward the forest is sparse. The tree that grows alone in an open space will be comparatively short and broad, since it does not have to compete with adjacent trees for sunlight, and its foliage is denser and wider to take full advantage of the sun.

Similar acclimations occur in animals. A sponge growing in a tide pool with constantly surging water tends to be squat and stiff to withstand the pounding surf. But a sponge of the same species living in calmer bottom waters of the ocean will tend to grow tall and thin. These responses to the environment are called **growth forms** (Figure 1.3) and are not directly inherited by offspring; however, the flexibility to adopt a growth form appropriate to the environment is genetic. Both the tall tree in the middle of the forest and the shorter one growing in the open will produce seeds with the genetic information that makes them capable of growing into tall, short, dense, or sparse trees in response to the external environment. Young sponges also have the inherent capability to grow in an appropriate form.

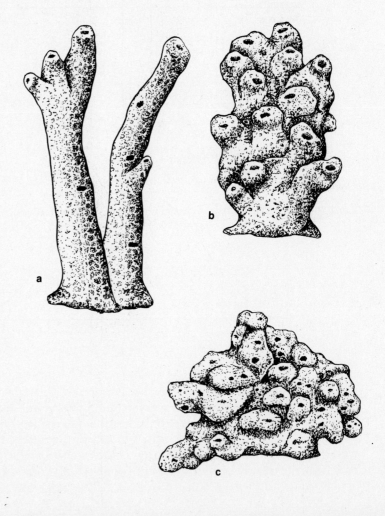

Figure 1.3 Growth Forms of Sponges. (a) From deep, calm water. (b) From shallow pools protected from the surf. (c) From rough, surging water. All of these sponges are different growth forms of the same species. The environment has caused the sponges to acclimate morphologically, but their potential to grow in a variety of forms is a genetic adaptation.

Organisms also possess many attributes they inherited from their progenitors. These are called **adaptations**; they are the set of inheritable anatomical, physiological, and behavioral characteristics of an individual that allow it to survive and reproduce. Virtually any aspect of an organism can be considered some sort of adaptation. If we examine the adaptations of cactus plants to their hot, arid environments, we find that they have no leaves and few appendages—this reduces the surface area from which water can be lost. They are built to store and absorb water for long periods of time; spines deter thirsty animals. Adaptations such as waxy needles of coniferous plants (pines, firs, and so on) reduce water loss. Many conifers live in a dry environment where much of the precipitation is in the form of frozen water—snow—which is unavailable to the plants except during the spring thaw. Other plants in areas subject to water shortage can fold their leaves to slow transpiration, shut down photosynthesis, or turn their leaves to avoid direct sunlight. Sea lions are insulated against cold ocean waters by thick layers of fat. There can be several ways to adapt to similar environmental conditions in the same or different habitats. Let us now examine some major environmental factors to which organisms must adapt.

LIGHT

Virtually all of the energy (the capacity to do work) that drives biological processes comes from the sun. Figure 1.4 illustrates the various types of radiant energy released from the sun. Much of this radiation is absorbed or reflected by gases and particles in the atmosphere. Of the radiation that does reach the earth, about 10% is ultraviolet, and the remainder is visible light and infrared radiation

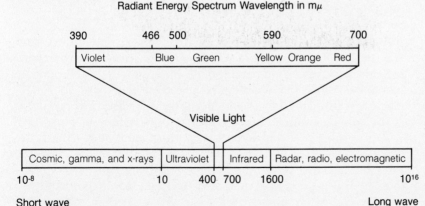

Figure 1.4 Radiant Energy Spectrum. Radiation emanating from the sun and reaching the earth comprises a variety of wavelengths. Most important to biological systems are ultraviolet, visible, and infrared radiation.

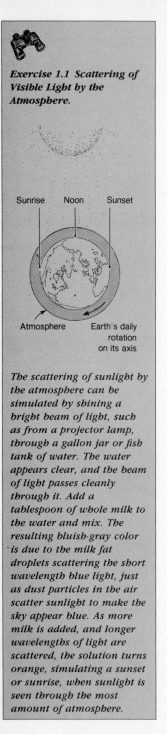

Exercise 1.1 Scattering of Visible Light by the Atmosphere.

Sunrise   Noon   Sunset

Atmosphere   Earth's daily rotation on its axis

The scattering of sunlight by the atmosphere can be simulated by shining a bright beam of light, such as from a projector lamp, through a gallon jar or fish tank of water. The water appears clear, and the beam of light passes cleanly through it. Add a tablespoon of whole milk to the water and mix. The resulting bluish-gray color is due to the milk fat droplets scattering the short wavelength blue light, just as dust particles in the air scatter sunlight to make the sky appear blue. As more milk is added, and longer wavelengths of light are scattered, the solution turns orange, simulating a sunset or sunrise, when sunlight is seen through the most amount of atmosphere.

in equal proportions. Visible light is the most biologically important of the numerous kinds of radiation that emanate from the sun.

Ultraviolet radiation (UV) plays a relatively minor role in ecosystems. Under usual circumstances it induces vitamin D synthesis in some animals, including humans. High doses of UV can be dangerous, however, and artificially generated UV is even used to sterilize medical instruments because it effectively kills microorganisms. The recent concern about the introduction of fluorocarbons into the atmosphere from spray cans involves the increase of UV radiation that would hit the earth. Fluorocarbons destroy ozone (O_3), which blocks the passage of UV through the atmosphere.

Infrared radiation is essentially heat. Although some heat energy is needed and used by plants and animals, it can be dangerous in excess (Exercise 1.1).

Why Are Plants Green?

Visible light is the ultimate source for virtually all the energy contained in biological systems. Light energy is converted into chemical energy by the process of photosynthesis, which occurs in green plants.

Photosynthesis, covered more fully in Chapter 2, can be briefly summarized as follows: Carbon dioxide in the atmosphere is taken in by the leaves, and sometimes stems, of a plant, and water is absorbed through the roots. Chlorophyll, a green pigment, helps to convert visible light energy to chemical energy, which is then stored in the plant matter. The intensity and duration of light, carbon dioxide concentration, water supply, temperature, and wind all affect the rate of photosynthesis. (How do you think each of these factors affects photosynthesis?)

A very interesting question is: Why are plants (or why is chlorophyll) green? Let's look at a green leaf.

Visible light hitting a leaf can act in three ways: It can be absorbed by the plant, be reflected by it, or be transmitted through it (Figure 1.5). Only the energy absorbed by the plant is used in photosynthesis. Of the light reaching the earth, only 1% to 2% actually hits plants. Since plants reflect some radiation and are not totally efficient at using all the radiant energy they do absorb, only 0.001% of the light reaching the earth is used in photosynthesis.

Any visible object absorbs some colors and reflects others; the object appears to be whatever color it reflects. Most plants are green, then, because they reflect green light. But why green?

The colors in the spectrum of visible light vary in their ability to provide energy for photosynthesis according to their wavelength; longer wavelengths such as red contain the least energy. A plant can

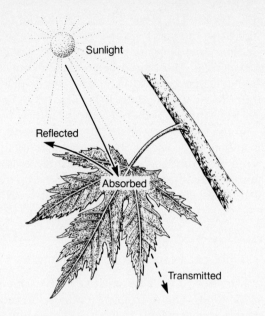

Sunlight

Reflected

Absorbed

Transmitted

Figure 1.5 Sunlight Falling on a Leaf. Sunlight falling on a leaf can be absorbed, reflected, or transmitted. Only the absorbed light can be used in photosynthesis.

theoretically be any color; if it reflected all (absorbing none) visible light, it would appear white. Since white plants would receive little or no energy, almost all the white plants that do exist are nonphotosynthetic fungi or parasitic flowering plants. If a plant absorbed all light, it would appear black. But a black plant would absorb a considerable, and perhaps lethal, amount of heat radiation, since visible light and infrared radiation are closely associated. Thus, plants compromise by reflecting some and absorbing other portions of the visible light spectrum. Green plants reflect most green and absorb most blue and red light. Darker green plants absorb more light energy, but also more heat. In dense forests, the shaded undergrowth of plants compensates for the minimal radiation by being dark green and absorbing the maximum amount of incoming energy. Dark green plants, then, are typically found in cooler, shady areas where the intensity and/or duration of solar radiation is low. The cooler environment makes heating up less of a potential problem. Ferns, mosses, and philodendrons, inhabitants of dense forests, are good examples of dark green plants. In hot environments plants tend to be light green since they can reflect a lot of light but still absorb a sufficient amount of energy without undue heating.

The digger pine is a good example of a plant adapted to a hot and dry climate. The digger pine lives at low elevations that are very dry and hot during the summer. These pines, unlike most pines, are light green in color, with sparse vegetation. Dark green pine trees with dense branches would soon overheat and die, so the dark pines are found only at higher and/or cooler habitats.

Red light is absorbed by water, so most of the light that reaches deep-water plants is blue and green. If deep-water plants were green in color, they would not use green light, and would be able to absorb only blue light. Since neither green nor blue light alone would provide sufficient energy, many deep-water plants, such as marine algae, are red to enable them to absorb the higher energy of the green and blue light together. The surrounding water makes overheating impossible. Plants in shallow water are in a radiation environment similar to terrestrial plants and are generally green.

Overheating is detrimental, but some heat is necessary. Heat is not only required for daily metabolic activities, but many seeds germinate and dormant bulbs sprout in response to increasing temperature.

Different Responses to Wavelengths

Organisms are not all equally sensitive to the same portions of the radiation spectrum. The red algae mentioned in the preceding section have accessory pigments that make them more sensitive to green light, as red light and some of the blue light is filtered out by the water. Color vision is well developed in most bony fishes, some reptiles and mammals, most birds, and some amphibians. Deep-water fish are most sensitive to blue light. Many insects can detect ultraviolet light, as can hummingbirds and at least one species of toad. Pigeons can also detect UV, as well as polarized light, which may help them in navigation and orientation. Bees are attracted to blue flowers because blue flowers apparently reflect more UV light than red flowers do, and bees are able to detect this UV much better than birds. In the tropics, where bird pollination of flowers is more important than that of bees, many flowers are red. In insect-pollinated areas of the world, blue flowers are dominant.

Photoperiod

Organisms also respond to the duration of light during the day—the **photoperiod**. At a latitude of about 40° N (such as at Washington, D.C.), the day length ranges from about 6 hours in the winter to 18 hours in the summer. The photoperiod serves as an important cue for timing plant and animal activity on a daily, seasonal, and annual basis. An excellent illustration of the effect of photoperiod is the migration of birds. We have all heard about the swallows' return to Capistrano (or the lesser known phenomenon of the return of the turkey vultures to Hinckley, Ohio) at about the same time each spring. Their predictable arrival is based on the lengthening photoperiod. Under artificial conditions, captive birds have been brought into reproductive condition and induced to show premigratory behavior (**migratory restlessness**) by a gradual lengthening of the photoperiod. Although captive birds will eventually come into breeding condition (enlarged gonads) even under a constant photoperiod, they will respond more quickly to a lengthening photoperiod. Thus, lengthening photoperiod itself does not cause gonadal development and subsequent migration, but rather serves as a "timer" to coordinate the bird's breeding cycle with the environment. Weather conditions affect the speed of migration, but they do not cause migration. The beginning of migration is based on photoperiod, which, unlike weather, is predictable. (The specific arrival and departure dates of birds, such as the Capistrano swallows, are not nearly as predictable as commonly believed, however.)

Deciduous trees lose their leaves in the fall as the photoperiod shortens. These trees would die if they were unable to drop their leaves and become dormant for the winter, but they would just as surely die if they dropped their leaves in response to an unseasonably early cold spell. You may be able to observe deciduous trees planted near street lights; many lose their leaves later than those away from street lights because street lights effectively lengthen the photoperiod for those trees. However, these leaves will eventually fall, demonstrating that photoperiod itself is not the cause of leaf drop, but rather serves as a cue for its timing.

Cool nights and dry weather also help to bring on the period of leaf drop that, to us, indicates autumn. The brilliant color displayed by many deciduous trees is due to the disappearance of the chlorophyll pigment, which has masked the yellow and red pigments underneath. In the eastern United States, the peak of fall colors occurs in the last week of September in the north and proceeds southward at about 800 kilometers per week.

How does a tree "know" when to drop its leaves? The blue pigment **phytochrome** exists in the leaves in one chemical form at night and another during the day. As the daylight period shortens, the pigment exists more and more in its nighttime form. When the level of nighttime phytochrome is high enough, leaves begin to turn colors. Phytochromes act through a chemical called **dormin**, which causes the trees and seeds to become dormant, and another chemical, called **abscisin**, which causes leaves to drop.

Some flowers bloom in the spring and some in the fall in response to day length. **Short-day plants** flower in the spring and fall and **long-day plants** in the summer. This is why we have lilies at Easter and poinsettias at Christmas. Some plants bloom at particular times of the day. In medieval towns, flowers were often planted in the town square in a circular pattern resembling a clock. By looking at the blooming plants in the flower clock, passersby could approximate the time of day.

The leaves of bean seedlings will droop at night and stand erect during the day. In complete darkness, however, they will show the same daily cycle, indicating that some plants, probably most, have an internal biological clock. The photoperiod simply keeps that clock accurate.

Photoperiod causes plants to respond, but it is not the changing day length that requires the plants to bloom or go dormant. Rather, this change is a cue that the season, and thus the environment, is changing, and it is to this major change that plants are actually responding.

Intensity

The intensity of light is also important. The rate of photosynthesis, for instance, depends, within limits, on the light intensity. There is a phenomenon among birds known as the dawn chorus. Near sunrise some bird species begin to sing; then others join in, then still others. After a short while, some drop out. The kind of birds singing and the strength of their song depend on the intensity of the sunlight. On a clear day, the chorus starts early and is loud; on an overcast day, it starts later and is more subdued. The order in which birds join in, however, stays the same. A few birds, such as the mockingbird, respond to very little light, and will sing even on a night with a full moon.

TEMPERATURE

Most climatic effects on organisms are due to a combination of temperature and moisture regimes, but often temperature alone is important.

The highest temperatures are generally in environments with the greatest amount of incoming solar radiation, such as the tropics. But cloud cover can prevent extremes of temperatures—during the day by partially blocking solar radiation and at night by inhibiting the reradiation of heat from the earth's surface. Recall that the hottest days and coldest nights are cloudless. The hottest areas in the world are the dry tropical deserts where temperatures may reach $60°C$* and the lack of rainfall makes them the most inhospitable places on earth. Some northern deserts, such as the Great Basin of

---

*As many Americans are not accustomed to using the Celsius scale, and most cannot remember the exact equation to convert Celsius ($°C$) to Fahrenheit ($°F$), here are two simple conversion equations that give *approximate* equivalents: $°F \rightarrow °C: °F/2 - 15 = °C$; and $°C \rightarrow °F: 2(°C + 15) = °F$.

North America, are cold deserts, with temperatures down to $-30°C$ and snowfall in the winter.

The coldest areas are at the poles, which receive the least direct solar radiation and experience months of little solar input (the six months of night). North American arctic tundra temperatures average between $-35°C$ and $+13°C$. Alpine tundras (above tree line on high mountains) are also very cold because as air rises it expands and loses heat, and thus the thinner atmosphere at higher elevations is colder (Figure 1.6). As one proceeds up a mountain, the average temperature drops about $6°C$ for each 1000-meter increase in elevation. Northward from the equator, the average temperature decreases by $6°C$ for every 777 kilometers. If increasing altitude and increasing latitude both produce similar temperature changes, it follows that changes in plant and animal life should also be similar, and this is the case.

The flora and fauna of alpine areas on mountain tops, for example, are similar to those in the northern arctic areas. However, the change in plant species with increasing altitude happens rather quickly and is easily observed, whereas the change in plant and animal communities with increasing latitude is more subtle, since it requires traveling over a greater distance. Let's examine some of these similarities in tundra regions. The vegetation of both the far north arctic tundras and high elevation alpine tundras consists primarily of lichens, mosses, sedges, and grasses. The plants are typically small, grow very slowly, and the shrubs and trees are much branched and stunted due to the effects of wind and cold. The plants tend to be **perennials** (persisting for two or more years) because seed germination and growth are very slow. Not only are the physical forms of vegetation similar, but 30% to 40% of the plant species found in the tundra of the Rocky Mountains is the same as that found in the arctic tundra. There are similarities among the animals also. If active during the winter, alpine and arctic tundra mammals tend to be white; hibernation is a common phenomenon among others.

Figure 1.6 Temperature Changes and Climatic Changes as Air Rises and Falls over a Mountain. As the prevailing winds push the clouds eastward against the mountains, the clouds must rise. As they do, they cool, and their moisture condenses out as rain or snow. Little moisture is left to be dropped on the lee side of the mountains. In fact, as the air drops, it warms and absorbs moisture. Thus, the lee side of the mountain is a rain shadow.

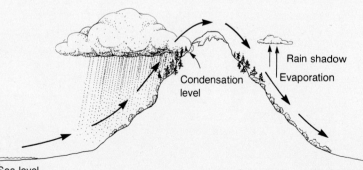

Rain shadow

Condensation level

Evaporation

Sea level

Life exists on earth between −70°C in the Siberian tundra to almost 100°C in hot springs, but the tolerance of most organisms is much smaller than that. Away from the equator, seasons occur because of the changing angle of incidence of the sun's rays as the earth moves in its orbit around the sun; in equatorial areas the constantly perpendicular rays produce temperature regimes that differ by only a few degrees throughout the year. Tropical organisms are, thus, not exposed to the extremes of temperature that temperate organisms must face. Seasonality in the tropics is dependent on wet and dry seasons rather than temperature differences.

Plants cannot migrate, so they must have other mechanisms to cope with temperature changes. Cacti are excellent examples of plants adapted to hot and dry conditions. Keep in mind that temperature and moisture are most often interdependent—adaptations or acclimations are rarely made to conditions of temperature or moisture alone. Cacti have few leaves, a waxy cuticle, and fine hairs or other surface coverings to reduce water loss due to **transpiration** (loss of water vapor through the plant's surface). These plants can expand to store water, which is not only used in metabolism, but also buffers the effect of temperature changes. The saguaro cactus grows tall and thin, allowing only a fraction of its surface to be hit by the sun. The prickly pear cactus grows so that its leaves are oriented edgewise to the sun during the hottest part of the day (Figure 1.7). Many plants become photosynthetically dormant in very hot or cold conditions. There are degrees of plant dormancy. Evergreen trees of

Figure 1.7 Prickly Pear Cactus. Stems are oriented to receive the least amount of sun during the hottest part of the day.

cold climates photosynthesize slowly throughout the year. Deciduous trees of more temperate areas lose their leaves and stop photosynthesis altogether. The perennial plants' entire above-ground body dies and is regenerated in following years from roots or bulbs; the annual plants complete their life cycle during the mild part of the year, producing seeds that overwinter and germinate to form a new plant the following year.

Thermoregulation—The Regulation of Body Temperature

Most animals are poikilothermic—often inaccurately defined as "cold-blooded." **Poikilothermic** actually means that the organism's body is physiologically incapable of maintaining a constant temperature, so it is roughly the same temperature as the environment, much like a plant. In fact, plants are also poikilotherms, but the term is usually reserved for animals.

Plants do produce some heat, as do all living organisms during metabolism, but few plants raise their temperature measurably above that of the environment. The skunk cabbage is a marvelous exception; it can raise its body temperature to 22°C and melt its way through a covering of snow or ice to bloom. Unlike most plants, its metabolic rate increases as temperature decreases. One advantage to this mechanism is that it is able to attract pollinators before other flowers bloom.

Like plants, animals have adaptations to modify or avoid the effects of extreme temperatures. Many reptiles, amphibians, and insects will expose themselves to the sun on a cold day to raise their body temperatures. Insects are most easily captured in the early morning when they are cool and sluggish (Exercise 1.2). Turtles have caused traffic accidents by resting on asphalt roadways in the evening to absorb the heat radiating off its surface. Many desert animals are nocturnal and avoid the day's heat in a burrow or under a rock. Bees can warm their hive by vibrating their wings and producing a small amount of body heat. Horned lizards can change color and thus selectively reflect or absorb radiant heat. Many invertebrates show a negative reaction to light, which keeps them out of the sun. Some frog species may pass the winter burrowed in the muddy bottom of a frozen lake, their metabolic rate so low that they absorb sufficient oxygen through their skin. Insects commonly enter **diapause**, a resting stage, during the hottest or coldest parts of the year. **Estivation**, or summer dormancy, occurs in many poikilothermic animals, such as turtles or frogs, which may burrow into mud during hot periods.

Poikilothermic animals are particularly affected by temperature, since their rate of metabolism is related to their body temperature, which is always close to that of the environment. Very cold or hot

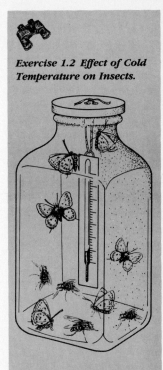

Exercise 1.2 Effect of Cold Temperature on Insects.

Obtain several flying insects such as flies, butterflies, or moths. Place them in a jar with at least a 2 liter capacity. Cover the jar loosely with a lid from which is suspended a thermometer. Place the jar in the refrigerator for a few hours or overnight. Remove the jar and record the temperature at which the insects become active and begin to fly. Do different individuals become active first? Why? Do some species become active before others? How is this adaptive to their natural environment?

days inhibit or stop the activity of these animals, and very hot days can bring on fatal stress.

The ability of poikilotherms to withstand freezing temperatures is generally due to behavioral mechanisms. Arthropods and salamanders, for instance, burrow under bark or decaying vegetation to avoid lowering their body temperature to 0°C, which would cause ice crystals to form in their body fluids, producing death. Some insects have glycerol in their body fluids, which lowers the freezing point several degrees. Physiological adaptations may also exist, as in some species of fish that live in antarctic waters at temperatures around −1.85°C. The water is not frozen because of its high salt concentration, and the fish have special proteins in their blood that prevent their freezing. Tuna and some sharks are able to warm their blood as much as 15°C warmer than the water because of a specially adapted circulatory system.

Birds and mammals are "warm-blooded," better termed **homeothermic**, because their body temperature is regulated internally and remains nearly constant. This adaptation has made them very successful, especially in cold climates, to which poikilotherms are not well suited. Since they produce heat internally, birds and mammals have evolved insulation in the form of feathers, fur, and fat. They may have special adaptations in the circulatory system of their extremities to reduce heat loss from their feet, hands, ears, and noses. The **rete mirabile**, a bundle of small arteries and veins at the base of the extremities, serves as a heat exchanger to reduce or increase heat loss from the appendages.

In very cold conditions some mammals, such as skunks and bears, go into a state of reduced activity for short periods of time, and may become active any time the weather warms. Organisms that cannot directly confront the environmental temperature must either migrate or go into a state of torpor. **Torpor** is a sort of "suspended animation" that allows an animal to drop its body temperature and conserve energy in order to survive a cold spell. This often occurs on a daily basis, with the organism being torpid at night and active during the day. Torpidity may extend through the winter. Such organisms do not hibernate as do ground squirrels, some bats, and groundhogs. True **hibernation** involves a severe reduction in body temperature (but always above the freezing point) for long periods of time, during which metabolism and respiration are extremely reduced. Among birds, a few, notably the hummingbirds, exhibit torpor during cold evenings, but only the poor-will of the southwestern United States and perhaps a very few related species and a few swifts undergo true hibernation.

Homeotherms are well able to reduce heat loss (Figure 1.8), but the need to lose heat in hot environments is solved with somewhat more difficulty. Evaporative cooling is one common solution to this

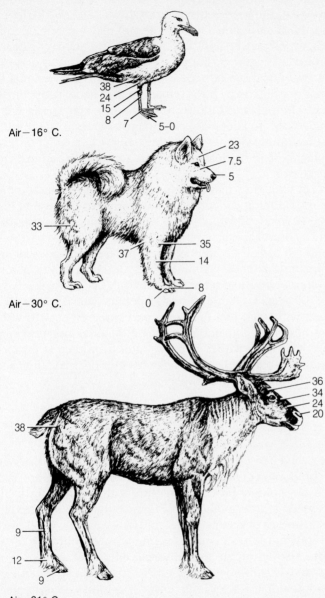

Air—16° C.

Air—30° C.

Air—31° C.

Figure 1.8 Adaptation to Avoid Overheating in Well-Insulated Animals. No animal can be totally insulated, since fur or feathers on the feet, legs, or nose would be a hindrance. These areas can thus be the sites of a considerable loss of body heat. A mechanism of heat exchange in the form of a capillary bed at the junction of the extremities and trunk allows the warm blood leaving the trunk to warm the cold blood coming into the trunk so that the extremities can become cold without the animal losing much overall body heat. Conversely, an active, overheated animal can lose excess heat by passing it on to the extremities. Jackrabbit ears, for instance, act as radiators of body heat. Numbers indicate temperatures in various body regions.

problem. Many mammals have well-developed sweat glands, and birds have accessory air sacs diverging from their lungs that help cool them by evaporation. Increased respiratory rates also aid in cooling by increasing water loss from lungs. The tail of a beaver is thought to be a cooling device, as blood vessels at the skin's surface lose heat to the water. Instead of sweating and losing precious water, a camel's body temperature rises as heat is stored in its body during the day; the body cools as excess heat is radiated to the cooler night air. Birds sitting on a nest in the sun may not be able to leave the eggs, which may be literally "cooked" on a hot day. Instead, the parent bird may pant or flutter its throat to increase evaporative cooling. You may have noticed the lack of bird activity on a sultry summer day when many birds remain in the shade. Some birds and mammals (herons and hippos, for example) may spend hot days walking in water to keep cool. Some birds defecate on their legs to increase evaporative cooling. Molting a heavy coat of fur or feathers and replacing it with a less dense coat adapts many animals to the summer season.

African elephants may be exposed to environmental temperatures of 43°C. Being dark gray in color and having no sweat glands, they face a considerable heat load. For an average-sized elephant to maintain its normal body temperature of 36°C it needs to lose body heat (enough to boil 22 liters of water per hour). It does so mostly through simple radiation from the body into the atmosphere. Elephants often wallow in mud or stand in water to cool themselves, but much of the heat loss is due to their large ears, which act as radiators. Elephants will often flap their ears and spray them with water to increase heat loss. On hot days, when water is not immediately available, elephants will even draw water from their stomach with their trunks to spray their ears. Elephants also decrease their activity and seek shade during the midday heat.

WATER

Life evolved in a watery medium, so it is no surprise that water is a basic component of life. The bodies of most organisms are primarily water, physiologic processes require water, and water provides habitats for an array of organisms. Minerals that are dissolved in the soil and in bodies of water are available to organisms. Water is essential as a solvent to the chemical cycling of matter through the environment. Also, through the physical effects of rainfall, runoff, erosion, and stream flow, water moves materials between ecosystems for immediate use or long-term deposition as sediments.

Precipitation

When a layer of air saturated with moisture is driven upward, it expands due to a lessening of atmospheric pressure. This expansion, which requires energy, causes the parcel of air to cool. Cooler air condenses the moisture, and precipitation occurs in the form of rain or snow. This is one part of the **hydrologic** cycle (Figure 1.9). Some of the water reaching the earth falls on the ocean or other

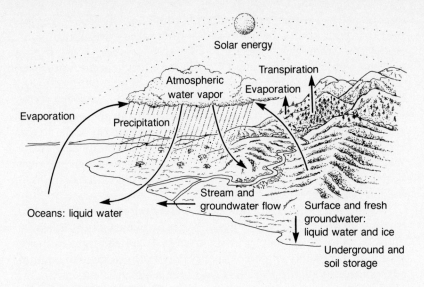

Figure 1.9 The Water (Hydrologic) Cycle. Other major considerations (not shown here) are human effects on the cycle. Damming rivers to form lakes, using water for agriculture, and pumping up groundwater all increase the evaporation rate. Much of the earth's surface water is polluted by sewage, runoff over agricultural or urban areas, and industrial use. The self-cleaning aspect of the hydrologic cycle, however, can reverse the effects of water pollution if the source of pollution is stopped.

bodies of water, and some falls on the land. Warmer air near sea level can hold more moisture, so surface water evaporates quickly and becomes atmospheric water vapor again. Of the water that falls on the land, some becomes runoff and flows to the ocean via streams and rivers; some seeps through the soil and rock layers to reservoirs of groundwater; and some of it returns to the atmosphere via transpiration and evaporation.

As the wind blows clouds against the sides of mountains, the air masses are deflected upward and cooled, causing the windward sides of mountains to receive more rainfall than the lee side. The clouds descend on the lee side and expand, retaining or absorbing water vapor. Thus, the lee areas tend to be much drier; they are in a rain shadow (see Figure 1.6). The Sierra Nevada, for example, annually receives a good deal of moisture as the winds blow inward from the Pacific Ocean. The western (windward) slopes at about 1500 meters elevation receive approximately 190 centimeters of precipitation annually, while the eastern (leeward) slopes at the same elevation receive only 50 centimeters. The Great Basin desert east of the Sierra results from this precipitation pattern.

The amount of precipitation and its seasonal occurrence over the year is of great ecological significance. As one ascends the slopes of a mountain range, such as the Rockies, the major vegetation changes from ponderosa pine, white fir, and Douglas fir to Englemann spruce, and to alpine tundra on the highest peaks. These changes in vegetation zones, often termed altitudinal zonation, are primarily due to temperature and precipitation. Higher and colder elevations receive more precipitation, and evaporation takes place more slowly than at lower elevations, which are warmer and drier. At the highest elevations snow falls rather than rain, effectively making the environment dry; the north and south poles, perhaps surprisingly, are very dry areas.

The distribution of rainfall around the earth is important in determining the type, distribution, and density of vegetation growth. Precipitation patterns are determined by global wind patterns, which, in the temperate zone, are primarily from the west. Some areas receive rain every day. The highest mountain on Kauai in the Hawaiian Islands is continually enshrouded in clouds and is the wettest place on earth; it receives 1105 centimeters of rain each year. Conversely, there are areas of the Chilean desert in which no rainfall has ever been recorded.

Again, the extremes of any environment are modified by organisms' adaptations. In wet areas most of a plant's body is above ground, while in dry areas much of it is in an underground root system that is widely spread or deeply penetrating to take maximum advantage of available moisture. **Succulents** like aloe and cactus are plants that store water in leaves, stems, or roots for times of drought. Bromeliads (Figure 1.10a) typical of tropical regions, often grow on other plants rather than in soil, and absorb moisture through their own leaves. Mangrove trees grow in salt water—impossible for most trees—because a physiological mechanism allows the trees to purify the water as it enters the roots (Figure 1.10b). The spines of some species of cacti point downward so that water drips off the spines onto the base of the cactus. Ocotillo, a desert plant of the southwestern United States and Mexico, sheds its leaves during a drought but begins to regrow them within 48 hours of a rainfall. This shedding and regrowth may occur five or six times a year.

Too much water can pose a problem for some burrowing animals, so they may build side chambers that do not flood as readily as the main ones, or simply come to the surface of the ground as earthworms do after a rainfall.

Metabolic Use of Water

All plants need water for photosynthesis. A little water is absorbed through the leaves, but most is soaked up by the roots. On hot, dry days when water is scarce and potential water loss high, transpiration is reduced to inhibit water loss. A reduction in transpiration necessitates a reduction in photosynthesis and consequently causes slower growth. Desert cacti may increase in height by only a centimeter or so each year. Conversely, in moist tropical areas, bamboo may grow as much as 20 centimeters or more per day. About 97% to 99% of absorbed water is transpired, as it requires about 500 grams of water to produce one gram of carbohydrate. A birch tree may transpire 350 liters in one day.

a   b

Figure 1.10. Root systems. (a) Bromeliads (members of the pineapple family)
grow on the bark of trees in moist, tropical areas and absorb water and nutrients
through their leaves. The plants also accumulate water at the bases of their leaves
for later use; this water may harbor dozens of different species of
microorganisms. Some frogs even lay their eggs there. (b) The red mangrove
grows in tropical and subtropical areas, where it forms dense thickets. The long
prop root system allows it to grow in deep water and move over mud flats as
mud accumulates. The pendulant aerial shoots will establish new plants; they also
absorb gases from the atmosphere. Mangroves are excellent soil stabilizers, and
they minimize shoreline erosion.

Aquatic animals are generally faced with the problems of fluctuating water quality (such as salinity, pollution, temperature) or
quantity (such as ponds and creeks drying up). Terrestrial animals,
however, are more often affected by the lack of available water and
have, like plants, adapted mechanisms to gain or retain moisture.
Roundworms, such as soil nematodes, have an external cuticle nearly impervious to moisture. Insects and other terrestrial arthropods
have a similar waterproof integument. Birds and reptiles excrete
uric acid in a paste form rather than liquid urea, which requires
considerably more water. Many desert birds and reptiles can exist on

the water in the food they eat. Some species of desert kangaroo rats never ingest water, but metabolize it from dry seeds. Gerbils, the dominant rodents of the deserts of North Africa and the Middle East, get their water by eating saltbush leaves. Although the leaves are succulent, they contain a high concentration of salt. Only by having one of the most powerful kidneys in the animal world, producing very salty urine, are gerbils able to eat the saltbush. A desert-dwelling cockroach can absorb water from the air.

Humidity

Absolute humidity is the amount of water in the air. Since the amount of water the air can hold is dependent upon air temperature and atmospheric pressure, we usually speak of **relative humidity**, the percentage of water vapor present in the air compared to how much the air could hold under the same conditions of temperature and pressure. Water loss and water requirements are both greater when the humidity is low. To avoid drying out, some arthropods, such as pill bugs, millipedes, and many insects, inhabit humid microclimates such as decaying logs and the undersides of rocks. Some animals, such as snails and slugs, are nocturnal because nighttime humidity is generally higher than daytime humidity. Since warmer air holds more moisture than cooler air, temperature, water, and humidity are closely related. Areas of dense vegetation, such as forests, are typically humid. In such areas, transpiration adds a great deal of water to the air, lower temperatures reduce evaporation, and the density of the vegetation reduces air flow.

ATMOSPHERIC GASES

The earth is composed of and surrounded by several atmospheric layers (Figure 1.11). The lowest layer, the **troposphere**, is the one in which life exists. It has a relatively constant composition of gases, a result of billions of years of volcanic activity, fires, and biological processes. The present atmosphere is approximately 78% nitrogen, 21% oxygen, and 0.03% carbon dioxide. Less than 1% consists of other gases, such as argon, neon, and sulfur dioxide.

The air's components are fairly equally distributed over terrestrial environments and are rarely in short supply, but the relatively recent high production of air pollutants can affect the air's quality. Increased sulfur dioxide (SO_2) concentrations can kill or sicken plants and animals because SO_2 in humid air becomes sulfuric acid (H_2SO_4). Combustion of **hydrocarbons** (organic compounds containing hydrogen and oxygen, such as gasoline and propane) in

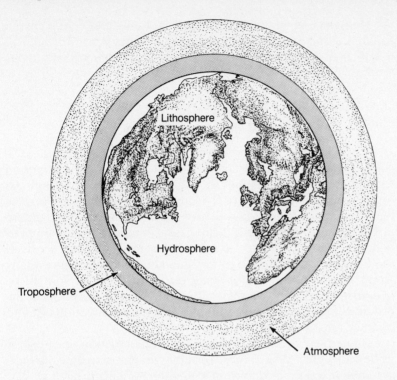

Figure 1.11 The Components of the Earth. The solid rock of the earth comprises the **lithosphere**; oceans and other bodies of water comprise the **hydrosphere**; the 12-kilometer high layer of air is the **atmosphere**, which includes the **troposphere**, a transitional zone. The thin **biosphere** is an amalgam of these spheres from a few kilometers above the earth to the bottom of the oceans and a few meters into the rock lithosphere where life exists.

factories, homes, and cars has increased the CO_2 concentration in certain areas. This is not totally harmful, since increased CO_2 content will slightly increase the rate of photosynthesis. (A popular myth that talking to plants helps their growth may contain a particle of fact, because if one spends a lot of time talking to, and simultaneously breathing on, plants, the CO_2 concentration around the plant may be slightly increased. However, the difference in growth between those plants and ignored ones would be negligible.)

Gases have a tendency to exist in bodies of water in an amount proportional to their abundance in air. The gaseous content of aquatic environments, however, does vary from habitat to habitat and within the same habitat because of the chemical and mechanical activities that occur in the water. The most important gas, oxygen, may only exist at levels one-thirtieth of that of the air above. The oxygen content depends upon many factors, such as abundance of plant life, water temperature, turbidity, water movement, rate of decomposition, and use of oxygen by animal inhabitants. A cold, fast-running stream is more oxygen-rich partly because of its continual churning than a warm stagnant pond; the two habitats will have quite different inhabitants for a number of reasons, and O_2 content is one of the most important.

Nitrogen is abundant in the air, but is often scarce enough in the soil to limit plant growth. Plants need nitrogen for growth, but cannot obtain it directly from the air. Nitrogen and other gaseous cycles are explained more fully in Chapter 3.

CHEMICAL NUTRIENTS

A **nutrient** is simply any substance necessary to sustain life. Organisms need numerous chemical elements and compounds in varying amounts for maintenance, growth, and reproduction. All organisms contain carbon (C), hydrogen (H), and oxygen (O), as these are the basic elements out of which all organic compounds are made. **Organic**, biologically defined, simply means "containing carbon." To speak of "organic" foods is biologically spurious, since all food contains carbon, and all organic compounds are potentially "food" (energy source) for some organism (or were formed by organisms, such as fossil fuels). From basic building blocks of hydrocarbon chains, an almost infinite variety of compounds vital to biologic functions is formed.

Numerous chemical elements are needed to form essential organic molecules. Most of these elements are the inorganic minerals available in a dissolved form in soil, water, or aquatic habitats; gaseous elements, such as nitrogen, are taken from the air. These elements, or the simple inorganic compounds in which they become incorporated, are either **macronutrients** (those needed in large amounts) or **micronutrients** (those needed in relatively small amounts) (Table 1.1).

Nitrogen and phosphorus are two of the most important macronutrients. Nitrogen is needed to form amino acids, which, in turn, form proteins. Phosphorus plays a crucial role in energy metabolism at the cellular level. Most commercial fertilizers contain primarily compounds of nitrogen and phosphorus and perhaps potassium. But even some elements needed only in minute quantities are indispensable, such as chlorine, which is necessary for photosynthesis. The absence or limited abundance of one or more of these nutrients can retard or prevent plant growth—the law of the minimum again. Animals need nutrients also, but much of what they receive they obtain directly or indirectly from plants. One interesting reversal of this role occurs in carnivorous plants, such as the Venus flytrap and pitcher plant, which trap and digest insects primarily for their nitrogen, as these plants often grow in nitrogen-poor areas.

Adding nutrients to the soil will increase plant growth in areas where they do not occur in sufficient amounts, but adding more than sufficient amounts will not necessarily increase plant growth proportionately. An excess of nutrients may even be harmful (recall the

Table 1.1 Macro- and Micronutrients

Macro-nutrients	*Example of Use by Plants*	*Example of Use by Animals*
Calcium	Part of cell wall	Skeletal tissue
Phosphorus	Energy transfer	Energy transfer
Nitrogen	Protein structure	Protein structure
Potassium	Enzyme activator	Protein synthesis
Sulfur	Component of some proteins	Hemoglobin protein structure
Sodium	Ionic balance (not needed by many plants)	Nerve impulse transmission
Chlorine	Photosynthetic reactions	Digestive juices
Micro-nutrients		
Sulfur	Most of these nutrients are required by	
Manganese	both plants and animals, but in differ-	
Iron	ent amounts; sodium can be consid-	
Cobalt	ered a macronutrient for animals and a	
Molybdenum	micronutrient for plants. Although list-	
Boron	ed as elements, they are typically found	
Silicon	as compounds in an ecosystem.	
Copper		
Zinc		
Iodine		

law of tolerance). Accumulated minerals may produce a high salt content in the soil, or they may run off into a nearby water source and instigate pollution.

In most natural systems, the input of nutrients is about equal to the output; little is lost, and little accumulates.

CURRENTS AND PRESSURES

Wind, water currents, barometric and water pressures, and gravity may not be as obvious as other environmental factors, but they are often powerful forces. Wind blowing against trees or shrubs may cause trees to bend over, tips of branches to break off, and buds to dry out. The gnarled, twisted, supine trees found at their altitudinal limits (**treeline**) are called **Krummholz**, and are the result of continual wind action. Other trees in windy areas exhibit **flagging** (Figure 1.12), in which the windward side of the tree has few or no branches, and the tree has the appearance of a flag on a flagpole. The

Figure 1.12 Flagging. This pine tree has taken on the appearance of a flag flying in the breeze because the consistent heavy wind has dried the buds and broken the branches on the windward side.

weight of accumulated snowfall can also cause trees to grow in a curved form. The flattened and stocky growths of coral, sponges, and sea anemones in tide pools are acclimations to the surging surf.

Wind

Wind, by drying buds or by breaking off branch tips, can cause branching by interfering with a plant's internal physiologic phenomenon called apical dominance. **Apical dominance** is the property that enables a plant to put most of its energy into growing upward rather than into branching. The terminal bud (at the plant's apex) produces hormones that flow down through the plant to inhibit lateral branching. When these apical buds are broken off, the plant hormones, called **auxins**, are no longer produced at the tip, and lateral growth can occur more rapidly. As a result the plant becomes more densely branched and bushier overall. Thus, the wind-swept sides of mountains and coastal slopes characteristically exhibit dense, shrubby vegetation. Salt spray along the ocean coast intensifies this effect. A more familiar example of apical dominance can be demonstrated around the house; we prune fruit trees and shrubs and pinch off growing shoots of house plants to cause increased branching for more fruit or an esthetic appearance (Exercise 1.3). In the northern tundra, where plants grow low because of the wind and slowly because of the cold, trees and shrubs form dense mats and are sometimes termed "cushion" plants. Wind also increases transpiration rates, causing plants to close their **stomates** (small openings in the epidermis of leaves through which gases pass) and reduce photosynthesis, contributing to their lessened growth rate.

Wind also affects animal life. Birds and flying insects may use wind to travel; or they may fly less to avoid injury or to avoid being blown in the wrong direction. Thousands of birds die each year due to navigational errors caused or exacerbated by the wind. On the other hand, wind has often assisted animals and plants to colonize new areas; for example, plant seeds are carried long distances in the air, and land birds could not have reached the Hawaiian Islands without assistance from the wind.

Some spiders "balloon" by spinning short strands of silk on which they hang and drift through the air. Strong winds, such as hurricanes and tornadoes, have been known to carry animals, even fish, to new habitats. There are even occasional news reports of a storm "raining fish" after a tornado scooped up the contents of a pond and dropped them elsewhere.

Wind also increases body heat loss and transpiration rates, but these are effects of temperature and moisture rather than physical air current, per se.

Water Currents

The effects of water currents are more apparent. Fish living in fast-moving water are sleeker and more muscular than their still-water counterparts(Figure 1.13). Some bottom-dwelling stream fishes have elongated pectoral and pelvic fins to help them hold onto the bottom. Insect larvae inhabiting streams, such as the larvae of dragonflies, caddisflies and stoneflies, are often equipped with

Exercise 1.3 Apical Dominance.

If a common fast-growing house plant (such as Coleus) is trimmed regularly by snipping the apical (tip) buds, two lateral shoots will arise at each place where a bud is removed and it will become densely branched. Compare this plant with one that is left untrimmed. In a natural environment what factors could cause branching?

Figure 1.13 Adaptations of a Stream Animal. This trout shows the typical torpedo-like body shape possessed by many stream animals. They are adapted to fast-flowing waters and can, for short periods of time, swim at a speed of 3 m/sec, although they prefer to avoid the current and spend much of their time in calmer pools.

hooks or hairs or other appendages for gripping rocks. Sponges growing in tide pools are shorter and thicker than those growing in the calmer open waters; tide pool barnacles have stronger shells than those in still water. Animals of calm water have no need for such adaptations.

Water currents are important to fish that migrate up or down rivers to spawn, such as salmon, steelhead, and lamprey. They orient to the current, a phenomenon known as **rheotaxis**.

Ocean currents, in combination with wind currents, account for the rainfall patterns around the earth. For example, as the westerly winds push moisture-laden air over the Pacific Ocean toward land, the warm air will drop moisture over cooler terrestrial environments such as the coastal Pacific Northwest and southern Chile. But where cooler ocean winds pass over warmer land, little moisture is dropped, and we find the coastal deserts of Baja California and similar areas in South America.

Air and Water Pressure

Barometric pressure does not seem to affect organisms to a great degree, mainly because it varies only a small amount. Pressure decreases with elevation, since the air is less dense at higher altitudes. Although a decreased amount of oxygen in high mountain habitats may have some effect, it is probably vastly overshadowed by the effects of temperature, rainfall, and wind. Birds tend to perch more as a low pressure center moves in because it is more difficult to fly in thinner air. That is why an abundance of perching birds often signals the onset of a storm, storms usually being associated with low pressure centers.

Pressure in aquatic environments is a much more serious problem. At an altitude of 500 meters above sea level, air pressure decreases to about one-half that of sea level pressure. At 500 meters below sea level, water exerts a pressure 50 times that at sea level. Thus, aquatic organisms need and have various devices to help them cope with water pressure and changes in depth, such as the swim bladder of many fishes, which controls buoyancy by filling or emptying with gas via a special mechanism in the circulatory system. In deep sea fishes this gas, 90% oxygen, may exert pressure in the air bladder twice that of a fully charged steel oxygen cylinder!

Gravity

Another factor that is easy to observe and that affects all organisms, but one that we do not generally think about, is gravity. We perceive most plants and animals as having a particular upright posture. Is it not surprising that when we plant a radish seed the leaves grow upward and the roots downward, no matter which way the seed was oriented? Have you ever questioned why an oak tree does not grow sideways? Plants grow upright because of the effects of plant hormones, auxins, that regulate growth. Auxins in the stem and root act in different ways to cause the plant to grow properly. A common demonstration of this property is to lay a potted plant, such as the common *Coleus*, on its side and observe how it bends upward within a few days. This response to gravity is called **geotropism** (Exercise 1.4).

The adaptations of terrestrial animals to gravity are generally too obvious to mention, but one interesting example is the well-developed system of valves in the neck of a giraffe that enables blood to reach the brain. Gravity has also limited the body size of terrestrial animals. Many of the largest dinosaurs were at least partly aquatic, because they would not have been able to support their weight on land.

The aquatic environment modifies the effect of gravity because water provides buoyancy. Thus, the largest animals, the whales, live in water. Various species of aquatic animals are essentially weightless at certain depths. *Nautilis*, the sole survivor of a group of chambered marine molluscs, has a curled, multicompartmented shell that is almost entirely filled with air; a 1400-gram *Nautilis* may weigh as little as one gram in seawater, and even the largest one may weigh only five grams in seawater. Although most aquatic creatures swim in what we would consider an upright posture, some fish maintain this posture by orienting their back to the light above. In an experimental situation in a glass aquarium, however, some fish will orient their dorsal surface to a light source even if the light originates from the side or below the tank. The fish will swim sideways or inverted, demonstrating that gravity is of little concern to some aquatic animals. Many animals have organs of equilibrium such as the **statocysts** of crayfish and jellyfish, which work in a way similar to the semicircular canals of the human inner ear to help the animal maintain its orientation. The air-filled bladders of the giant kelp, reportedly the

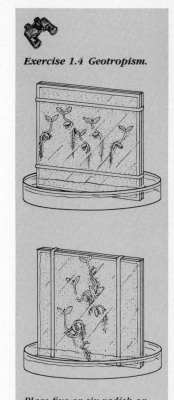

Exercise 1.4 Geotropism.

Place five or six radish or bean seeds on a wet paper towel, and sandwich the paper and seeds between two pieces of glass (10 × 10 centimeters works well). Hold the glass together with a rubber band, and set the sandwich vertically in a shallow container of water so that the paper towel continually absorbs moisture. As the seeds germinate, the stem and leaves will grow upward and the roots downward. Rotate the glass 90° every two or three days, and note how the stems and roots compensate. The stems and leaves show negative geotropism, *and the roots show* positive geotropism.

longest plant in the world (up to 700 meters in length), help to maintain the plant in an upright position as it stretches from the ocean floor to the surface.

FIRE

Fire is a common phenomenon in some terrestrial ecosystems, common enough for organisms to evolve specific or general adaptations to it. A common attitude towards fire has been to label it a destructive force, when in reality it has been, and still is, a natural phenomenon that often has beneficial effects. The "Smokey the Bear" extinguish-all-forest-fires attitude that prevailed for so long is being tempered with recent knowledge that fire is not necessarily detrimental, and, in fact, can be used as a management tool. Lightning is the most frequent cause of natural fires. If no effort is made to stop forest fires, they will often sweep through a forest. If the undergrowth of grass and shrubs and accumulated litter of twigs and leaves is not very dense, the fires are "cool" and burn only the undergrowth and litter, doing little damage to older trees. The forest is kept sparse, and the trees grow rapidly because of both lessened competition from undergrowth and the return of nutrients to the soil from the burned vegetation. But if fires do not occur for many years, or are extinguished when they do, undergrowth becomes dense, and litter accumulates. When a fire finally does take hold, it is a "hot" fire because of the abundance of fuel, and the trees burn along with the undergrowth. Forest managers now realize this, and fires are often allowed to burn themselves out and are even intentionally set.

There are grasslands around the world that remain grasslands only because of fire. If invading shrubs and trees are not burned, these grasslands will be replaced by other vegetation. Grass recovers from burning by growth from underground parts, but the invading plants cannot (Figure 1.14). The grasslands of Illinois have been almost entirely eliminated by agriculture, and virtually the only areas of native grassland that remain are along railroad right-of-ways. Frequent fires are caused by passing trains that create sparks that ignite vegetation adjacent to the tracks, maintaining these habitats. In recent years, a number of prairie areas have been reestablished by private and public concerns; these are intentionally burned regularly.

Some vegetation depends on fires for survival. Lodgepole and knobcone pine cones may remain on trees for as long as 20 years without opening to release their seeds. A fire causes mature cones to open. Shrub communities are also often fire dependent, not only to reduce competition from other plants, but to release seeds from

a

b

c

Figure 1.14 Maintenance of Grassland Habitat by Fire. Mesquite and other shrubs are a minor component of grassland, being contained by periodic fires. (a) In the absence of fire, shrubs gradually increase and replace some grasses. (b,c) Fire consumes both shrubs and grasses; however, the roots of the grass protect its growing buds, whereas the fully exposed buds of the shrubs are destroyed. (d) Grasses are capable of producing large crops of seeds much more rapidly than the shrubs. Shrubs, therefore, lose several years' growth following fire and are vulnerable again when fire recurs.

d

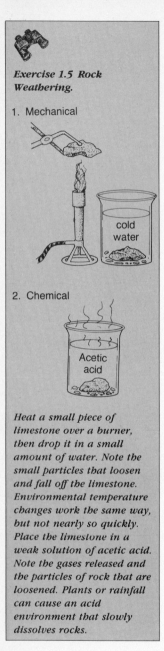

Exercise 1.5 Rock Weathering.

1. Mechanical

cold water

2. Chemical

Acetic acid

Heat a small piece of limestone over a burner, then drop it in a small amount of water. Note the small particles that loosen and fall off the limestone. Environmental temperature changes work the same way, but not nearly so quickly. Place the limestone in a weak solution of acetic acid. Note the gases released and the particles of rock that are loosened. Plants or rainfall can cause an acid environment that slowly dissolves rocks.

their coverings; some of these shrub species are readily flammable.

Animals may, of course, be burned out of their homes when their habitat is invaded by fire, but shortly afterward other animals are provided a new habitat. A grassland being invaded by shrubs is turned into a more dense and productive grazing land by fire. Kirtland's warbler, which needs bushy vegetation and short trees for nest sites, depends on fire to keep out large trees.

SOIL

Soil is both a physical and a biological factor. Soil is the result of the interaction of abiotic factors, such as geologic activity and climate, and biotic factors, such as vegetation type and rate of growth. Soil is the repository from which nutrients are drawn into the biological system, the site of water storage, an area of intense microbial action, and an anchoring place for plant roots. Soil is the intermediary between biotic and abiotic and organic and inorganic realms.

Soil Formation

Soil forms on geologic substrate such as sand, lava, or sedimentary rock. The action of wind, water, and temperature "weathers" the rock and slowly breaks it into smaller pieces. Wind blows smaller particles over larger ones, abrading their surfaces in the same way sandpaper abrades wood. Water movement erodes rocks, and when it freezes in crevices it expands and cracks the rocks. Water also dissolves some minerals from the rock, making them available to plants. Some of this decomposed rock is washed or blown away; the remainder becomes incorporated into the soil (Exercise 1.5).

Biological action occurs simultaneously. Hardy plants, such as lichens, begin to grow on the rocks. Their physical and chemical actions further erode the rock's surface layers. Other stalwart plants root in the weathered rock, and the root growth helps to fragment the rock. The plants also capture some of the minerals released from the weathering rock. The plants are then utilized by animals for food or shelter. When the plants and animals die or produce waste, decomposition occurs via the action of small organisms, such as worms and millipedes, and microorganisms like bacteria and fungi. Organic compounds are broken down into simpler ones and eventually into inorganic compounds, which again become available to plants. As more plants and animals invade the area and become established, they ultimately add their waste products and bodies to the soil, and the layer of soil becomes rapidly thicker. Worms and other soil animals mix the partly decomposed organic matter and the inorganic material with the weathered soil. The result is a continuum of soil components from the ground surface to the solid rock below.

Soil Layers and Soil Types

The layers of soil can be subdivided into litter, humus, leached humus, mixed organic and mineral matter, and bedrock, although the distinction between layers is not always clear. The weathering of rock is a physical process, but soil formation is mostly a biotic process, which combines organic and inorganic matter into characteristic horizons (Figure 1.15). Types of soils are classified by the distribution, presence or absence, and chemical composition of these layers. Hundreds of soil types are recognized, and the thickness of the **humus** layer of each is an important differentiating characteristic. Humus is partially decomposed organic material—the rich black earth that is generally considered indicative of soil fertility. A soil type may have less than 1% or more than 25% humus. The accumulation of humus, and soil in general, depends primarily upon the productivity of plants and the rate of decomposition. Plant growth in the hot deserts and cold tundras is slow, so organic matter is produced slowly. The dry climate and extremes of temperatures in both habitats retard decomposition, so organic matter is only slowly added to the soil. As a result, deserts and tundras have very thin soils. Coniferous forests grow rather slowly, and their waxy needles are resistant to decomposition, so soil forms slowly in these forests as well; **litter** (the top layer of nondecomposed matter newly added to the soil) is generally thick, however. Tall grass prairies, in contrast, are warm and receive abundant rainfall, so the plants grow rapidly. A good deal of organic matter, then, falls to the ground each year with the death of these grasses. The warm and moist conditions are also conducive to decomposition, so humus is quickly formed. Decomposition of organic matter does not take place as rapidly as does growth of plant matter, though, so litter accumulates and soil builds up before it can be reincorporated into living matter. In a tropical forest, on the other hand, decomposition occurs so rapidly that the decomposed matter is quickly reincorporated into the growing vegetation, and the soil tends to be thin.

Aside from the thickness of various layers and the chemical composition of the soil, its texture is also significant. Loose, coarse-grained soils allow water to percolate so rapidly that the soil is usually dry and leached of minerals. Compacted, fine-grained soils, conversely, hold water tightly and inhibit percolation; but this also results in a dry soil, and often invites erosion due to runoff. The best soils are those that provide penetration by water but have soil particles of a size that cohere to the water, slowing its downward prog-

Figure 1.15 (a)Typical Soil
Profile. (b)Soil Horizons. Soil
horizons are seen here in a
railway cut in Iowa. The top
layer is wind-deposited soil.
Lower layers were deposited
by the movement of glaciers.
The entire cut is about 12
meters high.

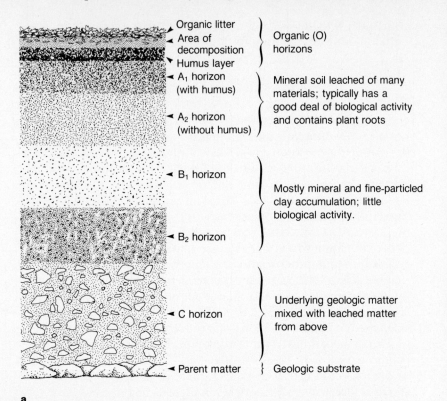

Organic litter

Area of
decomposition

Humus layer

} Organic (O)
  horizons

A₁ horizon
(with humus)

A₂ horizon
(without humus)

} Mineral soil leached of many
  materials; typically has a
  good deal of biological activity
  and contains plant roots

B₁ horizon

B₂ horizon

} Mostly mineral and fine-particled
  clay accumulation; little
  biological activity.

C horizon

} Underlying geologic matter
  mixed with leached matter
  from above

Parent matter

} Geologic substrate

a

b

ress and allowing plant roots to absorb the water and the minerals dissolved in it (Exercise 1.6).

Rate of Soil Production

Under ideal conditions soil formation may occur in only a few years. Composting household garbage and grass clippings can produce humus in a year or two. Deciduous tree leaves decompose rapidly under the proper conditions. Under other conditions, however, thousands of years may be required for an area to develop a stable soil. Soil may be deposited or eroded by water and wind; it may be leached of its nutrients by water running through or over it; and the vegetation growing on it may assist or impede its progress by physical or chemical means. The thickness of the soil and its fertility are affected by a wide variety of conditions.

Soil is thus produced and affected by physical and chemical processes and by the vegetation growing in it. At the same time, the soil, in conjunction with the climate, determines the kind and structure of vegetation that will grow in that area. Vegetation and soils are inseparable, as are most biological and physical processes of nature.

There are an infinite number of other phenomena in the biological world that are part of the environment of organisms. Those discussed in this chapter are only the major ones. They set the stage; the next chapter will discuss the players.

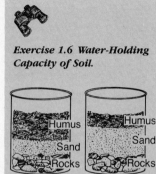

Exercise 1.6 Water-Holding Capacity of Soil.

Punch several holes in the bottom of two or more large coffee cans. Fill the cans with different types of soil horizons. Examples are shown here. Pour the same amount of water into each can. Which begins to drain first? Which holds more water? What soil types are these representative of, and what are the ecological implications of these soil types?

SUMMARY

Ecology is the study of the relationships between living organisms and the environment. The environment is composed of biotic and abiotic factors. The complex of these factors is an ecosystem. Both individuals and ecosystems have been molded by the processes of evolution.

Many environmental factors affect the lifestyles of organisms: Light, temperature, water, gases, nutrients, currents and pressures, fire, and soil are the major ones. Organisms acclimate to these factors during their lifetimes; they also have adaptations to these factors that evolved over many generations. Organisms are not adapted or acclimated to all conditions, however, so their distribution is limited to the habitats to which they are best adapted.

STUDY QUESTIONS

1. Define the following terms: acclimation, adaptation, ecology, environment, abiotic, humus, tropism, auxin, nutrient, stomate, ecosystem.

2. List five different ecosystems, ranging in size from very small to very large.

3. Explain the law of the minimum.

4. Compare acclimation with adaptation. Can acclimation be an adaptation?

5. Explain the importance of UV, infrared, and visible light to plants.

6. Why is the photoperiod used by so many plants and animals to begin or end seasonal activities?

7. How does temperature interact with or affect other environmental factors such as rainfall, soil, and wind?

8. Discuss the relative advantages and disadvantages of homeothermy.

9. Diagram the hydrologic cycle.

10. Discuss soil formation and how it is affected by biotic and abiotic factors.

Chapter 2

The Biological Environment and Energy Flow

Metabolism, growth, reproduction, locomotion, and all other physical and physiological processes require energy. The energy for green plants comes from the sun, and the energy for animals comes from eating plants or other animals, and thus indirectly from the sun. Since energy drives the biological world, it is important to look at the source of energy, its distribution, and its ultimate dissipation. Matter flow and energy flow are closely associated, but their modes of flow differ and we will consider them separately.

Living organisms exist in the biosphere, a thin shell around the earth about 1.5 kilometers thick. Within the biosphere there are numerous ecosystems, each of which contains abiotic components plus a **community**—all the living organisms of that ecosystem (or of a particular area or habitat within the ecosystem). Although ecosystems contain different kinds of organisms, they all are composed of the same basic functional units of plants and animals; that is, the roles filled in each ecosystem are similar. There are **producers**, which trap solar energy; **consumers**, which eat other organisms; and **decomposers**, which consume and degrade dead plants and animals and waste products. The roles in a community can be examined by tracing energy or materials; we will trace both, beginning with energy.

THE PRODUCERS

The producers are the green plants, which trap solar energy via photosynthesis (Figure 2.1). A simplified chemical equation for photosynthesis is:

$$6CO_2 + 12H_2O \xrightarrow[\text{Chlorophyll}]{\text{Visible Light}}$$

(From   (From soil
air)   or air)

$$C_6H_{12}O_6 + 6O_2 + 6H_2O$$
(Carbohydrates)   (Oxygen)   (Released via
transpiration)

Figure 2.1 Representative Producers. This cypress swamp of South Carolina contains a wide variety of green plants: dwarf palmetto, tupelo gum, bald cypress, algae, mosses, and lichens.

Photosynthesis is the method by which energy enters the biological system. Green plants convert radiant energy into chemical energy that is trapped in chemical bonds and made available when the bonds are broken. The plants capture but do not actually "produce" energy; rather, they produce organic compounds that are passed on to the consumers when they ingest the plants (or ingest other consumers that have eaten plants). Chlorophyll is a green plant pigment essential to photosynthesis, although other pigments may be involved (Exercise 2.1). Plants lacking photosynthetic pigments (such as the fungi) cannot photosynthesize and must derive their nutrients from other plants or animals. Green plants must, of course, utilize some of the food they produce for their own energy needs, such as normal metabolism, growth, and reproduction. Thus, an adjective commonly applied to green plants is **autotrophic** (self-feeding); they are also often called **autotrophs** (self-feeders).

The carbohydrate produced by photosynthesis is glucose, which can be combined to form starch or other foods that may be utilized as necessary. Glucose can be converted, or incorporated, along with minerals and other nutrients, into proteins, nucleic acids, pigments, hormones, vitamins, and the like. A plant obtains energy for its needs primarily from the stored carbohydrates via the process of **cellular respiration**, which is in some respects the reverse of the photosynthetic process:

$$C_6H_{12}O_6 + 6O_2 \rightarrow 6CO_2 + 6H_2O + \text{energy}$$

Exercise 2.1 Simple Pigment Chromatography.

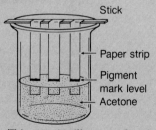

Stick

Paper strip

Pigment mark level

Acetone

This exercise illustrates how to extract various pigments from plant leaves. It can be used during discussions of photosynthesis, energy flow, and so on.

Materials: Beakers (500 ml) or the bottom halves of half-gallon milk cartons, pencils or sticks, transparent tape, filter paper cut in strips, 1 liter acetone, marking pens, centimeter rulers, green leaves, coin.

Procedure: Fill containers to the 1-cm level with acetone (don't stick plastic rulers in acetone). Transfer the plant pigment to a strip of filter paper (1.5 × 10 cm) by placing a leaf on the filter paper 1.5 cm from the bottom of the strip and rolling the edge of a coin over the leaf just hard enough to crush it. Place a

fresh part of the leaf under the coin and repeat the procedure in the same spot three times. Let the pigment dry a few minutes. Tape the top edge of the strip to a pencil or stick—you can tape three strips to one stick. Lower the paper into the acetone, making sure that the pigment mark stays above the liquid. The acetone will move up the paper, carrying various pigments different distances. Tape an identifiable sample of the plant to the top of the paper strip so the pigment print and sample remain together.

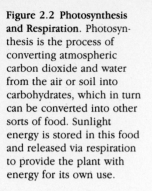

Figure 2.2 Photosynthesis and Respiration. Photosynthesis is the process of converting atmospheric carbon dioxide and water from the air or soil into carbohydrates, which in turn can be converted into other sorts of food. Sunlight energy is stored in this food and released via respiration to provide the plant with energy for its own use.

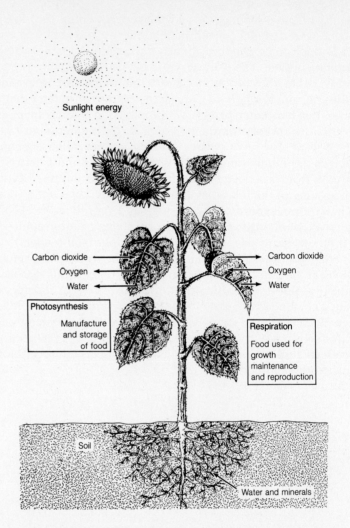

As energy is used by the plant, the raw materials of photosynthesis—carbon dioxide and water—are released. Plant respiration occurs whenever the plant is active. The net result of the opposite actions of photosynthesis and respiration is, of course, more photosynthetic than respiratory activity, or else plants would be unable to grow and reproduce (Figure 2.2).

Photosynthesis is responsible for the oxygen in the earth's atmosphere; perhaps up to 70% of atmospheric oxygen is produced by marine algae and tropical forests. Green plants are also responsible for removing up to 50% of the carbon dioxide produced by the burning of fossil fuels, and thus helping to cleanse the atmosphere of the by-products of human activity.

LAWS OF THERMODYNAMICS

To understand what happens to energy as it passes between organisms, we have to understand two physical laws: The **First** and **Second Laws of Thermodynamics**. The first law, stated simply, says that whenever energy is changed from one form (such as chemical, mechanical, heat) to another, no energy is lost or destroyed. The second law says that whenever energy changes form, some energy is converted to heat; that is, energy transformations are never 100% efficient. If we make an analogy to the automobile, we find that a car converts chemical energy in the form of gasoline into mechanical energy in order to turn the wheels. Heat is generated and must be removed by a radiator or other cooling system. Perhaps only 5% of the chemical energy is converted into mechanical energy which moves the car.

The implication of these laws of thermodynamics to the biological system is enormous because it means that energy is constantly lost from the system. As plants convert their stored chemical energy (respire) into other forms of chemical and mechanical energy, some is converted to heat. Since heat cannot be used as an energy source in the biological world, it is lost to the atmosphere. Thus, there is a continual and considerable flow of energy through the plant over its lifetime. Only a small part of that energy is contained in the living plant at any one time.

THE MEASUREMENT OF ENERGY

Energy, the ability to do work, is usually measured in calories in biological systems. A calorie (cal) is the amount of heat required to raise 1 milliliter of water 1° C. The caloric value of food often found in dieter's books (one apple = 110 calories, for example) is actually expressed in kilocalories or Kcal; (1 Kcal = 1000 cal). The caloric value of food or other organic matter can be determined by burning a small amount of the substance in an atmosphere of pure oxygen and determining the amount of heat emitted. Since the bulk of an organism or any of its component parts or products is made of carbohydrates, proteins, and fats, the energy contained in these compounds is what is measured (Exercise 2.2). Both proteins and carbohydrates contain approximately 4.2 Kcal per gram, while fat contains about 9 Kcal per gram. Fat is thus a more concentrated food source than carbohydrates and proteins, and is sought out by birds and mammals prior to migration or hibernation, and avoided by humans intent on losing weight.

PRODUCTION AND PRODUCTIVITY

Energy fixed (captured) by plants is termed **primary production**. The total amount of energy fixed by plants is **gross primary production (GPP)**; when the energy used in respiration (R) is subtracted, the result is **net primary production (NPP)**. Thus, GPP − R = NPP. NPP is the amount of plant material that can potentially be taken by a consumer, whether harvested by people or eaten by animals. When we harvest a crop such as corn, the amount of energy available to us, compared to the amount of energy required to produce that corn, is relatively small. Much of the corn plant is also inedible, so that energy goes unused, but it is still part of the NPP.

The rate at which energy is fixed is termed **primary productivity**. Productivity can be measured in units of either energy or weight (mass). In ecological studies, kilocalories per square meter per year (Kcal/m²/year) or grams per square meter per year (g/m²/year) are

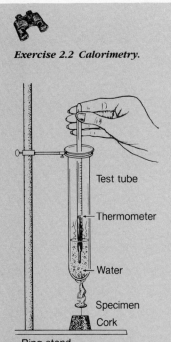

Exercise 2.2 Calorimetry.

Test tube

Thermometer

Water

Specimen

Cork

Ring stand

If an oxygen bomb calorimeter is available, its use can be demonstrated. If one is not available, here is a simple way to demonstrate the energy content of organic materials, although it is not nearly as accurate as a calorimetric instrument.
Materials: Ring stand
Balance
Test tube holder
Cork
Aluminum foil
Pin
Food samples such as peanuts, walnuts, cereals, crackers, sugar cubes, and so on.
Procedure: Wrap a cork with aluminum foil. Push a pin through the side of it at an angle, so the pin protrudes above the top by at least 2 centimeters. Weigh the specimen to be tested (peanut, wood, whatever) as accurately as possible and then place it on the pin. Fill a test tube with 20 milliliters of water and place it in the test tube clamp on the stand. With a match, set fire to the organic specimen. As soon as it ignites, place the burning specimen under the test tube. Place the thermometer in the top portion of the water (do not allow thermometer to rest on the test tube) and record the temperature. Continue to record the temperature until it stops rising or the specimen is completely burned. Determine the energy in the specimen by completing the following:
1. calories = (milliliters of H_2O) × (change in temperature in °C)
2. $\dfrac{number\ of\ cal}{number\ of\ g} = \dfrac{cal}{g}$
3. $\dfrac{Kilocal}{g} = \dfrac{cal}{1000\ g}$
Hints: Do not use too large of an organic specimen as the water will boil off. Try to reduce the heat loss to the environment by shielding the flame from wind and keeping the burning specimen as close to the test tube as possible.

commonly used. The NPP of different ecosystems varies enormously because of the climate, soil, plant composition, and so on. The NPP of one system may also vary considerably from year to year or season to season. The most productive ecosystems tend to be those with mild, stable climates, abundant sunlight, and a good nutrient and water supply. Coral reefs, estuaries, and tropical rain forests are such ecosystems. Conversely, habitats such as the open ocean, tundra, and desert, with harsh climates or lack of water or nutrients, have low productivity (Figure 2.3). Obviously, the higher the productivity, the larger the potential food supply for animals and, consequently, the higher the number and/or variety of animals.

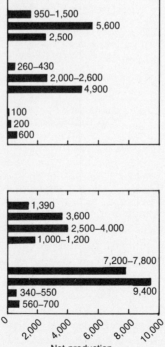

Figure 2.3 **Net Production of Natural and Agricultural Ecosystems**. A comparison of the net production levels of a number of natural and agricultural ecosystems. (The total net production of corn in the United States and of rice is calculated from grain yields.)

Figure 2.4 Use of Energy Trapped by Plants. The net primary productivity is distributed to other parts of the ecosystem through death and predation. For an agricultural crop, the bottommost arrow indicates what can actually be harvested. Perhaps only 1%–2% of the sunlight hitting a plant is used in the photosynthetic process. If the amount of sunlight assimilated contains 1000 Kcal, only about 100 Kcal will be incorporated as gross primary production and only 10 Kcal will become net primary production—a 99% loss from sunlight to consumers. Although there is a good deal of variability in these efficiencies, there is an average approximate loss of 90% of the energy between each of the levels: sun → plant (GPP) → plant (NPP) → consumers → consumer and so on.

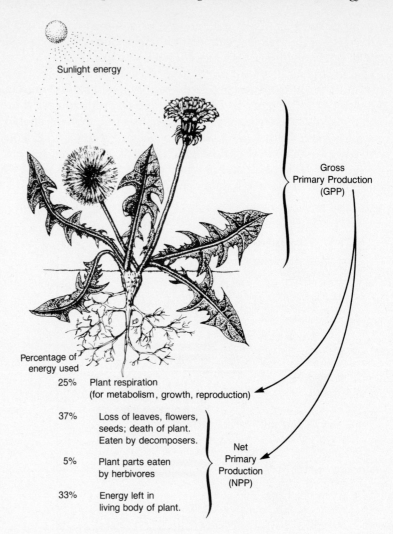

Sunlight energy

Gross Primary Production (GPP)

Percentage of energy used

25%	Plant respiration (for metabolism, growth, reproduction)
37%	Loss of leaves, flowers, seeds; death of plant. Eaten by decomposers.
5%	Plant parts eaten by herbivores
33%	Energy left in living body of plant.

Net Primary Production (NPP)

Fate of NPP

What happens to the NPP? As Figure 2.4 shows, some goes to the decomposers, some remains in the living plant, and the rest is eaten by herbivores. Not all that is consumed by herbivores, however, goes up the food chain to other consumers. Let's use a straightforward example in which grass is the producer and grasshoppers are the consumers (Figure 2.5). Grasshoppers cannot utilize all of the energy contained in grass for several reasons: (1) their digestive systems are not capable of assimilating the grass with total efficiency, and some is expelled as waste; (2) some leaves drop off the plant and are ignored by grasshoppers; (3) entire plants die before they

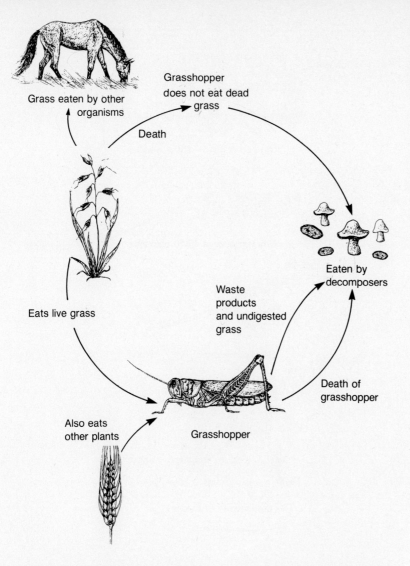

Figure 2.5 Use of Producer by Herbivore. There is really no such thing as a simple food chain in nature. To say that grasshoppers eat grass and pass the plant's energy to a higher food chain is a considerable oversimplification. First, not all the grass is eaten by the grasshopper, and then, not all of what is eaten is digested. Thus, much of the energy that is available to grasshoppers is in reality not used by them.

are eaten. Much of the energy contained in the grass, then, actually goes to the decomposers. Further, other organisms eat the grass and grasshoppers also eat plants other than the grass to obtain energy.

ECOLOGICAL EFFICIENCY

Because of respiration, unavailability, and incomplete assimilation, only about 10% of the NPP is passed on from the plants to the plant consumers, the grasshoppers. Similar processes are at work in the transfer of energy from the grasshopper to the grasshopper eaters (such as birds) and from the birds to the bird eaters (such as

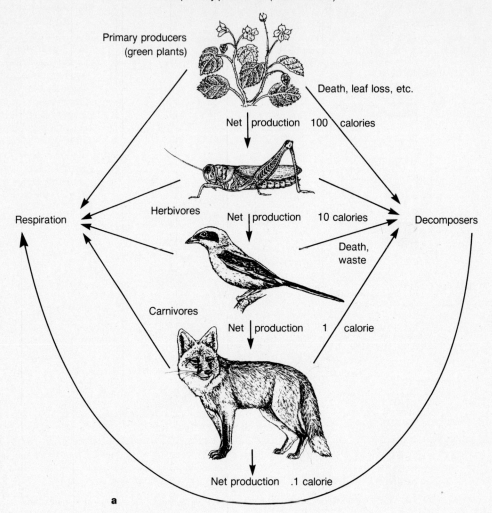

Gross primary production (1000 calories)

Primary producers (green plants)

Death, leaf loss, etc.

Net production   100   calories

Respiration

Herbivores   Net production   10 calories   Decomposers

Death, waste

Carnivores   Net production   1   calorie

Net production   .1 calorie

a

foxes). Respiration occurs in animals as it does in plants, and there is a continual loss of energy from the biological system so that the foxes receive only 0.1% of the energy fixed through photosynthesis (Figure 2.6a and b). This continued loss limits the number of consumers feeding sequentially on one another, since the amount of energy available in the highest consumer levels is so low that no organisms could exist on that energy source. Thus, nothing eats hawks or lions, leopards, killer whales, or wolves—not because they are big and mean, but because their level contains so little energy. Any organism attempting to exist by feeding on mountain lions would starve.

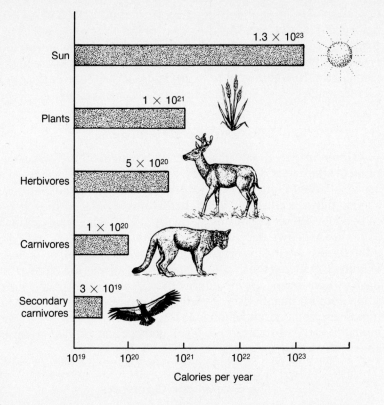

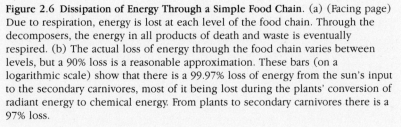

b

Figure 2.6 **Dissipation of Energy Through a Simple Food Chain**. (a) (Facing page) Due to respiration, energy is lost at each level of the food chain. Through the decomposers, the energy in all products of death and waste is eventually respired. (b) The actual loss of energy through the food chain varies between levels, but a 90% loss is a reasonable approximation. These bars (on a logarithmic scale) show that there is a 99.97% loss of energy from the sun's input to the secondary carnivores, most of it being lost during the plants' conversion of radiant energy to chemical energy. From plants to secondary carnivores there is a 97% loss.

To restate the causes for energy loss through the trophic levels we can say that: (1) energy is spent in "capturing" and ingesting food; (2) energy is required to break down the food in the digestive system; and (3) since not all of the food can be assimilated, some leaves the body as waste. Ecological efficiencies differ among feeding levels, among different organisms, and in different ecosystems. Poikilothermic animals are apparently more efficient at assimilating ingested food than homeothermic animals, but homeotherms seem to be more efficient in converting assimilated food into their tissues for growth. Producers in aquatic ecosystems are generally not harvested as efficiently by herbivores as are the producers in terrestrial ecosystems. For example, the range of ingestion efficiencies from plant to herbivore level in the open ocean is 5%–6%, while in a temperate zone marsh it may be 25%–30%. A good deal of research is needed yet to uncover the reasons for the variation in ecological efficiencies. In addition, at least 20 different definitions of ecological efficiencies exist, compounding the issue's complexity.

The Consumers: Herbivores

Heterotrophic organisms, or **heterotrophs**, are those that cannot photosynthesize and must rely on other organisms for food. These are also called consumers. The primary consumers are the plant eaters, more accurately called **herbivores**. Some of the NPP is utilized by these organisms, virtually all of which are animals; examples are cows, deer, antelope, rabbits, grasshoppers, and caterpillars—any organisms whose food is exclusively plant material (Figure 2.7). Being the first level of consumers, they are often called **primary consumers**. Of the energy they ingest (gross production), some goes for metabolism, growth, and reproduction. The remaining unrespired energy, which is incorporated into their bodies and available to the next consumer level, is about 10% of what was available to them from the producers.

The Consumers: Carnivores

The **carnivores**, or meat eaters, are the **secondary consumers** because they eat other animals rather than plants. They are two levels removed from the fixation of energy by plants, so they receive only 1% of the fixed energy. A third-level or **tertiary consumer** receives only 0.1%. Fourth-level consumers are very rare. There is so little energy left at the high trophic levels that a fourth- or fifth- or higher level consumer would use more energy searching for the rare (and thus low-in-energy) consumer at the next lowest level than it would get by eating that rare prey. High-level consumers live near the point of diminishing returns, and no consumer can exist by expending more energy than it takes in. Examples of typical carnivores are toads, bobcats, spiders, hawks, and snakes (Figure 2.8).

The Consumers: Omnivores

Omnivores eat both animals and plants, so they exist on an interface between consumer levels. Many animals generally considered to be carnivores also eat a good deal of plant material. Bears, coyotes, jays, raccoons, and people are examples; they may eat rodents, insects, berries, nuts, seeds, honey, and the like (Figure 2.9).

a   b

Figure 2.7 Representative Herbivores. The chiton (a) and grasshopper (b) are typical herbivores. The chiton is a mollusc that is highly adapted for living tightly attached to rocks and shells in surging water tide pools. The mouth is located on the ventral surface and a rasplike tongue scrapes algae from the rocks. Grasshoppers have chewing mouth parts and, in large numbers, may do severe damage to field crops.

Omnivory invokes an interesting problem. If eating lower down on the food chain provides more energy, then to maximize their energy resources animals should eat as much plant material as possible. But the digestive enzymes in an animal cannot effectively digest everything; certain materials pass through the digestive tract only slightly digested. For example, cows can eat and digest grass efficiently, but foxes cannot cope with the high cellulose content of grass because their digestive enzymes are inappropriate. Carnivores will eat vegetation when their normal food is scarce, however, and it may be digestible enough to prevent starvation.

Figure 2.8 Representative Carnivores. This toad feeds on insects, worms, spiders, slugs, and almost any other small animal that moves, even other toads. The salmon feed on small crustaceans and insects when young, and as they grow larger, feed on smaller fishes. Salmon rarely ingest vegetable matter. The grizzly bear will eat deer, fish, snakes, birds, squirrels, gophers, and the like. Although considered a carnivore, it will also eat grass, roots, berries, and other plant parts. The toad is actually much more a carnivore than the bear.

Figure 2.9 Representative Omnivores. Gulls are well known for their lack of discrimination among food items. They will eat fish, squid, insects, eggs, snakes, and all manner of live and dead foods. They are frequent inhabitants of garbage dumps, where they feed on scraps of food. Channel catfish eat worms, insect larvae, small fish, and most other small animals that live in or fall into the water, but they will also eat aquatic plants and decaying organic matter.

The Decomposers

The **decomposers** are actually a type of consumer. They eat the waste products of other organisms and the bodies of other organisms after death, breaking down the complex compounds into simpler ones that can be reused. Most of these creatures range in size from small to microscopic and are not very obvious, yet their role is exceedingly important in returning nutrients to the soil or water that would otherwise remain locked up in waste and dead organisms. The decomposers are a varied lot of creatures, and include leaf-eating millipedes, slugs, various types of insects, and other soil creatures such as isopods (Figure 2.10). The greatest number of decomposers are no doubt the bacteria and fungi, whose abilities to reproduce and chemically digest are well adapted for rapid decomposition, especially under ideal conditions of warmth and moisture and pH. Besides degrading large organic molecules into smaller inorganic compounds, decomposers help to produce soil. A home compost pile will produce soil most quickly if it is kept warm and moist and turned regularly to admit the oxygen the decomposers need for respiration.

a   b

Figure 2.10 Representative Decomposers. (a) Bacteria are found literally everywhere and are extremely important in breaking down dead matter so that it can be recycled. The concentration of bacteria in the topsoil or at the bottom of bodies of water is very high because that is where most dead matter ends up. (b) Mushrooms are only one of many kinds of fungi that absorb nutrients from dead matter, helping to recycle the matter.

TROPHIC LEVELS, FOOD CHAINS, AND FOOD WEBS

Each of the levels in the producer–consumer–decomposer arrangement is known as a **trophic** (feeding) level. The producers are the foundation of an ecosystem, since they trap the energy needed for all trophic levels. There are many examples of actual trophic level arrangements, such as algae–tadpole–fish–bird or plants–snail–bird–fox. These patterns are often called **food chains** because of their straight, linked arrangement (Figure 2.11).

Rarely are simple food chains found in nature; most often a **food web** exists. It is very unusual to find one kind of plant being the producer, one type of herbivore, and one kind of carnivore at at each level. Most often there are a number of producers and consumers; the consumers are most often omnivores and eat almost whatever is available. Their relationships are mixed and intertwined; an example is shown in Figure 2.12.

The evolution of food webs rather than simple chains helps to maintain the stability of a system. If only a food chain existed, then

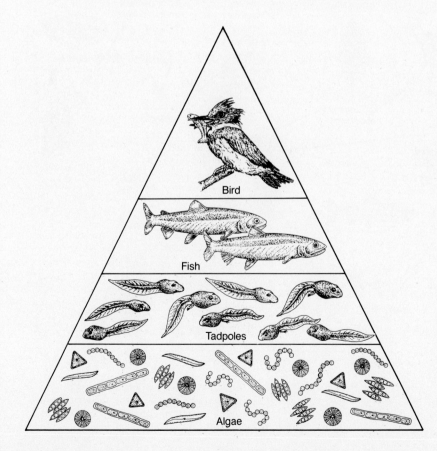

Figure 2.11 Food Chain. A simple food chain, rarely found in nature, but used here to exemplify the loss of energy, numbers of individuals, and biomass, as energy dissipates through the trophic levels.

the disruption of any one link could have a detrimental effect on the whole system. In the grasshopper example, only grasshoppers ate the sole producer, grass, and birds ate the only herbivore, grasshoppers. What would happen if the producers died off? If the grasshoppers went extinct? One of the problems with agricultural ecosystems is their simplicity. Clean, one-crop farming is more efficient and higher yielding than a mixed crop, but a drought, pests, heavy rainfall, winds, and other natural forces can destroy the producers in a very short time. The consumers subsequently succumb to lack of food and leave or starve. However, rarely do natural ecosystems collapse when inclement weather strikes. Some consumers may leave or die, but the fundamental operation of the system does

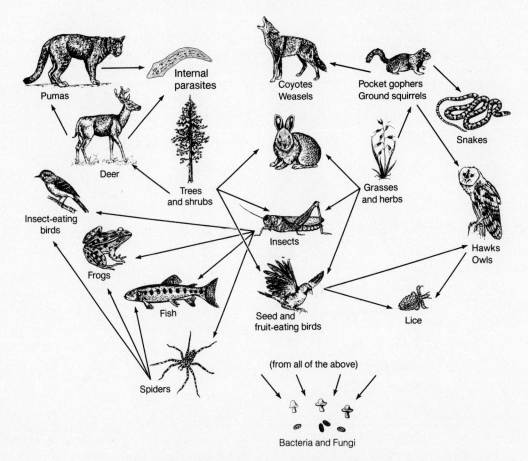

Figure 2.12 Food Web. Each trophic level is composed of a few to several species. Omnivores may be found in two or more trophic levels; the abundance of each species may change from season to season; and other factors change the relationships between species, making a food web a complex and dynamic entity.

not change. It takes avalanches, volcanoes, or severe floods to cause the demise of complex ecosystems. Essentially, the more kinds of organisms there are in a system, the less likely a change in the status of any one will have a major effect. Other factors are involved in an ecosystem besides food webs, of course, but these webs are the foundations of the structure of an ecosystem.

ECOLOGICAL PYRAMIDS

Food webs can be simply represented by three or four trophic levels; remember that each trophic level may contain one to many kinds of organisms and that many organisms may belong to more than one level. A progressive decrease in energy from the producer level upwards has other implications; this is often represented as a pyramid, as in Figure 2.13. Besides energy, there can also be a pyramid of productivity, a pyramid of numbers, and one of **biomass** (biological weight). It follows that if energy decreases up the pyramid, then so must productivity, numbers of individuals, and biomass. The pyramid of numbers can occasionally be inverted as when hordes of locusts destroy grasslands or crops or when bacteria feed on a live host or decaying organism. It is most often useful to think in terms of energy rather than numbers, weight, or productivity, however, and

Figure 2.13 Ecological Pyramids. P denotes producers and C, C_2, and C_3 represent, respectively, higher levels of consumers. These graphs give figures for decreases in energy, numbers, and biomass as one proceeds toward the highest level consumer. This decrease is often graphically represented in pyramidal shape. Although the numbers will vary and thus also the relative widths of the rectangles, the shape will remain pyramidal because of the constant decrease in energy, numbers, and biomass, no matter what type of ecosystem is considered. Decomposers are not included in these pyramids; they would, in fact, form their own pyramid.

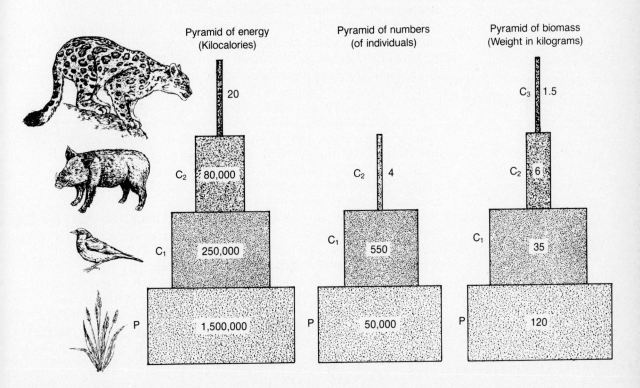

Pyramid of energy (Kilocalories)	Pyramid of numbers (of individuals)	Pyramid of biomass (Weight in kilograms)
20		C_3 1.5
C_2 80,000	C_2 4	C_2 6
C_1 250,000	C_1 550	C_1 35
P 1,500,000	P 50,000	P 120

we will often examine principles and processes in this light. The loss or gain of energy is very often a driving force in the evolution of an organism's anatomy, physiology, and behavior.

THE TROPHIC LEVEL APPLIED TO HUMANS

The principles of biological energy flow are relevant to the feeding of human populations: The lower the trophic level at which we eat, the more energy available to us. We have to consider, however, the relative digestibility of plant and animal tissue; animal tissue gives us more energy per unit weight, as it is more digestible. Vegetarians may suffer from vitamin deficiencies because sufficient vitamins, especially B_{12}, are difficult to obtain from eating plants alone. The best solution is to eat less meat, more high-protein vegetables, and other easily digested plants and their products. In underdeveloped countries, per capita meat consumption may average only 5 kilograms per year, while 100 kilograms per year is typical for more developed nations. Rice and other grains provide the major food source for many poor nations, while the more affluent ones use these grains to fatten beef cattle, which results in a loss to the food chain of 90% of the stored biological energy. The harvesting of fish, shrimp, and other sea creatures should perhaps be partially supplanted by the harvesting of marine algae and animals lower down on the marine trophic levels. We might also encourage eating more carp and catfish, which are lower level consumers than the more preferred tuna and salmon. Foods made from trash fish, petroleum by-products, and even newspapers are being experimented with as protein supplements. Whether they will be palatable enough to be acceptable as food is still questionable.

The industrialized countries of the world often invest many more calories of fossil fuel energy into producing or harvesting food than they obtain from the food. Twenty calories of energy are used to harvest 1 calorie of salmon, for example. What is occurring is simply an expensive trade of nonedible biological energy for edible energy. Not only is this economically costly, but it depletes dwindling energy supplies. Less energy-intensive and more productive means of agriculture and harvesting need to be implemented. Of course, the fundamental problem is the rapidly increasing human population. The lower on the trophic level we eat, the greater the number of humans that can be supported. But there is a limit to the rate at which energy can be fixed by biological systems, so there is a limit to the number of humans that the biosphere can support; with about half the world malnourished, we may be approaching this point.

SUMMARY

Nearly all the energy required for biological processes comes from the sun. It is trapped by green plants via photosynthesis and passed on through higher trophic levels, the majority of the energy being lost in the process. The rate at which energy is fixed is termed productivity, and the productivity of different ecosystems depends on many factors.

Organisms are classified as producers and consumers and organized into trophic levels with producers on the bottom. They can also be grouped into ecological pyramids by numbers, biomass, and energy.

STUDY QUESTIONS

1. Define community, producer, heterotroph, primary productivity, trophic level.

2. Explain the First and Second Laws of Thermodynamics as they pertain to energy flow through the ecosystem.

3. What environmental factors affect the productivity of ecosystems?

4. What are the most productive ecosystems? Why? How could they be made more productive?

5. Why are food webs a more accurate representation of nature than food chains?

6. How do trophic levels relate to food webs? Can organisms occupy more than one trophic level? Explain.

7. Give the equation for photosynthesis. What environmental factors affect photosynthesis?

8. How is biological energy measured and what units are used?

9. Explain the concept of ecological efficiency.

10. How can the earth's human population provide more food energy for itself by changing its diet?

Chapter 3

Biogeochemical Cycles

Energy, the intangible capacity to do work, is fixed from the sun; stored in the bodies of plants and animals; and used to perform physiological functions such as respiration, circulation, growth, and reproduction, and mechanical functions such as flying, burrowing, mating, and laying eggs. Energy drives these processes, but the processes themselves are performed via chemical reactions and/or physical movement, both of which require the tangibles we call matter. Energy is captured, stored, and transferred by matter, so matter and energy are nearly inseparable.

ENERGY FLOW AND THE FLOW OF MATTER

The Second Law of Thermodynamics, mentioned in Chapter 2, states that whenever energy is transferred from one form to another, some energy is lost as heat. This law also states that matter tends to move from a state of order to a state of disorder. The carcass of an animal begins to degrade to simpler chemicals shortly after death. Smoke from a fire disperses throughout the atmosphere. Pollutants dumped into a river rapidly disseminate. This phenomenon is known as **entropy**—the tendency for order to go spontaneously to disorder. Then why is there so much order in the world? Why are the bodies of organisms, as well as ecosystems, such highly ordered and finely-tuned entities? The answer is that individuals and ecosystems *are* tending toward entropy, but they are kept in an ordered state because of a constant supply of energy. Energy provides the means to maintain order by overcoming entropy. If energy is withdrawn from an organism (a green plant deprived of light or an animal of food), entropy will win out and the organism will die and decompose.

Energy flows through the ecosystem between trophic levels, dissipating as heat as it does so. Were it not for the continual energy input from the sun, all biological processes would soon cease. The sun, a giant fusion reactor, will ultimately burn out in 5 to 10 billion years, an essentially infinite amount of time. Matter, conversely, is not constantly supplied to the earth, but neither is it lost. Matter is recycled, the amount of material on earth remaining fixed. It may change, as grasslands burn or rock is weathered; it may move from place to place, as soil nutrients are washed away in runoff; or it may be stored for considerable periods of time, as in ocean mineral deposits; but none is ever lost or destroyed (the First Law of Thermodynamics again). There are two types of matter and energy flow systems, open and closed. In a **closed system**, such as the earth as a whole, energy is lost and gained but the amount of matter remains stable. Minor exceptions that occur are insignificant, since they are in the forms of incoming meteorites and outgoing satellites. In **open systems**, such as individual organisms or ecosystems, both energy and matter are lost and gained at varying rates.

If we consider matter in its fundamental form of atoms and molecules, it is essentially immortal; atoms and molecules combine to form complex chemicals and eventually degrade to simpler compounds. A molecule of oxygen (O_2) inspired from the atmosphere by an African antelope may be expired in a molecule of carbon dioxide (CO_2). The carbon dioxide may then be taken into an acacia tree via photosynthesis and incorporated as carbohydrates into the leaves of the tree. If the tree leaves are eaten by a giraffe, the digestive system of the giraffe will assimilate some of the leaves and pass some out as fibrous waste. In the process of decomposition, carbon dioxide is again released, either to be used by plants or to simply exist in the atmosphere for a long period until it becomes involved in some chemical change. The carbon atoms you may have ingested with a carbonated soft drink may once have helped to support the trunk of a giant redwood! The movement of matter is cyclic; it is used and reused but never lost (Figure 3.1). Energy flow tends to be linear, going from producers to consumers to decomposers, with some being lost as heat in the process of respiration.

Figure 3.1 Flow of Matter and Energy Through Ecosystem. Solid arrows indicate the flow of energy from the sun through the trophic levels. Curved arrows represent the loss of energy via respiration. Note that energy flow is not cyclic; it flows in only one direction. Dashed arrows indicate the flow of matter. Matter is never lost, as it is continually recycled. Note that it moves between the biotic and abiotic systems.

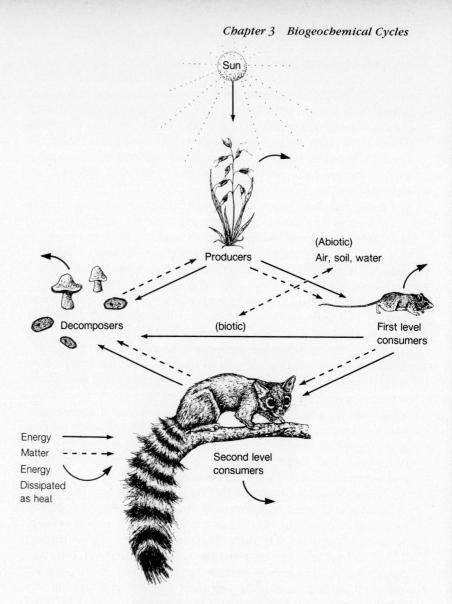

MOVEMENT OF MATTER

Geographic areas and ecosystems gain and lose matter. The oceans have become saltier over hundreds of millions of years because erosion of the land by water has washed minerals downstream to the ocean. Wind erosion has blown topsoil away from areas in the Middle East and turned them into deserts. Conversely, estuaries and marshlands receive organic matter in great abundance and are slowly becoming terrestrial environments. Many kinds of human activity also cause the movement of large amounts of material into or out of an ecosystem: logging, landfilling, building construction, dam erection, and agriculture.

In general, most mature ecosystems are stable; the amount of matter lost from the system more or less equals that gained. The movement of materials within a system (recycling), however, may differ greatly between systems. In a temperate forest, perhaps 25% of the organic matter present is tied up in the woody trunks of trees and is not recycled until the tree dies; in a tropical forest, 75% of the organic matter may be incorporated in wood. Conversely, 50% of the organic matter of a temperate forest may be present as soil, versus only 10% as soil in tropical forests. In a grassland, on the other hand, virtually all of the above-ground vegetation dies each year, and although some of it is recycled into next year's growth, much becomes soil. Soil thus builds up at a faster rate in tall-grass grasslands than it does in many other habitats; drier short-grass grasslands build up soil more slowly.

Whatever the habitat, the dead plant and animal matter that falls to the ground is ultimately decomposed and reused. Litter decomposition rates vary with the vegetation and environmental conditions. In a tropical rain forest, for example, decomposition may require only a year or 2; in a temperate forest, 4 to 6 years; in a cold coniferous forest, 15 years; and in tundra, up to 50 years or more.

ENTRANCE OF MATERIALS INTO BIOLOGICAL SYSTEMS

Materials enter a biological system from two major routes: the abiotic and biotic components of an ecosystem. Air, soil, water, rainfall, and bodies of water all contain chemicals that can be assimilated by organisms. The erosion of geologic material by water and the movement of particles by the wind make "raw" materials available (Figure 3.2). The other source of materials is previously existing biological matter that is decomposed into a once again usable form. The entrance or reentrance of matter into a biological system occurs primarily through the action of plants via photosynthesis and absorption through roots. Although animals do obtain minerals via the water they drink, their need to drink is based on the need for water, not on the need for dissolved nutrients; virtually all of their nutrients are derived by eating other organisms. Matter exits a biological system through death, wind action, runoff, emigration of animals, erosion, and so forth.

BIOGEOCHEMICAL CYCLES

The movement of every element or compound can be described by one of three kinds of biogeochemical cycles: **gaseous**, **sedimentary**, or **hydrologic**. **Biogeochemical** refers to the cycling of chemical

Figure 3.2 Erosion Erosion by water has caused gullies on a steep hillside previously denuded of vegetation and topsoil. Physical and chemical activity release the minerals in the soil. Soil and minerals are washed downward by rainfall and snowmelt and find their way into soil or river systems. They may be deposited at the outlet of the river or somewhere along the way. Soil deposited by a river is called **alluvium**.

elements and compounds through biological systems and the soil, rocks, and sediments of the earth. The gaseous cycles include elements and compounds whose most common form is gas; examples of these compounds include oxygen (O_2), nitrogen (N_2), and carbon dioxide (CO_2). The sedimentary cycles include elements and compounds whose most common form is solid or dissolved in water; examples are phosphorous, calcium, magnesium, zinc, and iron. The hydrologic cycle describes the movement of water. The water cycle is closely associated with the sedimentary cycle since many nutrients are dissolved in water (Exercise 3.1).

Gaseous Cycles

Figure 3.3 illustrates a typical gaseous cycle, that of carbon and carbon dioxide. Carbon is a basic constituent of all organic compounds and is closely associated with the storage and release of

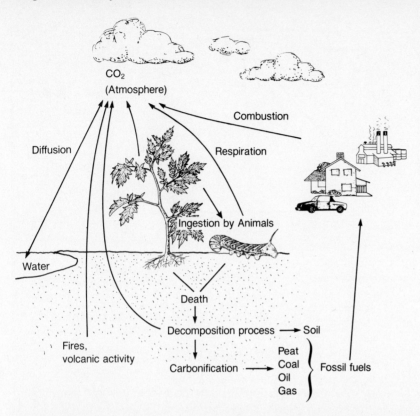

CO₂
(Atmosphere)

Diffusion

Combustion

Respiration

Water

Ingestion by Animals

Death

Fires,
volcanic activity

Decomposition process → Soil

Carbonification → Peat
Coal
Oil
Gas
} Fossil fuels

Figure 3.3 Carbon Dioxide/ Oxygen Cycle. The ultimate source of all carbon is carbon dioxide (CO_2) in the atmosphere. If the arrow directions were reversed, the diagram would represent the movement of oxygen, since plants take in CO_2 and release O_2, animals take in O_2 and release CO_2, and the processes of decomposition and combustion use O_2 and produce CO_2.

Exercise 3.1. (a) Measurement of Nutrients. *The level of nutrients in water or soil may be measured by various methods. The Hach Chemical Co., Ames, Iowa, manufactures kits that can be used to measure oxygen, nitrogen, phosphorous, and a number of other nutrients and dissolved gases. Or you may wish to devise your own tests; see M.A. Strobbe,* Environmental Science Laboratory Manual *(St. Louis: Mosby, 1972) or A.H. Benton and W.E. Werner, Jr.,* Manual of Field Biology and Ecology *(Minneapolis: Burgess, 1972). To demonstrate that organic*

matter contains minerals, a simple experiment can be performed as follows: Materials: newspaper or paper towel, filter paper and filter, beakers, phenolphthalein. Procedure: Burn newspaper, paper towel, or any other organic matter that leaves ashes. Put some of the ashes in water, stir, and filter. Add two drops of phenolphthalein to the filtered water. Phenolphthalein is colorless in acid and pink in basic solution. It will turn pink in the filtered water because the ashes are composed mainly of oxides of metals, which remained behind when the organic matter burned off.

Figure 3.4 The Nitrogen Cycle. Nitrogen gas cannot be directly used by plants, so it must first be converted by **nitrifying** bacteria into ammonia and then nitrates, which can be used by plants to make protein. This process is termed **nitrogen fixation. Denitrifying** bacteria break down the proteins of dead plants and animals, returning free nitrogen to the atmosphere. The terrestrial cycle is shown here, but a similar one occurs in aquatic environments, which contain nitrogen-fixing algae.

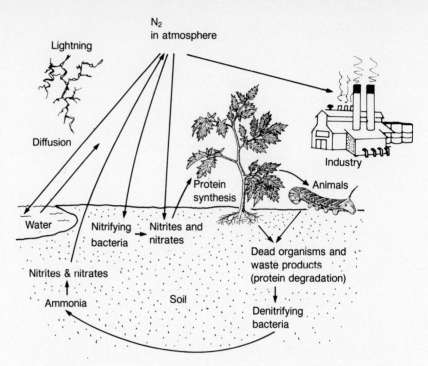

energy. The flow of carbon through the system to some extent parallels the flow of energy, since carbohydrates are produced with the fixation of energy, and CO_2 is produced when energy is released. The ultimate source of carbon is carbon dioxide in the atmosphere; CO_2 is incorporated into the biotic system through photosynthesis. Another typical gaseous cycle is that of oxygen, a product of the photosynthetic reaction. Its cycle is essentially a mirror image of the carbon dioxide cycle. Oxygen is used during combustion and decomposition; carbon dioxide is produced.

Nitrogen Cycle

The nitrogen cycle is also extremely important (Figure 3.4). Plants and animals require nitrogen (N_2) to form amino acids, which in turn are used to form proteins. Although 78% of the air is composed of N_2, it cannot be used by most plants and animals in that form. Instead, it must be converted (fixed) into other nitrogen compounds that can be utilized. **Nitrogen-fixing** bacteria found in soil convert N_2 into nitrites (NO_2^-) and then to nitrates (NO_3^-), which can be absorbed from the soil by plants (Figure 3.5). Some plants have nodules on their roots that harbor nitrogen-fixing bacteria. These bacteria incorporate much of the nitrogen into amino acids, which are released into plant tissues. Plants with these nodules are

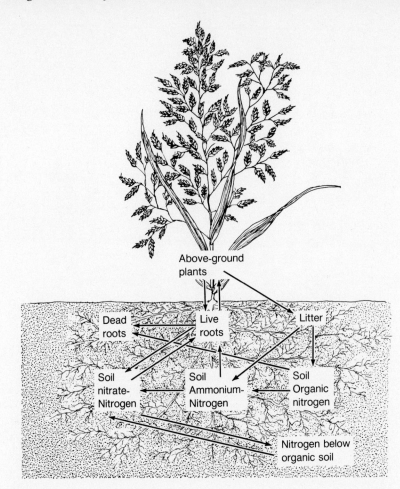

Above-ground
plants

Dead roots

Live roots

Litter

Soil nitrate-Nitrogen

Soil Ammonium-Nitrogen

Soil Organic nitrogen

Nitrogen below organic soil

Figure 3.5 A Nitrogen Cycle Model. This model shows a specific nitrogen cycle, that of nitrogen flow through components of a grassland ecosystem. The rates of flow from one component to another are affected by soil temperature, soil moisture, daily growth, death, decomposition, and nitrogen content. Nitrogen enters the system via rainfall, fixation by free-living organisms, and fixation by symbiotic organisms. Losses from the system are due to leaching, decomposition by bacteria, and movement in a gaseous form back to the atmosphere. Models like this one attempt to simulate the dynamics of actual ecosystems. There are instances where the application of nitrogen fertilizers has both increased and decreased the biomass of grass roots. If enough data are collected from actual ecosystems to provide an accurate model, then predictions can be made as to how the ecosystem will act when the environment changes or is changed. We may be able to reduce the use of fertilizers, increase the productivity of agricultural ecosystems, and reduce the effect of human disturbance by predicting the flow of nutrients.

thus able to grow in nitrogen-poor soils; examples are peas, clover, beans, alfalfa, redbud trees, and others in the pea (legume) family (Figure 3.6). Sometimes these plants are used to replenish nitrogen-poor soil via crop rotation. Some present research in genetic engineering is oriented toward giving other crop plants (corn, wheat) the ability to fix nitrogen, an ability that would greatly decrease the need for agricultural fertilizers. In aquatic ecosystems, blue-green algae serve as nitrogen fixers. Nitrogen can also be fixed in the atmosphere by the input of electrical energy, that is, lightning.

Denitrifying bacteria break down the proteins in the bodies of dead organisms or their waste products into ammonium, then nitrites or nitrates, and finally into nitrogen gas, which is released into the atmosphere from which it can be fixed and recycled again.

Although natural fixation produces sufficient nitrogen for natural processes, industrial fixation of nitrogen for use in fertilizer and other products exceeds that of the world's terrestrial ecosystems.

Figure 3.6 Peanut: A Legume. The peanut is a legume used for human food, for forage for farm animals, and to enrich the soil with nitrogen. Peanuts can be clearly seen growing from the branches. Nodules, which contain nitrogen-fixing bacteria, can be seen on the roots. Peanut oil is used as cooking oil, in soaps, in lubricants, and even in some plastics.

Sedimentary Cycles

Sedimentary cycles are often called "imperfect" cycles because minerals can be trapped for long periods of time (Figure 3.7). If we take calcium as an example, we may find that calcium-bearing rocks in a terrestrial environment are slowly broken down by leaching and erosion, releasing calcium into the soil. The calcium may be picked up by plants or run off into a watershed. The calcium (as a part of various compounds) may stay in the biological ecosystem for long periods of time; or it may move, via water, wind, or animals, to another ecosystem. It is continually being lost, although some is also usually entering the system so that the total amount of calcium remains fairly stable. Calcium also works its way downstream to rivers, estuaries, and ultimately the ocean. Along the way it may enter terrestrial, marshy, or aquatic ecosystems, but some of it eventually reaches the seas. The seas become "salty" because of this slow

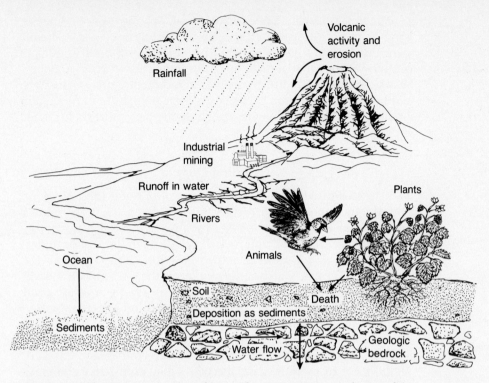

Figure 3.7 Sedimentary Cycle. Minerals enter the ecosystem via weathering of the underlying rocks, then entering the soil, and then perhaps entering plants and animals. Death and decomposition return minerals to the soil. Some may exit the ecosystem via runoff, and some enter with precipitation. Minerals flow through the soil and remain there for varying lengths of time. They may form sediments on, in, or under parts of the parent bedrock. The lateral movement of water through or over the sediments brings in or removes minerals. Deserts are notoriously poor in nutrients because water rapidly percolates out of the reach of plant roots. Some parts of the world have "pygmy" forests, noted for their stunted vegetation. This is due to an impervious bedrock under the soil, which not only prevents penetration of plant roots, but causes the nutrient-containing water to run off laterally, depriving the plants of water and nutrients. In a stable, mature ecosystem, input and outgo are nearly equal. In a young, growing system, there is a net accumulation of materials. Some examples of elements that undergo sedimentary cycling are calcium, phosphorus, iron, magnesium, potassium, and sulfur. Some, such as the sulfur in sulfur dioxide, may also be involved in a gaseous cycle.

but inexorable movement of calcium and other mineral salts from the land to the sea. When calcium reaches the ocean it may be utilized by marine animals for building shells (oysters, mussels, abalones), encasements (corals, marine worms), or skeletons (sponges, fish). The amount of calcium utilized may be enormous; the Great Barrier Reef of Australia, composed of enormous numbers

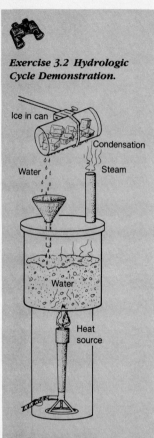

*A simple demonstration of
the hydrologic cycle can be
made with an inexpensive
apparatus such as that in
the sketch. Fill a can
halfway with water. Place a
candle or gas burner under
the can and heat to boiling.
Fill another can with ice
and heat over the first can.
Boiling water will vaporize,
condense on the upper can,
and drop back into a funnel
and then into the lower can,
simulating evaporation,
condensation, and
precipitation.*

of calcium carbonate skeletons of coral, is over 3000 kilometers
long. Some of these skeletons may be decomposed—starfish and
parrotfish will chip away at coral to reach the soft-bodied animals
within—or they may remain intact for thousands of years. Calcium
may be deposited on the ocean floor in sediments and remain there
until disturbed by ocean currents, geologic upheavals, or mining by
humans, and the cycle begins again. Similar patterns hold for other
minerals.

Hydrologic Cycle

The hydrologic cycle is the movement of water through biotic
and abiotic systems (see Figure 1.10). Water reaches the earth's
surface from precipitation as rain or snow. Some of this water percolates down through soil and rock to be stored in the soil or underground lakes called aquifers. Some water runs off the land and into
streams and rivers, which ultimately carry the water to the ocean.
Some water is picked up by plants, used in metabolic processes, and
returned to the air via transpiration. Transpiration plus evaporation
from the surface of the earth produce masses of water vapor called
clouds, which then release their water to begin the cycle again.

The movement and use of water in the hydrologic cycle is similar to that of gases in the gaseous cycle. Since excesses or shortages
of water tend to be local and short-term, the amount of water in an
ecosystem tends to remain almost the same from year to year. Besides being an essential chemical for physiological processes, water
also serves as a medium for the transport of minerals and a reservoir
for gases in aquatic ecosystems. Like the gaseous cycles, the hydrologic cycle is driven by the sun's energy, and immense amounts of
water are constantly being moved. (Exercise 3.2).

Water can be contaminated by natural processes, such as erosion
and runoff, or by human sources, such as sewage and industrial
wastes. The hydrologic cycle, distilling the water as it evaporates,
continually cleanses the water, and unless the contamination is severe or continual, many bodies of water can partially or totally recover because of the constant input of clean water. This process may
take a very long time, however.

Now, a new contamination problem has arisen—**acid rain**—and
there are increasing incidences of it. Sulfur dioxide gas is injected
into the air as a by-product of coal burning or as a result of a volcanic
eruption, and with precipitation, this falls as acid rain or snow. Acid
rain can kill fish in streams and lakes, damage vegetation, and inhibit the growth of soil bacteria. Acid snow, melting rapidly in the
springtime, can be devastating to the areas over which it runs off
because of the great accumulation of acid in the snow. Acid rain has
also been implicated in the rapid degradation of marble monuments

such as those in Venice, Italy, and of the Parthenon in Greece, and in the destruction of nylons on the legs of women walking in a sulfur dioxide polluted atmosphere (Exercise 3.3)

CYCLE IMPEDIMENTS

Under normal circumstances materials continue to move within and between biotic and abiotic systems, although there may occasionally be long-term storage of some matter. Manipulation of the environment by human activity has, however, altered the cycles of many materials so that excessive accumulation or depletion often occurs. The application of fertilizers, containing primarily nitrogen and phosphorous compounds, but often other substances as well, frequently leads to a buildup of some of these materials in the soil. Irrigation of farm crops may cause minerals to build up in the soil as the water evaporates or percolates downward. The retention of river water behind a dam prevents nutrients from flowing downstream; the fisheries industry at the mouth of the Nile River declined after the Aswan Dam was erected upstream, because the flow of nutrients needed to support the food chain was severely reduced. San Francisco Bay is becoming increasingly polluted as water is diverted upstream for irrigation, thus reducing the flushing effect. Further, many other bodies of water are accumulating materials due to increased erosion caused by land development or agriculture or by the disposal of industrial and residential wastes.

Some of these effects are obvious; others are more subtle and occasionally insidious. Sardine fishing off the west coast of South America has declined over the past few decades due to overfishing. The reduction in sardines reduced the considerable numbers of sea birds that greatly depended on sardines for food. This in turn reduced the amount of **guano** (bird droppings) left by the birds on their coastal nesting sites. Since guano is collected and converted into fertilizers, its increased scarcity has made fertilizer more expensive, thus having an economic, as well as biologic, effect on agricultural ecosystems. The scarcity of guano also has reduced the amount of nutrients that would wash into the ocean and be utilized by the producers, the marine algae. A reduction in the producers would reduce the width of the entire ecological pyramid.

A phenomenon that is often discussed is the "greenhouse effect." As fossil fuels and wood are burned at ever-increasing rates, coupled with decreasing forest lands, carbon dioxide concentrations are building up in the atmosphere. Carbon dioxide helps to trap heat that would normally be radiated off the earth's surface into the atmosphere. Like the glass over a greenhouse, the carbon dioxide causes the atmosphere below to increase in temperature. There

Exercise 3.3 Demonstration of Pollution by Sulfur Dioxide.

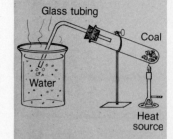

Materials: Test tube, one-hole stopper, glass tubing, bunsen burner, high sulfur coal, 200 ml beaker, phenophthalein.
Procedure: Place a crushed lump of coal in the bottom of a test tube. Stopper the test tube with a one-hole stopper through which is a piece of glass tubing. The glass tubing should be bent down into a beaker full of water. Heat the coal with a bunsen burner. As the air expands and the coal burns, air and gases liberated from the coal, including sulfur dioxide (SO_2), will bubble through the water. An acid indicator such as phenophthalein will detect an increase in sulfuric acid (H_2SO_4) and carbonic acid (H_2CO_3) due to SO_2 and carbon dioxide (CO_2) passing into the water.
Caution: Do not let test tube cool while the glass tubing is still in the water or the water will be drawn back into the test tube and crack it.

is speculation that an average increase of only 1° or 2°C per year will cause the polar ice caps to melt and subsequently inundate coastal cities around the world. Examples like these abound, making us question whether we should, or can, continue to move nutrients vast distances in great quantities.

BIOLOGICAL CONCENTRATION

With the increasing use of artificial chemicals—chemicals not naturally found in ecosystems—it is not surprising to find that some chemical compounds cannot be utilized, decomposed, and recycled as naturally occurring chemicals can. Occasionally they cause no apparent harm, but sometimes they have a toxic effect. The toxicity may be at a very low level, and organisms may be able to deal with and recover from it readily, or there may be an immediate deleterious effect. Often, there is an accumulation of the chemical in body tissues and the toxic effect may only slowly appear as the chemical concentration builds up.

Perhaps the classic case of biological concentration is that of DDT. DDT, an effective insecticide, is relatively resistant to metabolic breakdown and accumulates in the fatty tissues rather than being metabolized. Figure 3.8 shows how the addition of a low concentration of DDT can result in an increase in concentration of this chemical as it moves up the trophic pyramid to a level where it may be harmful or fatal to higher organisms. DDT has been particularly harmful to fish-eating birds such as the brown pelican, osprey, and bald eagle, because high DDT concentrations interfere with proper egg-shell production.

DDT is broken down in the environment only very slowly. It may persist for 2 to 15 years or more. Half of the DDT sprayed by crop-dusting planes never hits its target and may be carried far downwind. Being insoluble in water, it can be transported great distances via runoff from agricultural fields, in streams, and the ocean. Being soluble in fat, however, it is easily stored in the fatty tissues of animals and is carried with them as they move. DDT residues have even been found in Antarctic penguins, thousands of kilometers from the nearest application of the pesticide.

BIODEGRADABILITY

Cycles depend upon the **biodegradability** of a material—its ability to be broken down by the biological action of decomposers. Virtually every compound, natural or artificial, is degradable via the

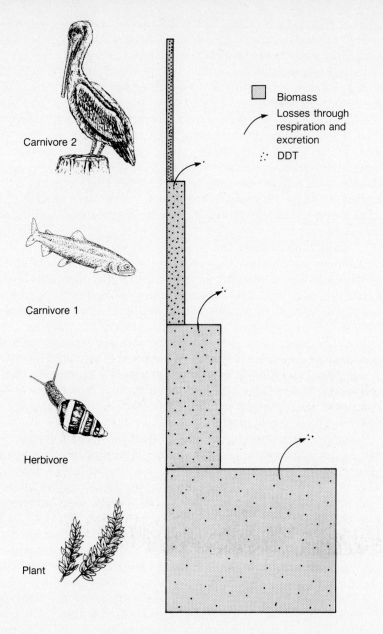

Carnivore 2

Carnivore 1

Herbivore

Plant

☐ Biomass

↗ Losses through respiration and excretion

∴ DDT

Figure 3.8 Biological Concentration of DDT. DDT residue is passed up a food chain in ever-greater concentrations. The top carnivores (here fish-eating birds) may eventually receive toxic doses. More commonly in birds, eggshell production is inhibited, and thin-shelled eggs, which frequently crack before development is complete, are laid. The number of brown pelicans was reduced considerably due to DDT.

processes of physical weathering, chemical action, and biological action, but primarily by a combination of all three processes. To say that something is degradable is meaningless, since everything degrades; it is really the length of time required for degradation that is significant. Skeletons of dead corals may take hundreds of years to

*Exercise 3.4 Decom-
position of Leaves.
Obtain a sample of tree
leaves (about a handful)
and pat dry to get rid of
excess water. Weigh the
sample and place in a mesh
bag (for example, of
cheesecloth). Place a
number of mesh bags in
different environments such
as in the shade of a tree, in
a sunny spot, suspended
from a tree branch, or
suspended in a pond or
stream. After a few weeks,
dry the samples in a drying
oven and weigh again. The
lightest ones have
decomposed the most, and
their matter has leached
through the mesh. Which
environments are most
conducive to decom-
position? Which parts of the
leaves decompose first?
Why? Experiment with the
leaves from different trees.
For example, compare
evergreen tree leaves with
deciduous ones.*

degrade, while a leaf in a tropical forest may be totally decomposed and reabsorbed by other organisms in a very few days (Exercise 3.4). Aluminum cans may persist for hundreds of years.

Soil organisms, especially bacteria, fungi, protozoa, earthworms, roundworms, and arthropods are responsible for most of the decomposition of the litter on the forest floor (Figure 3.9). The rates of decomposition depend upon environmental factors and the material being decomposed. **Aerobic** (oxygen-requiring) **bacteria** are more efficient than **anaerobic** (without oxygen) **bacteria**, so oxygen-rich environments are more conducive to decomposition processes; deeper layers of litter decompose more slowly, as does matter at the bottom of lakes. Moisture, temperature, acidity, and sunlight also affect the rate of decomposition. Moist, warm, and neutral pH conditions are ideal.

Leaves, animal flesh, and moist animal waste products tend to decompose rapidly since they contain chemicals such as cellulose, protein, and sugars which are easily reduced by the chemical activity of bacteria and fungi. The chemical action is often preceded by an initial mechanical decomposition by millipedes, worms, and the like, which may increase microbial decomposition rates by 20%. Other substances such as woody twigs and tree bark are decomposed much more slowly with the assistance of organisms like bark beetles that bore tunnels in the decaying wood. Experiments have shown that insecticides applied to a forest floor may decrease decomposition rates by 50% or more, demonstrating the importance of insects to recycling of nutrients.

Some lichens and other plants contain substances that inhibit bacterial growth and thus retard decomposition. The decomposition of coniferous needles, for example, produces an organic acid; the acidity of coniferous forest floors inhibits bacterial action and thus in part accounts for the slow decomposition of litter.

The ability of organisms to biodegrade is used in many industrial processes such as in producing wine and beer; making yogurt, cheese, acetone, some vitamins, and soy sauce; and in curing tobacco products.

Biodegradability is often used to describe man-made products as if this quality in itself made the product beneficial or at least harmless. This is quite misleading, as some of the most toxic pesticides degrade quite rapidly (a few days to weeks), and glass bottles, benign except perhaps as litter, may last 500 years or more. Phosphate-containing detergents may degrade rapidly in rivers or lakes, causing enormous growths of algae, which then die. As they are broken down by decomposers, an enormous amount of oxygen is used, depleting the oxygen in the water and resulting in fish kills.

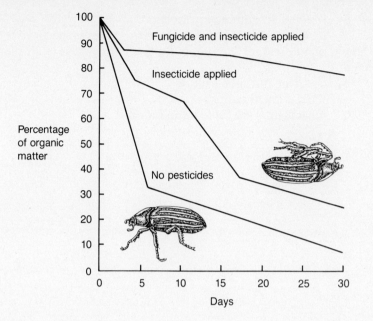

Figure 3.9 **Microorganisms and Decomposition.** The importance of arthropods and fungi to decomposition of organic matter can clearly be seen. Bags of organic litter were buried in desert soil. Some were treated with an insecticide, some with an insecticide and a fungicide, and some were left untreated. After 30 days, 90% of the litter in the untreated bags was decomposed, but only 70% in the insecticide treated bags and 15% in the insecticide and fungicide treated ones. This study underscores both the significance of microorganisms to the process of decomposition and the potential effects of pesticides on biological cycles.

Numerous examples exist, then, to demonstrate that everything decomposes, but where, how fast, and what it degrades to may be of great importance.

SUMMARY

Energy flows one way through an ecosystem and matter is recycled, but matter and energy are inseparable since matter contains energy. The movement of matter occurs via biogeochemical cycles: gaseous, sedimentary, and hydrologic. Sometimes matter becomes concentrated in organisms or stored in rock. The speed of recycling depends on the biodegradability of a material.

STUDY QUESTIONS

1. Define biodegradability, biological concentration, leaching, closed and open systems.
2. Compare the flow of energy to the flow of matter.
3. How does energy function to drive the biogeochemical cycles?
4. Diagram a gaseous cycle.
5. Diagram the hydrologic cycle.

6.   Diagram a sedimentary cycle.

7.   How are the three biogeochemical cycles related? Give an example of an element or compound that participates in all three cycles.

8.   Describe nitrogen fixation.

9.   What is "acid rain"?

10.   What factors are cycle impediments?

Chapter 4

The Origin of Life
and the Geologic Past

We have examined some of the fundamental processes that occur in the biological realm and some of the physical factors that affect these processes. What we now observe is the culmination of billions of years of geologic and biologic changes on the earth. The environments of historic and present time are quite different than they were eons ago. Enormous changes in the physiognomy of the earth and the composition of the air and water have resulted in equally large changes in the constitution and distribution of the flora and fauna. To better understand the present, we will briefly examine the past.

FORMATION OF THE EARTH

Most scientists generally accept the "big bang" theory of the formation of the universe. This theory states that all matter was once a large mass that exploded into billions of smaller masses which condensed into stars, planets, and other such bodies. Our solar system was only one of perhaps billions of solar systems formed. Revolving around the sun at an average distance of 155 million kilometers, the earth's components cooled and condensed, and the oceans formed. The atmosphere changed as volcanic eruptions belched enormous quantities of smoke, dust, and gases into the air,

Figure 4.1 Volcanic Eruption. Mt. St. Helens, Washington state, May 18, 1980. Volcanic eruptions release liquid lava, solid pieces of rock and ash, and great quantities of gases. Volcanic dust is extremely fine and may be blown around the earth several times before settling down. Water vapor is the most abundant gas erupted by a volcano, but sulfur compounds, nitrogen, ammonia, and methane are also present. Several square kilometers of ash and gases can be spewed forth, affecting both the earth's climate and the composition of the atmosphere.

as they still do (Figure 4.1 and Exercise 4.1). The earth formed perhaps 5 billion years ago; evidence seems to indicate that life did not arise until about 3½ billion years ago.

THE MYTH OF SPONTANEOUS GENERATION

Not having observed the origination of life, most of us simply presume that present life comes from preexisting life—a logical and correct presumption. But this lack of observation has caused some people to think otherwise, leading to the development of interesting myths. It was once thought, for example, that rotting meat created flies; it is obvious to us now that it is flies laying eggs in the meat that give rise to another generation of flies. Also, it was said that a flannel shirt placed in the corner of a barn would produce mice (which were seen when the shirt was lifted). Horsehairs from a horse's mane were supposed to turn into worms when they fell into a horse's water trough (Figure 4.2). These fables are examples of what is termed **spontaneous generation**—the origin of life from non-

Exercise 4.1 Model of a Volcano.
A working model of a volcano can be simply constructed. It will spew ash and steam in a manner similar to a real eruption. Materials: wooden base, staples, screen (or carpenter's cloth), plaster of paris, glass tubing, rubber tubing, flask with a one-hole stopper to fit, matches, heat source.
Procedure: Part I. Constructing the Volcano. Shape screen into a cone and staple to the wooden base. Insert glass tubing (see diagram) and tie it securely to the screen. Plug the small opening in the upper end of the tubing with paper to prevent the tubing from being clogged with plaster of paris. Cover the

screen with a layer of plaster of paris (make a thick creamy mixture). When the plaster is dry and the paper plug is removed, the volcano is ready for its first eruption.
Part II. Eruption. Place about one teaspoonful of ammonium dichromate around the edge of the crater. Lay a lighted match on the surface of the chemical. The chemical will ignite and burn very rapidly, causing ashes to fly up into the air and drop down the sides of the cone. Connect the glass tubing in the model volcano to a source of steam (see diagram). Place a clothespin on the rubber tubing for a short time; the steam pressure will build up (do not keep clamped any longer than 15 or 20 seconds), and when the clothespin is released, the

rush of steam will make the volcano appear even more realistic.

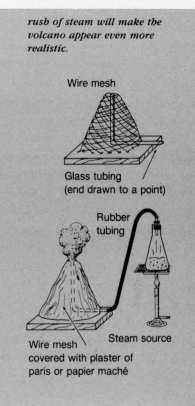

Wire mesh

Glass tubing
(end drawn to a point)

Rubber tubing

Steam source

Wire mesh covered with plaster of paris or papier maché

living matter. All the forms of life we know of now arise only from preexisting life of a similar form. Spontaneous generation does not now occur. But life did arise from nonliving matter at some time in the past.

THE ORIGIN OF LIFE

Evidence indicates that life—a property of matter that implies response to stimuli, energy use, growth, organization, and especially the ability to reproduce—was absent from the earth until about 3 to 4 billion years ago. Life arose from nonliving chemicals in an environment quite different from the one that exists now.

The early atmosphere of the earth consisted of methane, ammonia, hydrogen, and water vapor. (Our present atmosphere is primarily nitrogen and oxygen.) There was no free oxygen, since no plants—and thus no photosynthesis—were present. This primitive set of gases was experimentally exposed to electrical discharges by S. L. Miller in 1957. He found that organic molecules, including

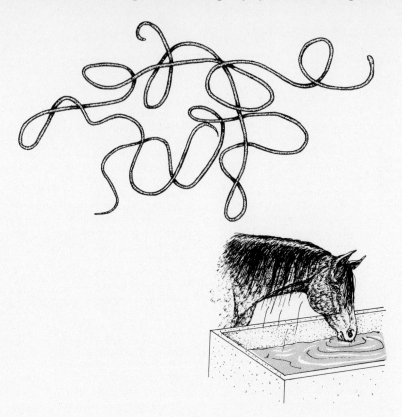

Figure 4.2 Horsehair Worms. It was once thought that when the hairs from a horse's mane fell into a watering trough they would turn into worms. Actually, a group of organisms called Gordian or horsehair worms live as internal parasites in arthropods or crustaceans. For a short period during the year, they emerge as adults to reproduce in water. Their long, thin shape and whiplike movement obviously stimulated imagination, and the story of horsehair worms arose.

amino acids—fundamental components of proteins—were produced. Later experiments demonstrated that other simple molecules could be produced with the input of various kinds of solar radiation or by volcanic activity. Various molecules eventually condensed into droplets possessing a definite chemical structure. These droplets absorbed (fed on, if you like) other molecules. Eventually the droplets became so large that they split into two or more smaller ones; the ability to reproduce evolved. This specific scheme is obviously open to argument, but the general scheme is probably accurate. The fact that life did arise and develop into an infinite array of forms is, of course, undeniable.

The earliest forms of life were very primitive microorganisms. Sedimentary deposits from the Precambrian era of 3.3 billion years ago show indications of bacterialike organisms. These earliest creatures, resembling bacteria and blue-green algae, photosynthesized and enriched the atmosphere with oxygen on which subsequent life forms depended. But it was only about 750 million years ago that recognizable higher forms of life began to appear and diversify. Thus, there was a great expanse of time during which very primitive organisms were the preponderant life forms. In the geologic strata

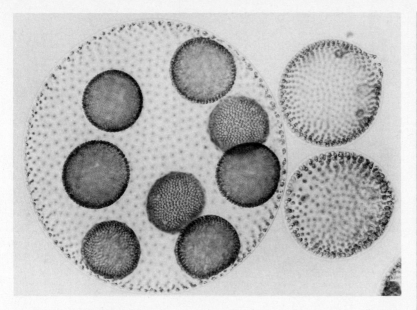

Figure 4.3 A Volvox Mother Colony with Daughter Cells. *Volvox* is one of the few multicellular protozoa and is considered to be similar in some ways to the first multicellular creatures.

Exercise 4.2 Prokaryotic and Eukaryotic Cells.

It might be instructive to compare the structures of the simplest organisms, the prokaryotes, with more advanced cells. Using either prepared slides or fresh material, examine bacteria and blue-green algae. Bacterial cells are rod-shaped (bacillus), spherical (coccus), or spiral (spirillus). Blue-green algae may be filamentous or globular or colonial, as well as other forms. Look at fresh or preserved eukaryotic animals or plant cells, whether single-celled, colonial, or multicellular. How are they more advanced? How are these advancements adaptive? Why are prokaryotic cells still so successful even without the "advanced" structure of the eukaryotes?

of the Cambrian period of about 600 million years ago we find a multitude of multicellular fossils: worms, sponges, jellyfish, and other such recognizable creatures. Of all the time that has passed since the earth's formation about 5 billion years ago, 88% of it is encompassed by the Precambrian era. Complex organisms are thus relatively new to the earth.

Life probably did not arise in one place at one time; most likely similar events occurred in many places over a period of time. From the bacteria and blue-green algae, which lack a nucleus and other structures typical of other cells, nucleated cells evolved. (Bacteria and blue-green algae are termed **prokaryotic cells**, in contrast to the more complexly organized **eukaryotic cells**—which are all other cells.) (See Exercise 4.2.) These cells ultimately aggregated to form colonial organisms. Recent evidence indicates that multicellularity may have originated independently in the plants, fungi, and animal groups. Division of labor among the cells evolved, followed by the appearance of multicellular organisms (Figure 4.3). Evolutionary changes since that time have been relatively swift.

The geologic time scale (Figure 4.4) can help put evolutionary events into perspective. What the scale does not do is impress upon

Figure 4.4 Geologic Time Scale. The major divisions of the geologic time scale and the major evolutionary events. Numbers indicate time in millions of years ago.

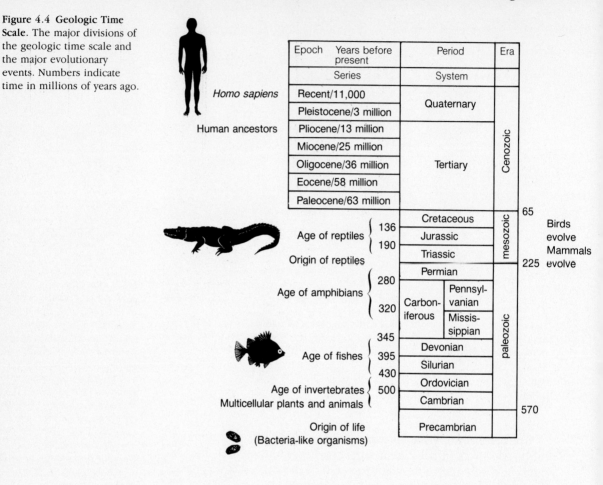

Epoch	Years before present	Period	Era
Series		System	
Recent/11,000		Quaternary	Cenozoic
Pleistocene/3 million		Quaternary	Cenozoic
Pliocene/13 million		Tertiary	Cenozoic
Miocene/25 million		Tertiary	Cenozoic
Oligocene/36 million		Tertiary	Cenozoic
Eocene/58 million		Tertiary	Cenozoic
Paleocene/63 million		Tertiary	Cenozoic

Homo sapiens — Recent/11,000, Pleistocene/3 million

Human ancestors — Pliocene/13 million

65

	Cretaceous	mesozoic
136	Jurassic	mesozoic
190	Triassic	mesozoic

Age of reptiles { 136, 190

Origin of reptiles

225 Birds evolve Mammals evolve

	Permian	paleozoic
280	Carboniferous — Pennsylvanian	paleozoic
320	Carboniferous — Mississippian	paleozoic
345	Devonian	paleozoic
395	Silurian	paleozoic
430	Ordovician	paleozoic
500	Cambrian	paleozoic

Age of amphibians { 280, 320

Age of fishes { 345, 395

Age of invertebrates { 430, 500

Multicellular plants and animals

570

Origin of life (Bacteria-like organisms) — Precambrian

the reader the overwhelming amount of time that has passed. Our short 70+-year lifetime does not allow us to comprehend the passage of centuries, let alone millions or billions of years. Humans have existed for fewer than 4 million years—less than 0.01% of the time that has elapsed since life began (Exercise 4.3). (If we compress the passage of geologic time into one year, we find that if life began on January 1, humans did not evolve until noon on December 31, they made tools at 11 P.M., and Columbus discovered America 20 seconds before midnight.) Our span on this earth has indeed been miniscule compared to the time that has passed. We must understand that today's environment is a result of steady evolutionary change over an almost infinite amount of time. As amazing as some plant and animal adaptations may seem—how could feathers, trees, and elephant tusks possible evolve?—their appearance seems less mysterious when one comprehends the enormous time span during which they have evolved.

EXPANSION OF LIFE FORMS

The first complex animals in the Cambrian and Ordovician eras (about 500 million years ago) were the aquatic molluscs (shelled animals such as clams, oysters, and snails), arthropods, and fish. During the Silurian period (about 400 million years ago) plants began to invade the land, followed by terrestrial animals. The first land animals were the arthropods such as scorpions and spiders, followed later by insects. Mammals, birds, and the flowering plants have been in existence only since the Jurassic period 160 million years ago.

A good deal of evolutionary change had to occur in order to free plants and animals from their dependence on aquatic habitats. Among the animals, shells, scales, or thick skin evolved as protection from abrasion and desiccation. Lungs evolved to extract oxygen from the air. Appendages were modified for locomotion over land. Plants evolved more sophisticated vascular systems to transport nutrients through their bodies and woody stems to support their bodies above the ground.

Spurts of evolution have occurred at times. When animals first invaded land, a totally new environment was open to them, and they proliferated. Birds, mammals, and flowering plants all showed a burst of evolution at about the same time; the animals and plants benefited from and took advantage of each other's existence. The plants took advantage of the mobility of animals to transfer pollen between plants and disperse their seeds. Many animals became adapted to eat plant parts, fruits, seeds, or nectar. At the same time some forms of organisms disappeared; some of these—the trilobites, for example—had been quite successful for millions of years (Figure 4.5).

A number of creatures have existed for millions of years essentially unchanged. The tuatara of New Zealand is a large lizard that looks very similar to its ancestors of 190 million years ago. Blue-green algae are remarkably similar to those that existed in the Precambrian era. Sharks, horseshoe crabs, and clamlike creatures called brachiopods (or lamp shells) are also very much like their earliest predecessors. They have survived with little change either because they are remarkably adapted to a variety of conditions or have lived in areas subjected to little environmental change.

There is a multitude of reasons for the emergence and disappearance of organisms. As the physical or biological environment changes, unfit organisms disappear and more fit ones survive. Slowly but inexorably, environments were and are filled with a variety of organisms capable of surviving and successfully reproducing. Reciprocal changes occur because as organisms change form or behavior, or their population size changes, the environment changes, because organisms are part of the environment.

Exercise 4.3 Visualizing Geologic Time.

To help comprehend geologic time, a graphical representation can be made. Obtain a strip of paper about 10 meters long (adding machine tape is ideal). Label one end of the strip 5 billion years ago— earth's formation. Measure 2 meters along the strip and label that point 4 billion years ago; do the same for the 3-, 2-, and 1-billion-year points. Label the 3.5- billion-year point as ''the beginning of life.'' In the last 2 meters, representing 1 billion years ago to the present, plot as many geologic, evolutionary, and historical events as possible, such as the evolution of fish, evolution of the flowering plants, and continental drift. Unroll the tape across the room. The small space that is occupied by the evolution of multicellular organisms and the miniscule space taken up by historical events should give some perspective of geologic time.

Figure 4.5 Trilobites. The dominant living invertebrates are the arthropods—animals with a flexible exoskeleton, jointed appendages, and well-developed nervous systems. This large group has more known species by far than any other and includes spiders, crabs, mites, insects, millipedes, and the extinct trilobites. Trilobites are the dominant fossils in rocks formed during the early Cambrian. By the Ordovician period their numbers were at a peak, but they died out sometime during the Permian period, after existing for perhaps 300 million years or longer.

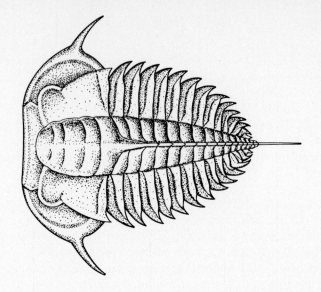

CONTINENTAL DRIFT AND OTHER GEOLOGIC CHANGES

A glance at a map of the world shows continents whose shapes seem to fit together like pieces of a jigsaw puzzle. This led some scientists to believe that all continents were once one land mass and slowly drifted apart. But it has only been within the last 30 years that sufficient evidence has been gathered to make their theory acceptable to the scientific world. Evidence has been gathered by examining rock types and strata on the continents, and by mapping in detail the ocean floor, directions of magnetic fields, and the distribution of fossil flora and fauna.

Continental drift can be explained fairly simply. Ridges in the mid ocean floor are spewing out **magma** (molten rock), causing the ocean floor to spread. The continents act as flat plates and are pushed apart by the expansion of the ocean floor at the rate of about 2 centimeters per year. In a few cases one plate has slid under another, causing a buckling of the land's surface, resulting in mountain ranges. The country we now call India has slid partially under Asia, causing the Himalayas to be formed. Continental drift, more recently termed **plate tectonics**, began about 200 million years ago (Figure 4.6).

The original single land mass, Pangaea, split into a northern land mass (Laurasia) and a southern one (Gondwana) sometime in the Mesozoic era. These land masses were separated by a large body of water called the Sea of Tethys, although some land bridges may have connected them for a short time.

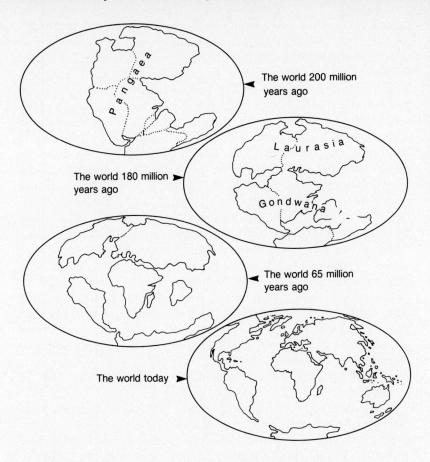

The world 200 million years ago

The world 180 million years ago

The world 65 million years ago

The world today

Figure 4.6 Continental Drift. About 200 million years ago, all continents were one land mass. As the floor of the sea expands due to the extrusion of lava from the earth's core, the continents are spread apart. These movements are called continental drift. Besides geologic evidence, there is abundant biological evidence to indicate close physical contact between continents in the past.

An understanding of continental drift and other large geologic changes, such as the uplifting of mountains, the formation of islands, and the appearance and disappearance of land bridges, such as the isthmus of Panama, sheds some light on the present-day distribution of organisms. Why, for instance, are there cases of similar fauna on continents separated by a vast expanse of ocean? The sequoia trees of California have many characteristics in common with the metasequoias of China. Fossils of similar insect-eating mammals (moles, hedgehogs, and shrews) have been found in both North America and Asia. And there are three living species of lungfish: one in Australia, one in Africa, and one in South America. The ancestors of lungfish seemed to have evolved in Gondwana, and their descendents now remain on the continents that once formed Gondwana.

To reverse the question, why are some flora and fauna so different on different continents? In the case of mammals, most evolved

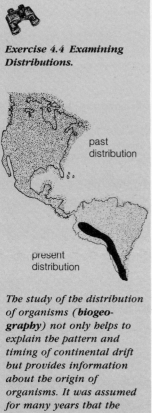

Exercise 4.4 Examining Distributions.

past distribution

present distribution

The study of the distribution of organisms (biogeography) not only helps to explain the pattern and timing of continental drift but provides information about the origin of organisms. It was assumed for many years that the present high diversity within groups of organisms in a geographic area identified that geographic area as the center of origin of that group. We know now, however, that that idea is oversimplified and often erroneous.

Analyze the literature on a group of organisms such as lizards, pine trees, orchids, molluscs, or horses, and try to determine their time of origin, center of origin, and dispersal as related to continental drift, land bridges, and the like. Draw a map of their distribution at various geological times including the present.

after Pangaea split into northern and southern positions, and many evolved after Gondwanaland had begun to separate. Thus, we find that marsupials, a relatively primitive group of mammals, are restricted to South America and Australia—an indication that these continents were still attached by their southernmost tips for some time after Africa had drifted off, since Africa has no marsupials. The one exception is the opossum, which reached North America via the Panama land bridge. Evidence indicates that much of the fauna and flora of North America has derived from Eurasian or South American stock. Camels provide similar evidence for the connection of the two continents. Camel fossils are common in the deserts of the southeastern United States. Camels apparently evolved in North America, migrated north and westward over the Bering land bridge from what is now Alaska to Asia and southward to North Africa. G. G. Simpson, a well-known paleontologist, has categorized the paths of dispersal of organisms three ways. **Corridors** are land connections that allow the free movement of organisms from one land mass to another in either direction. **Filter bridges** are land connections that only some organisms are able to move across, others being filtered out by unfavorable climate, insufficient or inappropriate food sources, and so on. **Sweepstakes routes** are essentially chance dispersals to isolated areas, such as the colonization of the Galápagos Islands by tortoises. Dispersal mechanisms are discussed in detail in Chaper 7.

Organisms spread from their center of evolution to various parts of the world via these mechanisms, inhibited or assisted by continental drift, the emergence and submergence of land bridges and mountain ranges, and climatic changes. Mammals, for instance, originated and diversified in Eurasia before spreading to other parts of the world. Extensive dispersal of plants during the tertiary period over the Bering land bridge accounts for the similarity between much of the Eurasian and North American flora.

Of course, evolution has also occurred within the boundaries of each continent to give each land mass a distinctive set of organisms. Isolation from other continents and geologic and climatic changes have formed the habitats and their inhabitants as we see them today. Chapter 8 discusses these habitats in detail (Exercise 4.4).

Conversely, knowledge of past and present flora and fauna reveal much about the abiotic components of the past. Fossilized pollen taken from various levels of prehistoric swamps tells us, for example, what kinds of trees once grew in that area (Figure 4.7). Knowing the present requirements of the trees, we can intelligently speculate what the weather was like when these grew. The oldest living organisms, bristlecone pines, exist in the White Mountains of the southwestern United States. Most are about 1000 years old, but a few exceed 4000 years in age. Study of the width of the tree rings,

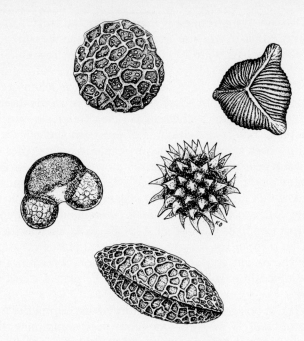

Figure 4.7 Pollen Grains. Pollen grains are found in a wide diversity of shapes and sizes, often having wings, spines, and other projections. Fossilized or recent pollens can thus be identified readily.

which indicate the age of the tree and its rate of growth, combined with soil layer and pollen studies, has provided a good deal of reliable information about the climate of the White Mountain area over the past 6000 years.

Animal fossils also provide us with information about past climates. In the Eocene epoch, about 50 million years ago, tropical birds, such as trogons, occurred in Asia, Europe, and North America; parrots were present in places like Nebraska and France. Presuming that these birds were as suited to tropical regions then as now, France and Nebraska probably had tropical climates; other kinds of evidence support this supposition.

FOSSILS

Much of the information about life in the geologic past has been gleaned from fossils. Essentially, **fossils** are organisms or parts of organisms that have been buried in sediments of sand, soil, or the like. Fossils are most common in sedimentary rock—rock formed by the deposition of sediments carried by wind, water, or air. In the process of fossilization, after the organism is buried in sediment, the softer body parts readily decay. The harder parts, such as bones and shells, persist and are gradually replaced, grain by grain, by the minerals dissolved in the water. The fossil record is biased because those organisms with hard parts—corals, molluscs, vertebrates—are

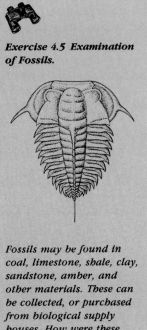

Exercise 4.5 Examination of Fossils.

Fossils may be found in coal, limestone, shale, clay, sandstone, amber, and other materials. These can be collected, or purchased from biological supply houses. How were these fossils formed? Why do some organisms fossilize more readily than others? What other modes of fossilization are there besides entrapment in sediment?

found relatively more frequently than soft-bodied creatures, such as worms and jellyfish. Aquatic flora and fauna are fossilized more often, since they have a greater chance of falling into sediment than terrestrial organisms.

Fossilization can occur in other ways, such as when organisms are trapped and buried in some material that prevents their decay. Volcanic eruptions, such as the recent Mt. St. Helen's in Washington state, spew out enormous amounts of fine-particled ash and water that readily cover plants and animals unable to escape. Molten lava may burn and nearly vaporize trees as it moves around them and may leave holes in the new lava surface where the trees once were; examples of this can be seen on the island of Hawaii where volcanoes spew lava every few years. Small creatures may be trapped in tree sap that fossilizes into amber. Asphalt pools, such as the famed La Brea "tar pits" in Los Angeles, trapped and preserved numerous fossils of the recent geologic past. A few woolly mammoths have been preserved in excellent condition in the glacial ice of Siberia (Exercise 4.5).

Fossils can be aged directly or indirectly by dating the material in which they were preserved. One of the most common methods is the carbon-14 test. Living organisms contain both C_{12} and its radioactive isotope C_{14}. When an organism dies, C_{14} decays to C_{12} at a known rate. Determining the proportion of C_{14} to C_{12} gives the age of the organic materials with considerable accuracy, up to 50,000 years of age. For older specimens, a similar method using potassium and argon is used.

RECONSTRUCTING THE PATTERNS OF THE PAST

The fossil record, carbon testing, morphological analysis, biochemical testing, distribution, and even the behavior of living groups of animals have all been used to reconstruct the past scheme of evolution and the relationship of extinct to living forms. There are enormous gaps in our knowledge, to be sure, and theories are being revised at an overwhelming rate. But the disagreements among paleontologists, anthropologists, and others who study the life of the past, are generally over sophisticated details such as rate of evolution of particular groups, the reasons for the persistence of some groups and the disappearance of others, whether or not some dinosaurs were homeothermic (some probably were), and whether evolution occurred rapidly or gradually. The overall pattern, however, is generally agreed upon and most new evidence fills in the gaps and substantiates what is already known. Figure 4.8, for instance, shows a simple scheme for an evolutionary tree that most scientists would accept.

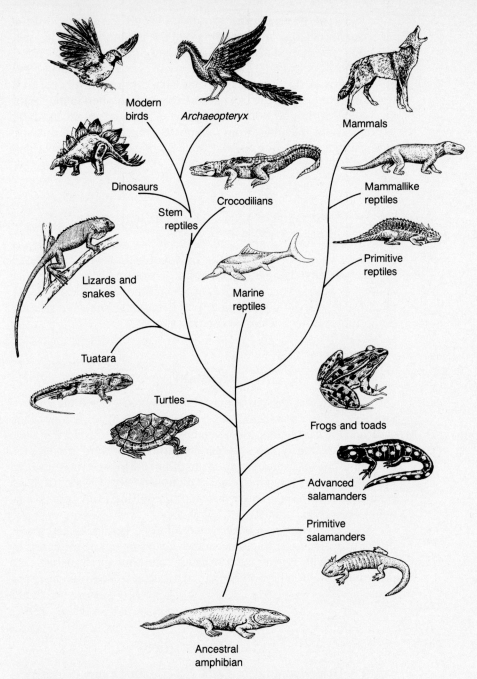

Figure 4.8 Evolutionary Tree of the Tetrapod Vertebrates. The tetrapods are the vertebrates excluding the fish. Essentially, fishlike ancestors gave rise to amphibians, amphibians to reptiles, and from reptiles rose both mammals and birds.

Occasionally, a rare and fortuitous event occurs that bolsters both our knowledge and confidence in the geologic past. The coelacanth, a lobe-finned fish that was thought to be extinct for 70 million years, was caught alive off the coast of South Africa in 1938. Since that time, scientific expeditions have captured dozens of additional specimens and the fish have been photographed in their natural habitat. Although none have survived capture, much has been learned about the physiology, chemistry, and ecology of this species. Studies on "living fossils" such as these tell us a great deal about the past.

ECOLOGY AND THE GEOLOGIC PAST

Tremendous changes have occurred in climate, topography, and types of organisms over the past half-billion years. We know much about what happened, but not nearly so much about why. Just as the past helps explain the present, examining ecological processes of today may help us explain past biological occurrences. The remaining chapters discuss those processes.

SUMMARY

The earth was formed about 5 billion years ago and life evolved about 3½ billion years ago from nonliving matter. The geology of the earth and the earth's atmosphere both have changed considerably since then and continue to do so. The earliest forms of life were primitive microorganisms; the larger forms of life did not arise and diversify until 750 million years ago. One major evolutionary step was the transition from aquatic to terrestrial environments, requiring many new adaptations of plants and animals.

The continents were once joined into one land mass and slowly drifted apart. The distribution of some organisms can be explained by continental drift. Fossils, carbon dating, morphological analysis, and present distribution have all been used to reconstruct evolutionary patterns.

STUDY QUESTIONS

1. Define continental drift, sweepstakes route, prokaryotic, spontaneous generation.
2. Compare eukaryotic and prokaryotic cells.
3. Describe the mechanism of continental drift and how Pangaea split into the different continents.

4. What tests could you perform to demonstrate that spontaneous generation is not a valid concept?

5. Describe the atmosphere of the earth before life evolved.

6. Why do you think primitive organisms existed essentially unchanged for nearly 1½ billion years before multicellular organisms evolved and radiated into many forms?

7. What do you think accounts for "spurts" of rapid evolution?

8. What are the possible reasons, in addition to continental drift, that flora and fauna are so different on different continents?

9. In what ways can living organisms reveal facts about the geologic past?

10. How are fossils formed, why is the fossil record incomplete, and why do some groups fossilize more readily than others?

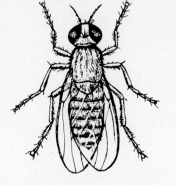

Chapter 5

Evolution

EVOLUTION AS A BIOLOGICAL PRINCIPLE

"Nothing in biology makes sense except in the light of evolution" (Theodosius Dobzhansky, 1973). There are many principles, axioms, and rules in the biological sciences, but none is so pervasive as evolution. To understand evolution is to understand the development of organisms, their anatomy, physiology, behavior, and role in the natural world. The development and functioning of ecosystems can be explained via the evolutionary process, as can the operation of any other biological system. In turn, the workings of evolution shed light on observations of living systems.

Chapter 4 gave us a broad overview of the events of the past. Enormous changes occurred over an almost incomprehensible amount of time. Organisms and their environment are quite different now from what they were during the Cambrian or Triassic periods. As the physical geology of the earth, the composition of the atmosphere, and the climate changed, organisms adapted to the new conditions. This, of course, is evolution. This chapter will examine the process of evolution, which has been described by some scientists as the very "backbone of biology."

Ever since Charles Darwin published the *Origin of Species,* the concept of evolution has been controversial. Its teaching has been challenged in the courts several times, most notably in 1925 during the "monkey trial," when John T. Scopes was prosecuted by the State of Tennessee for teaching evolution in high school. This controversy and the resulting legal and moral conflicts are mainly due to a clash of fundamentalistic, theistic beliefs and secular, scientific information. Most unfortunate is the fact that these conflicts stem primarily from the scientific ignorance of most antievolutionists.

The phrase "theory of evolution" is frequently misinterpreted as meaning that the existence of evolution is theoretical. There is little doubt in the minds of the vast majority of scientists and informed laypersons that evolution occurs. Evolutionary "theory" refers to the explanations of how evolution works. Prodigious evidence exists to document the disappearance of dinosaurs and the proliferation of birds, mammals, and flowering plants about 150 million years ago, as well as the appearance and disappearance of many other organisms. That dinosaurs existed, disappeared, and were replaced by other vertebrates is not a theory; a tremendous amount of evidence shows that to be the case. But how and why the dinosaurs disappeared, and how and why other organisms arose is not so clear. Numerous explanations for the dinosaurs' dominance and later demise have been proposed, all of which have some validity. That is evolutionary theory—the whys and hows of the evolutionary process, not the ifs.

BEFORE DARWIN: LAMARCK'S THEORY

A French naturalist of the late eighteenth and early nineteenth century, Jean Baptiste de Lamarck, developed a theory to explain evolutionary changes. Lamarck's essential argument was that the environment caused organisms to adapt certain forms and functions. And although the environment does cause changes in organisms, the mechanism Lamarck proposed turned out to be incorrect.

Lamarck said that offspring would inherit any characteristics that were acquired by the parent during the parent's lifetime—the **Theory of Acquired Characteristics**. Giraffes, then, according to his theory, have long necks because previous generations of giraffes stretched their necks reaching for tree leaves, and the slightest elongation was passed on to the next generation. Although this sounds somewhat plausible, it has no basis in fact. Experiments have demonstrated, for example, that cutting the tails of mice has no effect on the tail length of their descendents. And we, of course, would not expect silicone injections or implants done to increase a woman's bust size to have any effect on the bust measurements of the woman's daughter. Lamarck's theory predicts that it would.

DARWINISM

Charles Robert Darwin (1809–1882) is considered by most to be the father of evolutionary theory. Although indications of the concept of evolution can be found as far back as ancient Greece, it was not until the eighteenth and nineteenth centuries that the concept began to crystalize. Darwin's naturalist grandfather, Erasmus Darwin, speculated that all living things could have come from a "single filament" of life. But it was Charles Darwin who collected sufficient evidence to propose the idea to the scientific world.

Darwin was a self-taught naturalist. His only formal schooling was two years at the University of Edinburgh Medical School, which he quit to pursue a divinity degree at Cambridge, completed in 1831. In late 1831 he began a journey on the H.M.S. *Beagle* that lasted nearly five years. The journey took him along the coast of South America and to a number of islands, notably the Galápagos Islands of the South Pacific. The remainder of his life was devoted to interpreting the information and cataloging the specimens he collected during the voyage. He wrote several books on both geology and biology, beginning work on his most famous book, the *Origin of Species*, in 1856. Charles Lyell, a geologist, and Joseph Hooker, a botanist, urged publication of the book, but it took the instigation of Alfred Russell Wallace (1823–1913), who, independently, had come to the same conclusions as Darwin, to convince Darwin to present his theory to the scientific world. Wallace and Darwin jointly presented their papers in the *Journal of Proceedings of the Linnean Society* in England in 1858; the *Origin* was published the next year.

The theory of evolution was accepted by most scientists of Darwin's time and was ferociously defended by some, notably Thomas Huxley. But it also caused a good deal of consternation among the general public, igniting a controversy that still smolders today.

Although the *Origin of Species* contains an overwhelming amount of information from which Darwin drew his conclusions, almost nothing was known about the mechanisms of genetics. He postulated that the best adapted organisms survived and reproduced more successfully than less well adapted ones. But why were some more well adapted than others? Organisms of the same species differed slightly from each other. But how did these differences arise? Darwin himself realized that he could not fully explain the workings of evolution without understanding heredity.

Gregor Mendel (1822–1884) was an Austrian monk who worked out some of the fundamental principles of genetics using garden peas; he published his results in 1866. Unfortunately, it was not until 1900 that the importance of his work was recognized. Many other scientists, too numerous to mention, also expanded evolutionary thought by filling in gaps and honing the rough edges of the theory. The modern theory of the workings of evolution is often called neo-Darwinism—it incorporates Darwin's ideas and genetics, forming a strong set of tenets.

Volumes have been written on evolution, and still its explanations of many observed natural phenomena are not fully satisfactory. Although details of the process are beyond the scope of this book, you should receive here a good foundation for further reading. An understanding of ecology is essential for comprehending the evolutionary process, which, condensed to its basics, is actually very logical and simple.

EVOLUTION DEFINED

The word *evolution* evokes images in the ill informed of humans evolving from amoebas or chickens hatching out of snakes' eggs. These images are not only incorrect, but ignore the ponderous complexity of evolution and its inextricable link to all life processes. More exact and appropriate definitions are available, but for our purposes we will consider evolution to mean simply "change over time," or, better, as Darwin said it, "descent with modification."

As mentioned earlier, 95% of all organisms that have ever lived are extinct; what we see today are the descendants of ancestral forms. Treelike "seed ferns" flourished for 150 million years but disappeared over 200 million years ago. Sea scorpions, which are related to spiders and reached a length of nearly 2 meters, are no longer **extant** (the opposite of extinct) (Figure 5.1). Modern plants and animals (flowers, birds, mammals, moths, and so on) did not appear in abundance until about 135 million years ago. Estimates vary, but it is likely that extant species number only 1.5 to 2 million, while the total number of species that have ever existed is 1.5 to 2 billion or more. So species have changed over time and, for the most part, in a very gradual fashion. Before Darwin's time it was common to explain these changes by assuming that catastrophes regularly

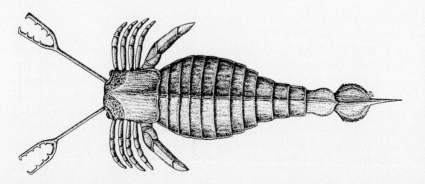

Figure 5.1 Sea Scorpion. The sea scorpions (or eurypterids) lived from the Cambrian to the Permian periods from 500 to 250 million years ago. From their anatomy it appears that they could swim, using their legs and tail as paddles. They were very common for millions of years but began to die out as the fishes became abundant, suggesting replacement by the latter.

Figure 5.2 The Opossum. The opossum is somewhat of a ''living fossil,'' as it has existed virtually unchanged since the Cretaceous period over 100 million years ago. Sharks, oysters, the horseshoe crab, and the maidenhair tree are similarly ''ancient'' since they look like their ancestors of very long ago.

exterminated all life, and were followed by a new creation. As more evidence mounted, however, it became obvious that a ridiculously large number of catastrophes would have to have occurred, and, additionally, all groups did not become extinct all at once. Some have persisted for hundreds of millions of years—others for much less than a million (Figure 5.2).

THE PROCESS OF EVOLUTION

The workings of the process of evolution can be explained by examining four basic mechanisms: overproduction, variation, competition, and natural selection. These mechanisms are the core of evolution; to understand them is to comprehend the mode of evolution.

Overproduction

Every environment is limited in the number of individuals it can support. Limited resources restrict the size of any biological population. A forest can provide for just so many deer, and a limited number of cattails can grow in a marsh, for example. Of the offspring produced, the proportion of individuals that reach maturity from ''infancy'' is never 100% and is often much less. Some will survive and some will not because organisms always produce more individuals than can possibly survive.

Variation

Identical individuals are common among asexually reproducing species such as bacteria and protozoa, and in asexual vegetative plants that reproduce via above or below ground runners (**stolons** or **rhizomes**). Individuals produced via asexual reproduction have the same genetic makeup and are identical in all ways to the parent and each other—they are **clones** (Figure 5.3). Clones are produced via **mitosis**—cell division that produces identical "daughter" cells.

Sexual reproduction, which occurs in many organisms in one form or another and is the only form of reproduction for many species, results in offspring that are different, however slightly, from each other. We can readily observe that our human friends all look different; these differences occur in plants and animals also, but they are not as apparent to us because we are not accustomed to looking for differences among individual mangrove trees or ground squirrels. We even have a difficult time discerning individuals of a different human ethnic group if we have not had sufficient exposure to that group. But the differences are there—among humans, spiders, water lilies, salamanders, and slime molds.

The range of variation in which any variable character is usually distributed can be represented by a bell-shaped curve (Figure 5.4). The most common variation is the one that is intermediate in quality. Thus, if you measure the heights of a species of shrub, you will

Figure 5.3 "Clone" Example. Many varieties of fruits and vegetables are reproduced asexually by taking roots, shoots, or branches from one plant and using them to produce a number of offspring. Thus, one plant with the most desirable characteristics can produce a field or orchard full of plants with exactly identical characteristics because they are all genetically the same; they are clones.

Figure 5.4 **Range of Variation**. See text for explanation.

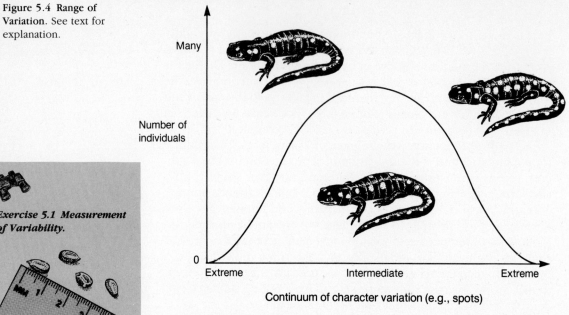

Continuum of character variation (e.g., spots)

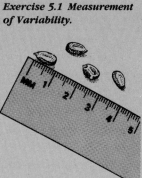

Exercise 5.1 Measurement of Variability.

Obtain a sizeable number (100 or more) of biological specimens—beans, corn kernels, mealworms, and acorns are all appropriate, but many other things can be used as well. Measure one or more of the variable characteristics—length, weight, circumference, speed of movement, color, or whatever. Plot these measurements on a graph with the vertical axis representing the number of individuals and the horizontal representing the range of variability. What kinds of curves are derived? Do curves differ between different species; that is, are some wider or taller than others? Why?

find that tall and short shrubs are much less common than ones of medium height. If you ascertain the amount of cholesterol in the blood of a group of people, you will note that those with high and low amounts are fewer than those with intermediate amounts. You can easily derive one of these curves by measuring the heights of your classmates, their shoe sizes, range of hair color, and so on. (Some physical features will produce a different curve for men and women, however, because men are, on the average, larger.) (See Exercise 5.1.)

Not all characteristics vary, of course: All rainbow trout have two eyes and only two, and a daddy longlegs has eight legs. Any variation in characters such as these is so deleterious that such a variant could not survive and thus does not exist. Can you imagine a robin with three wings? This is not to say that these variations do not arise; "monsters" do occur, but since they are almost inevitably unsuited to survive, they do not persist in nature. Many, many characteristics do vary, however, and these are our focus here.

Variation in the living world is due to three mechanisms: **environmental plasticity**, **genetic recombination**, and **mutation**. Environmental plasticity was discussed in Chapter 1 in the context of acclimation and growth forms. Individual acclimations to the environment are not inherited, so these changes cannot be passed on to future generations and are not part of the fundamental process of evolution.

Genetic Recombination

This type of variation is genetic and does produce changes that appear in the next generation. Chromosomes carry the genetic material, the genes, and it is the arrangement and changes in this genetic material that are important. In mitosis the number and kind of chromosomes are the same in the parent as in the daughter cells, but in meiosis, which occurs in the sex cells, the number of chromosomes in the daughter cells is one-half that of the parent cell (Figure 5.5). This **haploid** cell is a **gamete**—a sperm or egg (various other names can be given to gametes, depending on their role and the organisms in which they are found).

There can be different forms of the same gene (**alleles**) on a pair of chromosomes—one for short fur and one for long fur, for example. But only one of these genes will enter a gamete, since only one of each chromosome pair does so. Thus, the gamete carries only half of the genetic information of each parent. When a gamete from a male meets a female gamete and **fertilization** (fusion of cell nuclei) occurs, a **zygote** (fertilized egg) is formed. This zygote is the beginning of a new individual. Since this new individual obtained half its genetic information from one parent and half from the other, it will be different from each of them. Whether it is intermediate between its parents, resembles one more than another, or resembles neither very much, depends on the genes it receives and their interaction. The trait for brown eyes in humans is dominant to blue eyes, so a blue-eyed father and brown-eyed mother (or vice versa) would produce brown-eyed children, assuming the brown-eyed parent had only brown-eye alleles (was **homozygous**). (In reality, eye color inheritance is much more complex than this.) Sometimes, however, dominance is incomplete, and some offspring inherit a characteristic intermediate between the parents (Figure 5.6).

Another source of variation due to genetic recombination is **crossing over**. In the process of meiosis, chromosomes from the mother (maternal chromosomes) and the father (paternal) line up alongside each other. Sometimes they overlap each other and pieces are exchanged. If this occurs, each chromosome is part maternal and part paternal, creating more possible variation in gametes.

But no organisms have only one or two traits. If we consider an organism with only ten traits, each with two alternative alleles, the number of possible different gametes is 1024. If only 6.7% (a realistic figure) of the genes in humans have alternative alleles (gene pairs are **heterozygous**), any one human could potentially produce 10^{2017} different gametes (Exercise 5.2). That number is considerably greater than all the grains of sand in the world and all the stars in the

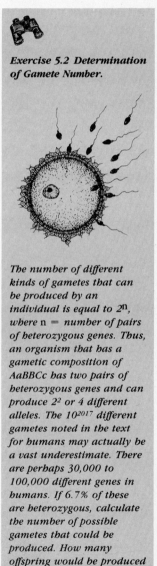

Exercise 5.2 Determination of Gamete Number.

The number of different kinds of gametes that can be produced by an individual is equal to 2^n, where n = number of pairs of heterozygous genes. Thus, an organism that has a gametic composition of AaBBCc has two pairs of heterozygous genes and can produce 2^2 or 4 different alleles. The 10^{2017} different gametes noted in the text for humans may actually be a vast underestimate. There are perhaps 30,000 to 100,000 different genes in humans. If 6.7% of these are heterozygous, calculate the number of possible gametes that could be produced. How many offspring would be produced by the random fusion of pairs of gametes?

Figure 5.5 Meiosis and Mitosis Compared. In mitosis the resulting daughter cells are identical to the parent cell; they contain the same chromosomes and genes. In meiosis the resulting cells have only one-half the number of chromosomes as the parent cell. Note the differences in these haploid cells (gametes) and note that the fusion of any one of them (in fertilization) with a gamete from the opposite sex will produce an individual that is different from each of the parents. Besides the rearrangement of chromosomes, crossing over exchanges chromosome pieces and, thus, genes.

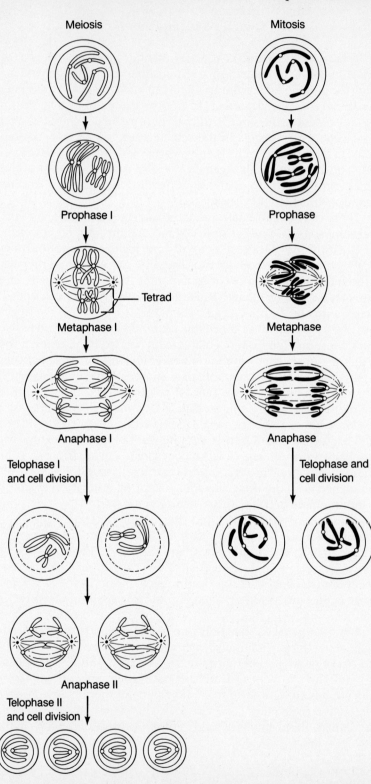

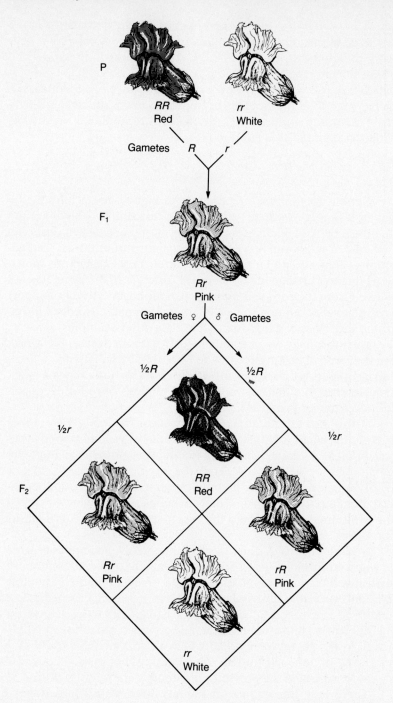

P

RR
Red

rr
White

Gametes   *R*   *r*

F₁

Rr
Pink

Gametes ♀   ♂ Gametes

½*R*   ½*R*

½*r*   ½*r*

F₂

RR
Red

Rr
Pink

rR
Pink

rr
White

Figure 5.6 Incomplete Dominance.
Although some traits have dominant and recessive alleles, resulting in the traits' expression as either bald or fuzzy, black or white, this or that, many traits show incomplete dominance—a blending of traits—which gives rise to characteristics intermediate to those of the parents.

sky. If any one of the 10^{2017} different eggs has the potential of being fertilized by any one of 10^{2017} different sperm, the number of potentially different offspring is $10^{2017} \times 10^{2017}$, a boggling figure. The entire human race will probably never produce that many individuals. Thus, it is essentially impossible to produce two identical humans (except for multiple births from the same fertilized egg). The same reasoning applies to all other sexually reproducing organisms; variation via meiosis is enormous.

Mutation

Genetic recombination produces variation without altering the genes; it simply rearranges them in a process similar to shuffling and dealing a deck of 52 different cards (How many different deals can be had?). A mutation, however, may change chromosomal configurations (Figure 5.7a). Or a new gene (allele) arises (a "point" mutation) due to a change in the chemical structure of the gene; it is not derived from the parents (Figure 5.7b). **Albinism** (lack of pigment) is a relatively common mutation among animals. **Dwarfism** in humans, and a loss of wings in fruit flies, are other examples.

Mutations may occur spontaneously with no discernible cause, but their rate of occurrence can be increased by radiation or chemicals. In the fruit fly, about 1 in 100,000 gametes undergoes a mutation. In corn, the gene for seed shape mutates once for every one million gametes produced.

Mutations have no direction; they occur randomly. Like the result of tinkering casually with the mechanism of a finely tuned Swiss watch, these random changes are not likely to produce an improvement; most are deleterious. Estimates vary, but perhaps 90% to 99% of all mutations are either harmful or neutral—very few are beneficial. But, importantly, the few that are beneficial are the raw material for evolution, as they make the organism better adapted than its forebears.

Nonscientists often, and mistakenly, perceive mutations as large changes in form—a lizard sprouting feathers or a fish being born with a well-developed pair of lungs. Large mutations are extremely rare and probably always harmful. Most mutations, beneficial or otherwise, are very small, and many are likely not to be detected at all. They may simply increase the rate at which a vitamin is synthesized, lower the temperature at which an enzyme reaction occurs, make the underside of a leaf a bit more fuzzy, or allow a marine worm to withstand the pressures of deeper depths in the ocean.

There is no direction to mutations, no certain course toward which changes are directed. We can see patterns and trends in the past, but we cannot predict future trends. Mutations that arise that make the organism better adapted will be maintained and those that

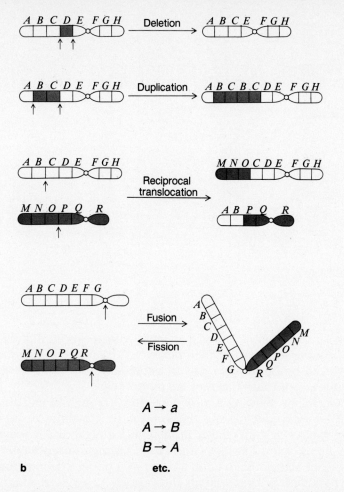

Figure 5.7 Mutations.
(a) Chromosomal Mutation. Changes in the arrangement and/or number of genes on a chromosome can result in variation in some characteristic(s). These changes occur spontaneously or may be caused by x-rays or a chemical **mutagen** (mutation-causing agent). (b) Gene Mutation. In this case, the genes are not shuffled around, but change from one form (allele) to another.

do not will disappear. The environment does not cause certain mutations (for example, a cold environment will not induce a gene for thick fur to arise); it does, however, determine which of the genes that do appear are adaptive and which are not.

Variation, then, arising through changes in or rearrangement of chromosomes and genes, is genetic and thus inheritable. It places in the environment an assemblage of individuals only slightly, but surely, different from each other. Genetic changes cause alterations in the **genotype** or genetic composition of an individual. The physical expression of that genotype is the **phenotype** (Figure 5.8).

Competition

We discuss competition in Chapter 10, but we can define it now as "the striving for limited resources." If resources are limited, not all individuals will receive sufficient amounts of all requisites. Some

Figure 5.8 Genetic and Physical Makeup.
(a) Genotype. The genotype (genetic makeup) of an organism is the organism's complement of genes, which determines the expression of all characteristics. The bands on the chromosome represent the location of genes on the chromosome of *Drosophila melanogaster,* a fruit fly. (b) Phenotype. The phenotype (physical makeup) of an organism is the visible expression of those genes. Here are four different phenotypes of *D. melanogaster.*

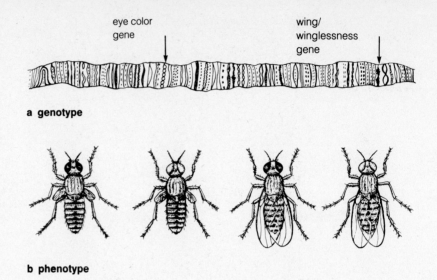

eye color gene

wing/ winglessness gene

a genotype

b phenotype

organisms are better adapted than others, albeit ever so slightly, and will be more successful in their pursuit for resources and in production of offspring. Another way of expressing this higher level of adaptiveness is to say that they are more "fit."

Organisms may be better competitors through having a faster growth or reproductive rate, the ability to photosynthesize at lower temperatures, having thicker insulation or a higher body temperature, being able to swim or run faster, having a more stable root system, or being able to build a better nest or dig a deeper burrow. The best competitors are, then, the most fit.

Natural Selection

The essence of the evolutionary process is survival of the fittest (and, conversely, reduced survival of the less fit). "Survival" is not totally accurate, since organisms that survive but do not produce offspring leave as many descendents as those who die (this is **genetic death**). What survival of the fittest really means is that the better adapted individuals produce more offspring, thereby passing on more of their genes to future generations. Less fit individuals pass fewer or no genes on. Because the genetic constitution is what is passed on through the gametes, **natural selection** can be defined as the "differential reproduction of genotypes," meaning that some genotypes survive and reproduce better than others (Exercise 5.3).

The environment, however, does not act directly on the genotypes, but on their physical expression, the phenotypes. (Think of

Exercise 5.3 Natural Selection.

It might be assumed that deleterious genes will disappear from a population rapidly and beneficial ones increase quickly. Here is an exercise to examine those ideas.

You will need a tray with an edge and a supply of 50 red and 50 black (or two other colors) beads (or marbles). Let the red beads represent dominant genes and the black beads recessive. Shake the tray to mix the beads. Tip the tray to allow the beads to fall to one edge and become aligned—representing fusion of gametes. Each pair of beads will then represent an individual organism. Assume that each pair of red beads (homozygous dominant) and each pair consisting of one red and one black bead (heterozygous) represent an individual adapted to the environment. Each pair of black beads (homozygous recessive) represents an individual with a deleterious trait. Remove the pairs of black beads. Mix the bead population again (a new "generation") and again remove the pairs of black beads. Record the data for each generation as shown at the left: How many generations did it take to remove all the black beads (get rid of all the deleterious genes)? Modify the exercise by making heterozygotes unfit also. How many generations does it take to remove the deleterious genes? How would you change the exercise to make hetero-zygotes more fit than the homozygotes? What other changes could you make?

Generation	Number of Pairs Red	Number of Pairs Red/ Black	Number of Pairs Black
1	•	•	•
2	•	•	•
•	•	•	•
•	•	•	•
•	•	•	•

the blueprints of a house as the genotype and the house itself as the phenotype. If the house falls down in a high wind, the phenotype is acted on by the environment, but the genotype will also disappear as those blueprints are destroyed with the house.) Natural selection, then, operates by allowing the better adapted individuals to produce more offspring. Ultimately the more fit become more abundant and the less fit rarer or even extinct. New variations are always arising, so the better fit are always replacing the less fit (Figure 5.9). The environment "selects for" some individuals and "selects against" others. Since the environment is always changing, so are the selective pressures, and evolution will occur more rapidly in faster changing environments (within limits, dependent on such things as reproductive rates, generation time, and mutation rates).

Darwin understood the role natural selection plays in evolution and said, "Can it be thought improbable . . . that . . . variations useful in some way to each being in the great and complex battle of life, should occur in the course of many successive generations. If such do occur, can we doubt (remembering that many more individuals

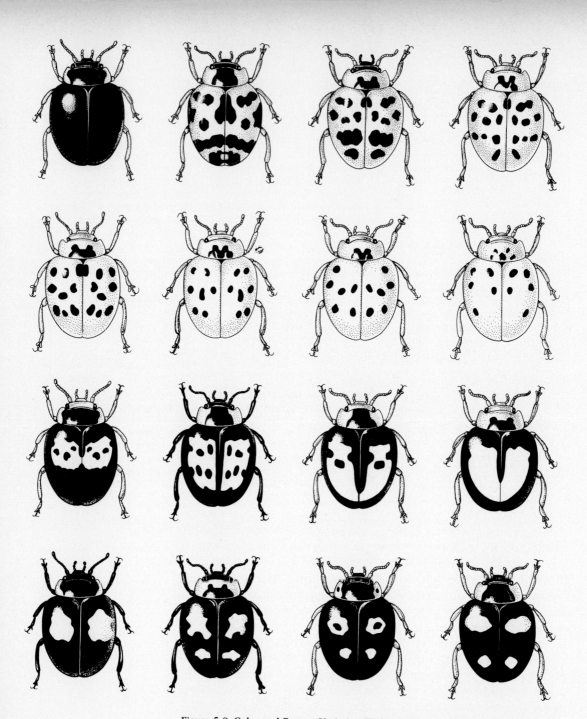

Figure 5.9 Color and Pattern Variation Within a Species. The wing covers of the lady beetle species *Harmonia axyridis*, which occurs in Asia, vary considerably depending on where the individuals are found. The almost entirely black one is dominant in west-central Siberia, and the black spots on yellow phenotypes are found further east. The bottom row have red spots on a black background and are found only in the far east (Japan and Korea). New variations are always arising in a population. Those that are best fit survive to reproduce. In this case many variations happen to be adapted to a variety of environments.

are born than can possibly survive) that individuals having any advantage, however slight, over others, would have the best chances of surviving and procreating their kind? On the other hand, we may feel sure that any variation in the least degree injurious would be rigidly destroyed. This preservation of favourable individual differences and variations, and the destruction of those which are injurious, I have called Natural Selection, or the Survival of the Fittest." (*Origin of Species.* New York: New American Library of World Literature, A Mentor Book, 1960, pp. 87–88).

EVOLUTION IN ACTION

To put all these processes together, let us consider a hypothetical example. A population of rabbits lives in a grassland. If we measure the length of their hind feet, we find the expected variation in foot length distributed in the characteristic way (Figure 5.10).

Few have long or short feet; most have an intermediate length foot. How is this adaptive? Consider what a rabbit uses its feet for: running and digging burrows. Those with short feet and those with long feet are not as adept at going about their business as those with intermediate length feet. Cottontail rabbits of the eastern United States, for example, have a hind foot length of 87–104 millimeters; those with a foot length near 87 or 104 are not as well adapted as those whose feet measure around 95 millimeters. Note also that you will not find rabbits with feet measuring less than 87 or more than 104 millimeters; these would be so unfit that they apparently cannot

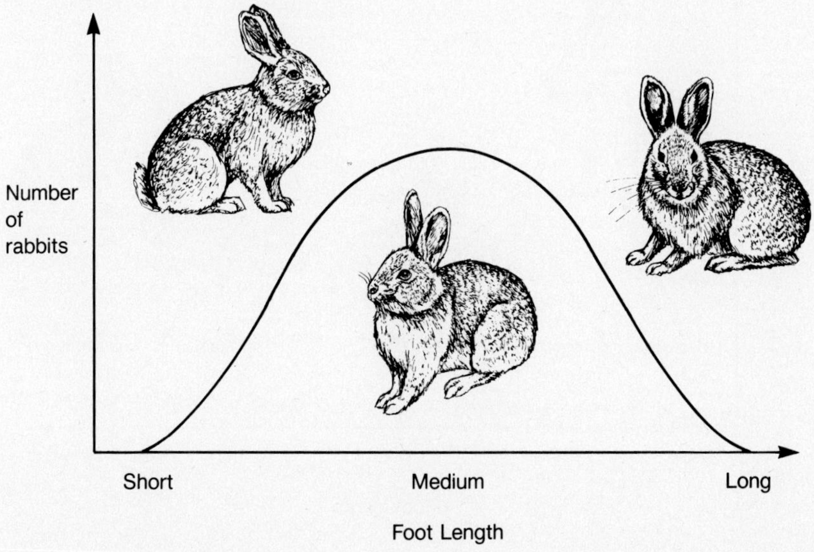

Figure 5.10 Variation in Foot Lengths in Rabbits. See text for explanation.

exist at all. Note, however, that "very short" and "very long" footed rabbits are produced, though they are selected out very quickly and thus will rarely, if ever, be captured by a scientist studying these rabbits.

Stabilizing Selection

Under conditions of a stable climate and ecosystem, natural selection acts as a stabilizing force, keeping the rabbits' feet in the same range of length. Remember—and this is significant—variability continues to occur in the population, and rabbits in a wide range of foot length are continually being born; if they all survived, the foot length range would be much greater than 87–104 millimeters. But they do not all survive. The environment allows certain ones (the "medium-footed" rabbits) to survive and reproduce optimally, and others to survive and not reproduce; still others do not survive at all ("very long-" and "very short-footed" ones). This is called stabilizing selection (Figure 5.11). The rabbits have been divided into these five groups based on the effect of environmental pressures on their survival and reproduction. Medium-footed rabbits are "selected for," short- and long-footed are "selected against" (reduced reproduction), and very long- and very short-footed rabbits are "selected out" (die). These categories are not clear-cut since they occur along a continuum; the distinction between selected out or against is often not made, as the results of both are frequently the same in the long run.

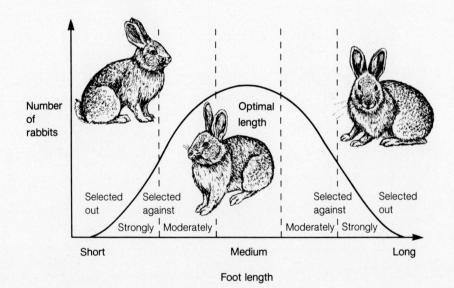

Figure 5.11 Stabilizing Selection (Rabbit Foot Length). See text for explanation

Put simply, the medium-footed rabbits can run faster and dig burrows deeper or quicker, making them more able to avoid predators and reduce the effect of bad weather. They are then more fit and will have more young; therefore, succeeding generations of rabbits will consist mostly of medium-footed individuals. Put another way, future generations will most commonly possess the genotype that results in the medium-footed phenotype. Thus, rabbits fit for the environment (as far as foot length is concerned) are selected for, and the unfit disappear. Although this type of natural selection, stabilizing selection, does not appear to be evolutionary (because changes in the adult population are subtle and not readily apparent) it definitely is! We do not see the individuals with the extreme (short- and long-footed) traits because they are rare and do not survive as long as the more common individuals with intermediate foot length.

Directional Selection

When the environment changes, selective pressures are altered, and what was once fit may no longer be, and what was deleterious may become neutral or even beneficial.

To continue our rabbit example, let's say that over a period of several hundred or thousand years the climate changes in such a way that the rainfall increases considerably. What was once a grassland becomes a marshlike environment. Burrows are not feasible to use, and escape from predators depends on the rabbits' ability to run across a wet or watery surface. It is logical to assume that the foot length that was well adapted to the grassland is not as well adapted to the marsh. Longer (and wider) feet provide more surface area, reduce sinking into the wet surface, and make locomotion easier.

So how, in our example, does the population of rabbits become long-footed in response to the development of the wet habitat? Long-footed rabbits exist already, but in low numbers. Now, as the environment changes in their favor, natural selection selects them and selects against medium-footed rabbits. Thus, the rare long-footed rabbits we used at the beginning of our example are now the norm, and medium- and very long-footed are the extremes. Short- and very short-footed rabbits are no longer found because they are totally unfit, but very long-footed rabbits, although low in number, exist now, whereas they could not do so in a grassland habitat (Figure 5.12).

This is evolution in the classical sense; the environment changes, and the appropriate organisms are "filtered" by the process of natural selection moving the curve of variation in a certain direction. A gradual change in one or more characteristics produces

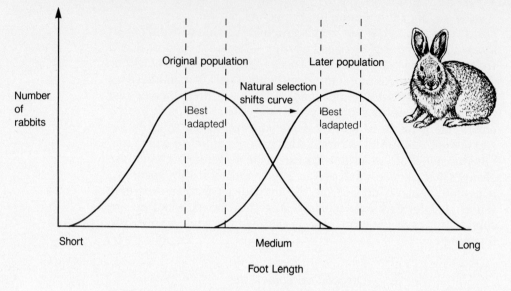

Figure 5.12 Directional Selection (Rabbit Foot Length). See text for explanation.

a somewhat different organism. Of course, natural selection works not only on foot length, but also on hair color, weight, blood proteins, resistance to freezing, immunity to diseases, the shape of the genital structures, and many other characters. The sum effect of all the variable features in an individual determines its fitness. This is not a chance process, as creationists often imply; random variations do occur, but the elimination of some and the perpetuation of others is totally dependent on the individuals' inherited ability to cope with environmental pressures.

Other Examples

One of the classic cases of evolutionary change is that of the peppered moth. This species, found in England, includes a "peppered" form (white with black spots) and a melanistic (blackish) form. In 1848 the melanistic form comprised 1% of the population and the peppered form 99%. In 1898 the melanistic form made up 99% of the population. How did this change occur?

It occurred because in the part of England where this study took place, the trees had become blackened with coal soot from nearby industries. Before 1848 light colored lichens covered the trees. The peppered moths, resting on the bark, were difficult for bird preda-

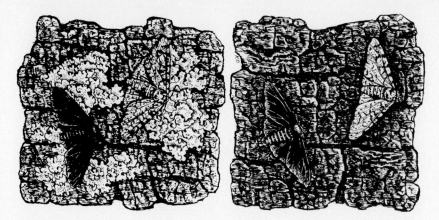

Figure 5.13 Directional Selection in *Biston Betularia*. (Left) On a "normal" lichen-covered tree bark, the peppered (light) form of the moth blends in with the background and the dark form stands out. (Right) On the soot-covered tree, the peppered moth contrasts with the bark and the dark form is much less obvious.

tors to detect; the dark forms stood out and were eaten. As the industrial soot covered the trees and killed the lichens, the tree barks darkened, and the peppered moths became more obvious while the dark forms blended in. Thus, predation pressure (natural selection) shifted to the light forms and allowed the dark forms to increase—directional selection again. In recent years, since air pollution has diminished, the shift has been back toward the peppered moth form (Figure 5.13).

The evolution of resistance to pesticides by insects is common. A population of mosquitoes may be continually exposed to DDT, for example. A few mosquitoes may possess a genotype that provides them resistance to the pesticide. These will survive while the nonresistant ones will not. Ultimately the entire population will consist of DDT-resistant individuals. This is usually what has happened when it is discovered that a previously effective pesticide no longer works as well as it once did (Figure 5.14). In a similar way, pine mice exposed to endrin (an insecticide-rodenticide) for 11 years developed a 12-fold tolerance to the chemical.

LARGE CHANGES, OR MACROEVOLUTION

Virtually all changes in an organism happen in small, almost undetectable amounts, usually over a long period of time. But given sufficient time (and organisms have been evolving for 3½ billion years, a very long time), small changes accumulate, and present-day

pesticide
resistant
forms

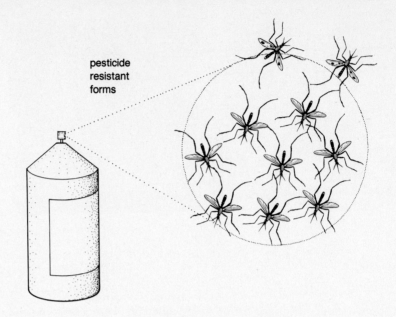

organisms are usually quite different from their ancestors. Thus, it is no surprise that birds evolved from reptiles—not as a chicken popping out of a snake's egg, but as an aggregation of many, many small changes (**microevolution**) over a period of 150 million years. By a similar process, algae gave rise to plants called psilopsids, which eventually gave rise to ferns.

We now see only the present results of evolution. The earliest ancestors and most intermediate forms are known only from fossils, and the fossil record is incomplete. Thus, the evolutionary history of an organism or group appears to have happened in large, discrete steps. In all probability, however, a large number of intermediate forms arose and disappeared in a continuous line of forms that differed only slightly from one another.

However, Stephen Jay Gould and Niles Eldridge, well-known contemporary scientists, argue convincingly that large changes can occur, that is, that evolution can happen in discontinuous steps. One of their examples is that some structures, such as the external cheek pouches of kangaroo rats and pocket mice, do not lend themselves to gradual evolution. What good would partly developed cheek pouches be, they ask? They make a plausible case for this **macroevolution**, reconciling it with Darwinism by postulating that small changes in early embryological development can result in large changes in the adult form.

EVOLUTIONARY RELATIONSHIPS

If one or more groups evolved from another group, we would expect them to share some similarities (Figure 5.15). We can deduce relationships of organisms via these similarities if we make the distinction between **convergent evolution** and evolution from a common ancestor. In convergent evolution, organisms resemble each other because they evolved similar features independently. Birds and insects have wings, for instance, but even a superficial examination reveals that the wings are very different in structure. They have the same function but a different derivation; bird and insect wings are **analogous**. The wings of birds, the wings of bats, the hands of humans, and the flippers of porpoises all work differently, but they are constructed on the same basic plan (Figure 5.16) and share a similar ancestry; they are **homologous**. There are many other similar examples from which we can draw geneologies ("family trees"), because the more similarities two organisms share, the more closely they are related (Figure 5.17). The fossil record has also been extensively used to determine the relationships of both extinct and living forms (Figure 7.1).

ARTIFICIAL SELECTION

Artificial selection is simply the selective breeding of domestic animals or plants to produce varieties that serve best the needs of humans. Cows, horses, dogs, cats, pigeons, wheat, corn, tomatoes, peaches, roses, fungi, and many other organisms have been artificially selected for desirable characteristics. Dogs are bred for show, for work, or for hunting. Fruit trees are selected for their growth rate,

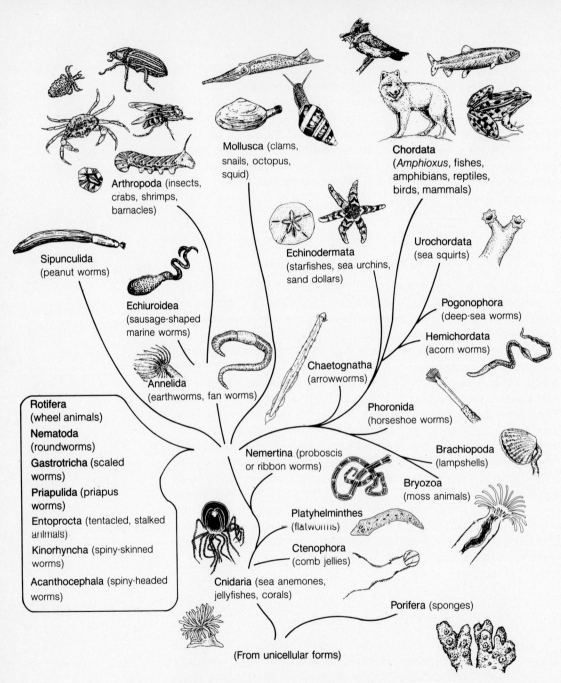

Figure 5.15 Major Groups (Phyla) of Animals. The evolutionary process has produced many changes over time. This tree shows the implied pattern of modifications leading to existing major groups of animals.

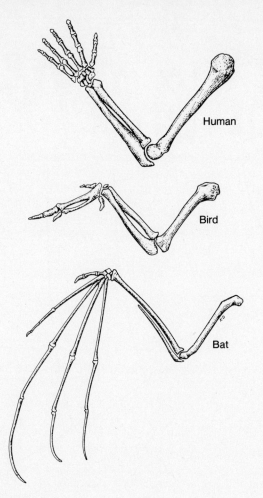

Figure 5.16 Homology in Forearms. The forearms of vertebrates are all derived from the same basic pattern, since they all evolved from a common ancestor. This is not convergent evolution, but homology. The bird and the bat wing are both constructed for flight, although the fingers of the bat are long and thin to support the wing membrane, and the bird's fingers are fused to support feathers. The similar human hand is used quite differently. All three have the same bones, modified for different functions.

fruit production, and resistance to disease. Molds were selected for their ability to produce penicillin in large amounts.

This selective pressure is not natural, since humans do not use the same criteria as the natural environment to select for or against individuals. Indeed, many varieties of pigeons, cats, dogs, and cattle seem quite unfit for anything except display or food production.

EXTINCTION

Descendant plants and animals are somewhat different from their predecessors, who frequently have become extinct. Sometimes, however, the entire species becomes extinct and leaves no descendants. If the environment changes so rapidly that there is no

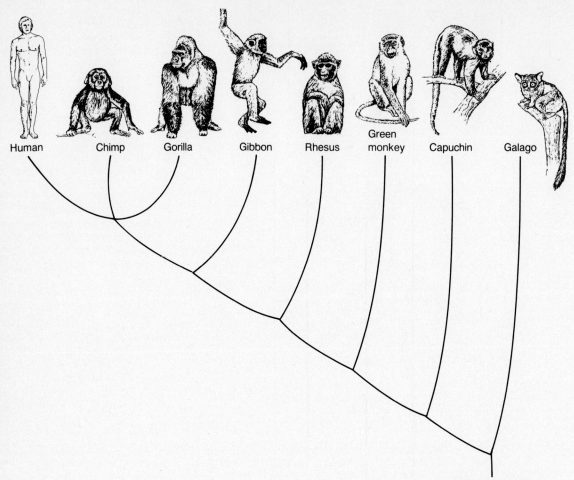

Human   Chimp   Gorilla   Gibbon   Rhesus   Green monkey   Capuchin   Galago

Primitive mammal

Figure 5.17 Human Evolution. Humans and the other primates evolved from a common ancestor. The similarity of humans to living primates is unmistakable; skeletal structure, muscles, physiological processes, DNA, blood proteins, and so on are quite comparable. For instance, in a sequence of 146 amino acids in the hemoglobin molecule, there is only one difference between man and chimpanzee; differences from other animals are much greater. Note that monkeys and humans are contemporaries; humans did not evolve from monkeys.

chance for evolution to occur, then all the organisms may be unfit and disappear. In our rabbit example, for instance, a marsh could be created by a sudden geological force that causes a subsidence of the landscape, flooding it in a few weeks or months. Rabbits can produce perhaps only two litters a year, and even if the appropriate genotypes were present, they would not be present in sufficient numbers to reproduce quickly enough to maintain the population; the entire population would disappear.

In many cases the genetic composition does not possess any genes that would provide appropriate variations to allow an organism to adapt to a changing environment. Consider a hypothetical example of a fish living in a pond. Although the fish produces many offspring with a wide range of variability, it is highly unlikely that any offspring would be produced that could live on land. This is too large of a change to occur. Not only would the respiratory system of the fish have to change, but also its skin, fins, mouth, feeding habits, and so on. The movement to land could not occur unless there arose a set of large, beneficial mutations—extremely unlikely. Thus, if the pond dried up over a period of years, the fish would simply die. Many extinctions can be explained by an environment changing too rapidly for appropriate genetic variation to arise and be incorporated. Sometimes the changes are made by humans. For example, the introduction into Hawaii of rats, mongooses, and exotic birds has resulted in the extinction of at least 23 native bird species in the past 200 years; at least another 28 are in danger of extinction.

COEVOLUTION

Because the evolutionary response of any species is dependent on selective pressures imposed upon it by external environmental factors, both biotic and abiotic, we often find situations in which two or more species evolve in response to each other.

Female butterflies of a particular species lay their eggs only on the immature flowers of a species of lupine in Colorado. The hatched larvae feed on the flowers. So much damage is done to the flowers that they produce only 37% of their potential number of mature flowers which would, of course, go on to produce seeds. In another population of lupines where these butterflies are rare, 75% of the flowers produce seeds. So the butterfly considerably reduces the potential reproduction of lupines. How do the lupines react?

In the attacked population, the lupines begin and end their flowering nearly a month earlier than the unattacked population. The plants begin to flower before the adult butterflies emerge, thus reducing the amount of potential damage. If they flowered at the same time as the unattacked population, the flowers would presumably be eliminated by the selective pressure of the butterflies. (From D. E. Breedlove and P. R. Ehrlich, "Plant-herbivore coevolution: Lupines and Lycaenids," *Science* 162 (1968): 671–672.)

EVOLUTION CONTINUES

". . . whilst this planet has gone cycling on according to the fixed law of gravity, from so simple a beginning endless forms most beautiful and wonderful have been, and are being evolved." So ended the *Origin of Species*, C. Darwin, 1859.

SUMMARY

Evolution refers to change over time, and evolutionary theory refers to the explanations of how evolution works. Darwin and Wallace brought the concept of evolution clearly into the scientific

Figure 5.18. The Tennessee Anti-Evolution Act.

THE TENNESSEE ANTI-EVOLUTION ACT

Chapter 27, House Bill 185 (By Mr. Butler)

Public Acts of Tennessee for 1925

An Act prohibiting the teaching of the Evolution Theory in all the Universities, Normals and all other public schools of Tennessee, which are supported in whole or in part by the public school funds of the State, and to provide penalties for the violations thereof.

Section 1. Be it enacted by the General Assembly of the State of Tennessee, That it shall be unlawful for any teacher in any of the Universities, Normals and all other public schools of the State which are supported in whole or in part by the public school funds of the State, to teach any theory that denies the story of the Divine Creation of man as taught in the Bible, and to teach instead that man has descended from a lower order of animals.

Section 2. Be it further enacted, That any teacher found guilty of the violation of this Act, shall be guilty of a misdemeanor and upon conviction, shall be fined not less than One Hundred ($100.00) Dollars nor more than Five Hundred ($500.00) Dollars for each offense.

Section 3. Be it further enacted, That this Act take effect from and after its passage, the public welfare requiring it.

Passed March 13, 1925.
W. F. BARRY, *Speaker of the House of Representatives.*
L. D. HILL, *Speaker of the Senate*
Approved March 21, 1925
AUSTIN PEAY. *Governor*

---

world. The process of evolution involves (1) overproduction, (2) variation, (3) competition, and (4) natural selection. Natural selection is the process by which the most well-adapted individuals are selected by the environment to reproduce at a higher rate than the less fit individuals, thus increasing the propagation of genes of the well-adapted types. Evolutionary changes occur slowly and in small steps, but large changes may also occur. Many plants and animals coevolve adaptations. Directed selection by human intervention is artificial selection.

STUDY QUESTIONS

1. Define selection, homologous, convergent evolution, phenotype, genotype, environmental plasticity, alleles.

2. What is the Theory of Acquired Characteristics?

3. Differentiate between stabilizing and directional selection.

4. Name and briefly describe the four major mechanisms of evolution.

5. What are the causes of variation in organisms?

6. Natural selection is the "differential reproduction of genotypes," yet selective pressures act upon phenotypes. Explain.

7. If vegetative (asexual) reproduction does not produce different individuals, and thus does not allow evolution to occur, why do you think it is so common in the biological world?

8. Why are variable characteristics always distributed in a bell-shaped curve?

9. Discuss genetic recombination and its role in evolution.

10. If the environment were unchanging, do you think evolution would still occur? Justify your answer.

Chapter 6

The Species Concept

Over billions of years of evolution, plants and animals have diverged into many different types, yet they all retain some common features, indicating mutual origins somewhere in the past. All organisms are made of cells, contain DNA, and share many chemicals and metabolic processes. Yet most organisms are quite distinct—trees and spiders, for example, are unmistakably different organisms.

As Darwin observed, the great diversity of environmental conditions led to a great diversity of life forms. Since all these life forms derived from a common ancestor, their relationships can be represented by the branches of a tree. (See Figure 5.15.) As a tree grows, new buds form new twigs; some twigs become branches, some do not; some branches become larger, some die. But in some way, all the twigs and branches are related to each other, even those on the opposite sides of the tree, because they share a common origin.

Like the branches of the tree, all organisms are related to some degree. Knowledge of organisms, anatomy, physiology, behavior, and other characteristics help scientists determine the degree of relationship. The more similar creatures are, the more likely they are to be related. Flora and fauna are grouped into categories based on the closeness of their relationship. Each organism belongs to a set of categories, or classification levels, the broadest containing many other organisms, the narrowest naming only one type (Figure 6.1). The science of classifying and naming is called **taxonomy**. Although people have always categorized animals and plants in

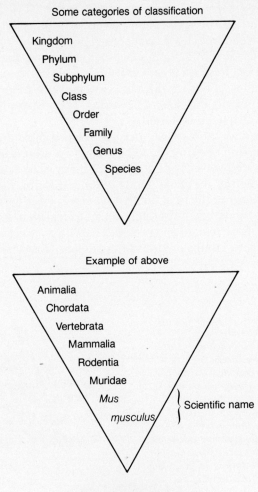

Some categories of classification

Kingdom

Phylum

Subphylum

Class

Order

Family

Genus

Species

Example of above

Animalia

Chordata

Vertebrata

Mammalia

Rodentia

Muridae

Mus

musculus   } Scientific name

Common name: house mouse

Figure 6.1 A Diagrammatic Scheme of Classification.
(a) Categories of Classification. Arranged in a rank order with the largest (most inclusive) groups on top, descending through smaller ones, this is called the **hierarchical** scheme of classification. Further, each category can be divided into many other categories: subfamilies, suborders, tribes, divisions, and so forth. (b) Classification Example. Note the bottom two names, *Mus* and *musculus*. The genus (*Mus*) is capitalized but the trivial name or specific epithet (*musculus*) is not; both are italicized. The genus and specific epithet together comprise the scientific name and refer to a specific type of organism.
The scientific name is usually derived from Latin or Greek and is often descriptive. *Anas platyrhynchos* is the scientific name for the mallard duck; *Anas* is Latin for duck and *platyrhynchos* is a Greek word meaning flat- or broad-billed. "Mallard" may be easier to pronounce, but common names tend to be colloquialized and thus vary in different geographic regions. Scientific names are standardized and can be used by scientists internationally. If a German friend wrote to you and said he saw a Schleiereule outside his window, would you know it was a barn owl—a species found both in North America and Europe? If you were both familiar with scientific names, you could instead discuss the *Tyto alba*. Appendix A lists the scientific names for all of the common names mentioned in the text.

some way, it was not until a Swedish naturalist, Carolus Linnaeus, published his *Systema Naturae* in 1758 that the modern system of classification began.

Taxonomy is often subjective because of the tremendous number and variability of organisms. There is often disagreement among taxonomists as to how many, what kind, and what degree of differences comprise a separate genus, or class, or other category. Differences between species have largely been based on anatomical features such as skull length, weight, intestinal configuration, color of skin, and so forth. In recent years data from chromosome counts, blood proteins, ability to withstand extremes of temperature, behavior, nest construction, time of flowering, and many other characteristics have also been included.

Taxonomy is somewhat artificial; relationships are really on a continuum, and drawing a line between two groups is often based on an educated guess. For example, for many years scientists debated as to whether the giant panda belonged to the raccoon family (*Procyonidae*) or the bear family (*Ursidae*). (It is currently placed in the raccoon family.) Although not an exact science, taxonomy is becoming more refined and provides us with an acceptable method for referring to groups or kinds or organisms (Exercise 6.1).

Exercise 6.1 Using a Key.
The identification of organisms is made simpler through the use of a "dichotomous" key. Dichotomous simply refers to the way the key works; there are usually two different choices—is the flower of the plant red or blue, for example. By making enough careful choices—are the hind legs longer or shorter than the forelegs, is the skull wider than 100 millimeters, and so on—you will finally end up with a list of characteristics long enough to identify the organism to some level of classification. Keys may go to order, family, genus, species, or almost any other category. They can be used for any organismic group anywhere in the world. Most are based primarily on anatomical features, but habitat,

behavior, or physiological characteristics are sometimes included. Since keys are rarely comprehensive—a key to the flowering plants of North America would be imponderably huge and nearly impossible to use— keys are typically restricted to small sets of organisms or to a limited geographical area. Thus, we find keys to the fishes of Illinois, to the flowers of southern Arizona, and to the birds of the West Indies.
Learn to use keys appropriate to your area, beginning with ones for relatively easy organisms such as trees. Once the idea of keying is learned it is universally useful. For a greater challenge, one of the best ways to become familiar with the identification of a group of plants or animals is to invent a key. Begin simply; it's harder than it sounds.

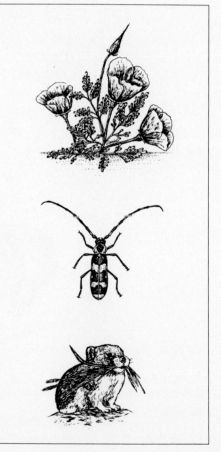

THE CONCEPT OF SPECIES

The use of similarities to define relationships is limited, though, and even closely related organisms that look and act very much alike may be considered definitively different. We may call these different groups of organisms species. A **species** is a population (or group of populations) of organisms, all of which are capable of interbreeding and producing viable young. ("Species" is both singular and plural; "specie" is not a biological term—it means coined money.) A population whose individuals are incapable of interbreeding with those of another population typically belongs to a different species. This gives us a natural, biological test of a relationship between two "types" of organisms. Mushrooms and termites are obviously very different organisms, incapable of interbreeding, but what about very similar birds such as the eastern and western meadowlarks? Their physical appearances are nearly identical, and the only sure proof that they are different species is their lack of interbreeding (although their songs are different, which first gave scientists a clue that there were more than one species) (Figure 6.2).

Western meadowlark

Eastern meadowlark

Song

high
pitch
low

length

length

Range

Figure 6.2 **Eastern and Western Meadowlarks**. The Eastern and Western Meadowlarks are virtually identical in physical form, behavior, and habitat, but their songs are quite different, which enables them to recognize their own species where the two forms overlap. The two species are found separately throughout the United States, Mexico, and southern Canada; they overlap only in the central United States and part of Canada.

Because of the widespread variation within a species, there may be many differences between different populations of the same species. Thus, we can find **clines**, which are gradual changes in a characteristic from one area to another. We can, for example, observe a decrease in the size of the eggs of meadow frogs from the northeastern to the southeastern United States, and populations of the yarrow plant show a decrease in height from the California valley to the high Sierra. Figure 6.3 shows clinal variation in body size for house sparrows in North America. Within a species there may also be discretely different morphological types with no intermediate forms. Snow geese have a white and a gray form, the screech owl has a red and gray phase, and the red-backed salamander has red-backed and gray-backed forms.

Bird species that are distributed over a wide latitude are clinal with respect to clutch size. The European blackbird for example, lays one to three eggs in the Canary Islands, three to four in Spain,

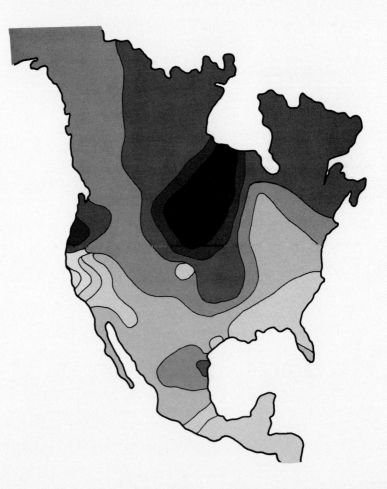

Figure 6.3 House Sparrow Clinal Variation. House sparrows have larger body sizes in the northern portion of their ranges. The darker areas indicate the larger bodied birds' range. Larger bodies mean lower heat loss per body weight, so the larger, more northerly birds use proportionately less energy and can withstand lower temperatures than the smaller, more southerly birds.

five to six in Germany, and five to seven in Russia. The usual explanation for this variation is that the southern latitudes are milder, so the birds there are more likely to raise successfully a higher proportion of their young and/or have the opportunity to lay two or three clutches during the year—thus, fewer eggs are necessary per clutch.

TYPES OF CLINES

Clines can be based on variation in color, shape, size, metabolism, or various other characteristics. Three of the most common types of clines—those concerning body size, appendage length, and color—have been designated "rules," and apply mainly to homeotherms.

Bergmann's rule states that the farther away from the equator one goes in a species range, the larger the body size becomes. As bodies become larger, their volume increases relative to their surface area, over which heat is lost. Thus, less heat (energy) is lost per body weight, and larger animals are better able to withstand colder temperatures (Figures 6.3 and 6.4).

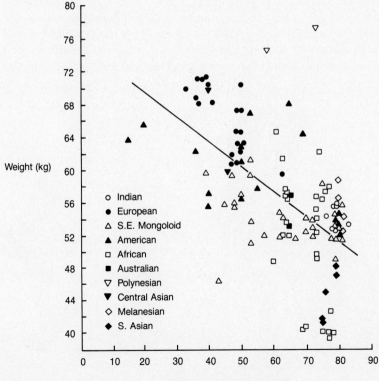

Figure 6.4 Bergmann's Rule. There is a negative correlation between the average annual temperature of the habitat and human populations. In general, people in colder habitats are heavier and taller.

Allen's rule, based on the same reasoning about heat loss, states that homeotherms farther from the equator have shorter appendages than do those animals in the warmer areas. Shorter appendages reduce the surface area for heat loss (Figure 6.5).

Gloger's rule says that organisms in warmer, more humid regions are darker in color than those in colder, drier regions. This may be related to temperature regulation or protective coloration, but the explanation is not clear.

These are by no means strict rules, and there are numerous exceptions. Some darkly pigmented insects, for example, are found in both humid and cold areas, and light-colored ones in warm and dry areas. Local conditions can also serve to modify a cline.

THE GENE POOL

Over the course of evolution, genes for many characteristics have arisen and been incorporated into each individual. But each individual does not possess the same genetic makeup (genotype). Since individuals exist as part of a population of individuals, we can also refer to a population of genes. The total of all the genes within a population of organisms is called the **gene pool**. Hypothetically, any organism within the population is free to mate at random with any other organism; this random mating is called **panmixis**. All of the genes are, in a sense, shared by the members of the population. And any new beneficial mutation that arises may eventually spread throughout the population after a number of generations. Thus, the

Figure 6.5 Allen's Rule. The black-tailed jackrabbit found in warmer open areas at lower elevations has much longer ears than the snowshoe hare of the higher elevations and colder climates. Shorter ears, among other adaptations, reduce heat loss.

entire population evolves, not as a unit, but through changes in the individuals in the population.

One can see differences in groups of the same species in different parts of the geographic area over which a species exists. Although the groups are of the same species, and thus capable of interbreeding, they exhibit geographic differences in the gene pool. The most striking example is among humans; we recognize several races: Eskimos, American Indians, Mongoloid, Negroid, Caucasian, Polynesian, and many others. Among plants and animals the differences may be very slight or very distinct. These geographical variants of the same species are known as races; if the races are given scientific names, they are called **subspecies** (Figure 6.6). The designation of subspecies is extremely subjective, as clear distinctions

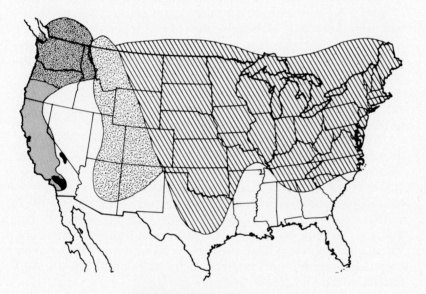

Figure 6.6 Subspecies of *Hoplitis producta*, a **Leafcutting Bee**. Barriers to easy dispersal between populations, and the influences of the local climate, initiate the evolution of different forms of the same species, perhaps eventually leading to new species. Shown here are the ranges of subspecies.

between races are rare. For example, depending on the taxonomist, there are somewhere between 7 and 30 races of Canada geese.

SPECIATION

Speciation is the formation of new species; it is one result of the evolutionary process. Via inheritable variation and natural selection, new "kinds" of organisms arise. As the variations are spread through the gene pool, all the organisms change similarly, and, though ancestors and descendants may be quite different after many generations, individuals of the population share a common gene pool and thus the same species. For a population to split into two or more species, geographical separation is usually necessary.

Consider an area of grassland occupied by a species of mouse (Figure 6.7a). Any individual mouse can potentially meet and mate with any other mouse of the opposite sex in that population. As long as that can occur, different populations cannot arise. If, however, through some geologic force, a river (or mountain, or chasm) appears and divides the population into two populations of the same species, we have **geographical isolation** since the mice from one side of the river cannot contact the mice on the other side (Figure 6.7b). They are then known as **allopatric** populations.

Now the two populations no longer share the same gene pool; they each have a separate one. Within each gene pool, inheritable variations continue to arise and spread through the pool. The populations continue to change, but in slightly different directions. Even though the environments may be the same on both sides of the river, variation is random, and natural selection thus has different "raw material" to work on. After all, there are a variety of ways to cope with the same environment, and there is no reason why both mouse populations cannot be somewhat different but still equally well adapted to their environment.

As differences continue to arise and become incorporated in each population, the populations diverge; their morphology, physiology, and/or behaviors become different. If the environments on either side of the river are different, different selective forces will act on them, and the divergence will occur more rapidly. If the barrier between the populations now breaks down (if the river dries up or is diverted, for example), the populations again share the same geographic region: they are **sympatric** (Figure 6.7c).

During the time of separation, each population has accumulated some changes. If the sum of these changes is not too great, the mice will again form one gene pool, interbreed, share the accumulated variability, and overwhelm the differences between the two populations. No new species has arisen. If, however, sufficient changes

Figure 6.7 Speciation. (a) A Population of Mice. In a large, flat grassland, mice of the same species can freely move, intermingle, and interbreed. Any new beneficial variations that arise via mutations will ultimately be shared by future generations. (b) Two Allopatric Populations. Separated by a barrier of flowing water, the population is now split in two. Each will develop its own set of variations, and the two populations will slowly diverge. (c) Two Sympatric Populations. Allowed to intermingle again, the two populations may again merge as one species or retain their respective identities and thus become two species.

a

b

c

have taken place so that, although sympatric, the two populations cannot interbreed, they are then two different species. Speciation has occurred. This is usually called **allopatric speciation**.

Parapatric Speciation

Although geographical isolation, which makes populations allopatric, is fundamental to the origin of most species, it is not always necessary. If gene flow among individuals is slow or inefficient, as among some wind or water pollinated plants, widely separated individuals on each end of the gene pool could lose the capability to interbreed. This is common in clines, since one end is often distant from the other. In the meadow frog, a cline (of several features) is distributed from Quebec to Florida. Adjacent populations, those of Vermont and Maine, or Georgia and Florida, interbreed fairly readily and produce viable offspring. In experimental cross-breeding of populations with greater separations (Florida and North Carolina, for example), fewer viable offspring are produced. And at the ends of the cline (Florida and Quebec), interbreeding of the two populations produces few or no viable offspring. The most distant populations, although the end points of a continuous gene pool, are, therefore, virtually separate species. This is known as **parapatric speciation** (Figure 6.8).

Sympatric Speciation

Parapatric speciation requires some limitation of gene flow even though strict allopatry does not occur. Geographic isolation produces reproductive isolation, but if reproductive isolation continues to occur in a sympatric situation, then speciation can also occur when populations or individuals are in close proximity. The most common way for sympatric speciation to occur is via **polyploidy**, a multiplication of the haploid number of chromosomes, so that the organism is diploid, triploid, tetraploid, or of greater multiples. (Recall that adult organisms, with few exceptions, are **diploid**, that is, they contain pairs of matched chromosomes. As a result of meiotic divisions, gametes (sex cells) are produced that are haploid, containing only one chromosome from each pair.)

Polyploidy occurs during meiosis when the paired chromosomes fail to divide and the resulting gametes are diploid. A polyploid individual's gametes cannot fuse with a "normal" haploid gamete so it is reproductively isolated and thus creates a new species. In addition, polyploids can only interbreed with others of the same chromosome number (diploids and triploids cannot cross). Polyploidy is very rare in animals, but more than half of all flowering plant species have arisen through polyploidy.

Figure 6.8 Parapatric Speciation. Although gene flow can occur between any two populations, such as A and B or E and F, flow between populations becomes increasingly more difficult with distance. It can be experimentally shown that greater decreases in the viability of offspring are found when individuals from greater distances are crossed. When A and G are crossed, it is possible for no viable offspring to be produced. Although all these populations share the same gene pool and are thus theoretically the same species, we should recognize that A and G are, nevertheless, distinct entities. Conversely, it would be presumptuous to consider A through G as separate groups; they comprise one large population whose extremes are reproductively isolated.

ISOLATING MECHANISMS

Separate species, in order to be called such, must not be able to interbreed with each other. Obviously, a fern and a fox cannot produce viable offspring. But it may not be as apparent that a coyote and a wolf, or a sugar maple and silver maple, also cannot interbreed successfully. Many organisms that are separate species look virtually identical to the untrained eye, and in many cases only careful scrutiny can discern that they are different (Figure 6.9). But these organisms maintain their separate identities; for one or more reasons they are together unable to produce offspring capable of living, developing, or reproducing. These reasons are grouped under five categories known as isolating mechanisms.

Figure 6.9 Discerning Different Species. There are several species of planthoppers (plant sap-sucking insects) that look virtually identical. The only way that they can be distinguished with certainty is by examining male genital organs. The top figure shows what the insects look like from a dorsal view (right wings are removed). The two figures on the bottom show the differences in the genitalia (genital structures) of two species, indicating that they are, in fact, two different species and are probably isolated by mechanical mechanisms.

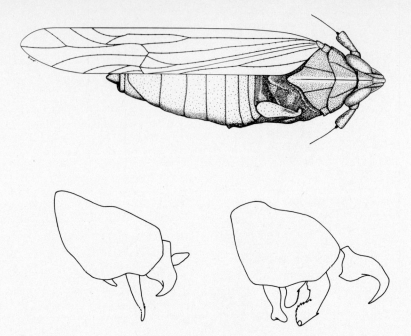

Ecological Isolation

If species make use of different portions of the environment, they are unlikely to encounter each other very often. The hermit thrush prefers the dense interior of a deciduous forest while the Swainson's thrush favors the forest edge. Of two species of garter snakes, one is mainly aquatic and one terrestrial. Some species of oaks are restricted to different soil types in the same area. Species of parasites are often restricted to specific hosts.

Temporal Isolation

Species may have breeding cycles that take place at different times of the year. The brown trout breeds in the spring, while the rainbow trout breeds in the fall. Although they inhabit the same streams, only one species at a time is physiologically capable of spawning. Two species of wandering jew, a common houseplant, bloom at different seasons and thus cannot cross-pollinate.

Two Florida dragonfly species patrol their territories at different times of the day. The two species are similar in size and appearance and reproduce during the same season, so the temporal separation of the territorial flights may be the only reproductive isolating mechanism between them.

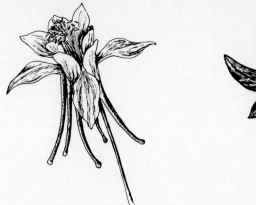

Figure 6.10 **Mechanical Isolating Mechanism.** Two species of columbine flower, *Aquilegia pubescens* and *formosa*, possess two very different flower structures. *A. pubescens* has yellow flowers with long, thin spurs adapted for pollination by hawk moths with their long, slender proboscises (tongues). *A. formosa* has red and yellow flowers with short spurs about the length of a hummingbird's bill. It also has a greater nectar supply than *A. pubescens*. Since these flowers are adapted to different pollinators, there is little chance they will receive pollen from other flower species.

Behavioral Isolation

Many organisms respond to only a particular color, pattern, behavior, or song given by the opposite sex and will not respond to signals by other species. For example, fireflies (lightning bugs) flash at different rates; two species of narrow-mouthed toads give calls of different duration and frequency; the bright colors of male ducks and the distinctive courtship rituals of many male birds ensure that only **conspecific** (same species) females respond to them.

Two species of wolf spider look alike, live in the eastern United States, and inhabit the same layer of dead leaves on the forest floor. When females of both species were experimentally anaesthetized, the males of either species would attempt to mate with the females. Awake females, however, would respond and mate only with a conspecific male. The females sense sounds made by the males as they approach; males of different species make different sounds and a female responds only to the proper "love song."

Mechanical Isolation

Often the physical inability to pass on or accept gametes prevents interbreeding. Being built quite differently, an oriole and a boa constrictor could not possibly mate. The size or structure of reproductive organs prevents fertilization. Experimental studies on five species of dragonflies showed that the males were unable to distinguish among females of the different species. But when the males attempted to mate with a nonconspecific female, they were usually prevented from doing so because the appendages used by the males to clasp the females did not fit properly on the female's thorax and thus the species were reproductively isolated. Figure 6.10 shows the mechanical isolating mechanisms of two species of closely related plants.

Genetic and Developmental Isolation

Even if for some reason an organism of one species is able to mate with (or pollinate) an individual of another species, the differences in their genetic makeup usually lead to (1) inability of the gametes to fuse and form a zygote, (2) death of the zygote, (3) abnormal or arrested development of the embryo, (4) production of nonviable offspring, or (5) viable but sterile offspring. In a cross of two species of sea urchin, for example, only about 1% or less of the eggs are fertilized (fertilization percentages within the same species range from 75% to 100%). A cross between a horse and a burro results in the sterile mule because horses and burros contain different chromosome numbers.

In most cases, two or more isolating mechanisms exist to produce a reproductive barrier. In the previous dragonfly example, mechanical isolation is important because those species look very similar to each other. Other dragonflies are isolated because they are distinctively colored and can recognize their own species. Some breed in different habitats (ecological isolation), and some fly at different times of the day (temporal isolation) (Exercise 6.2a–e).

HYBRIDIZATION

Isolating mechanisms usually work well between sympatric species. But in allopatric species, which do not interbreed simply because they are geographically separated, isolating mechanisms may not be as effective in the event that individuals of the different species do come into contact. If the two species meet, a hybrid form may arise. Hybridization often produces individuals that resemble an intermediate between the parents or may closely resemble the species of either parent. Readily occurring crosses between Jeffrey and ponderosa pine, and between various species of oaks, and between the mallard and pintail ducks, demonstrate that the concept of species needs elucidation (Figure 6.11). There have been a number of cases in which hybridization occurred so frequently that scientists finally decided that the two species "hybridizing" were only geographic races of one species. Thus, the eastern Baltimore oriole and the western Bullock's oriole have been lumped into one species, the northern oriole.

Populations of the same species may become sympatric after a period of allopatry and may hybridize if they have not diverged sufficiently. If the differences in the populations are small, hybridization could produce viable offspring and simply add genetic variability to the population, a beneficial result. Thus, hybridization between similar populations would be advantageous and lead to the merging of the populations. However, if the populations have di-

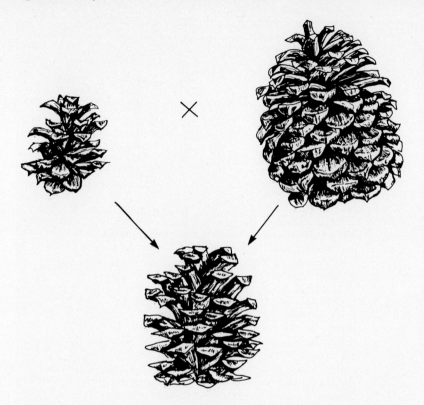

Figure 6.11 Hybrid. The ranges of ponderosa and Jeffrey pines overlap in some areas of the western United States. Ponderosa pine is found at lower elevations than Jeffrey, but they overlap at mid-elevations. In the area of overlap, they hybridize. A ponderosa pine has small cones that can be almost enclosed in one's hand. The Jeffrey pine has cones twice that size. The sketch shows cones from trees at elevations from 350 to 2000 meters, the range of species overlap being about 1000–1500 meters. A typical ponderosa pine cone is shown at top left, and to its right is a typical Jeffrey pine. The other is an intermediate, produced by hybrids in the zone of overlap.

verged considerably, the hybrids might be unfit or sterile, and the parents would have wasted energy in their production, a definite detriment. Hybridization between dissimilar populations may be disadvantageous, and those individuals who participate pass few or none of their genes on. Selection thus acts to prevent hybridization, isolate the populations, and ultimately form new species. Most often, it is a combination of some of these isolating mechanisms that serves to isolate species.

"SPECIES" REDEFINED

The definition of a species is not as clear as we first defined it—that is, any populations that are reproductively isolated are separate species. Populations that share the same gene pool on either end of a cline, and allopatric populations that have never been in contact but that may not have developed totally successful isolating mechanisms, do not clearly fit this definition. Scientists are often divided on the issue of whether a group of individuals should be one or more species (those that opt for fewer species are called "lumpers" and those for more, "splitters").

Exercise 6.2 Demonstrating Isolating Mechanisms.
A number of field or laboratory investigations can be done to demonstrate the existence of one or more isolating mechanisms between two or more different species. In lieu of field or lab work, data can be derived from the scientific literature. The following are brief suggestions:

a. *Ecological. Watch the foraging behavior of two species of birds in a habitat and determine what part of the habitat is used (height of vegetation, portion of tree used, species of vegetation, utilization of forest edge or center of forest, and so on). Or you may wish to examine the distribution of two plant species with regards to their location in the shade or sun, soil moisture, slope and so on.*

b. *Temporal. You can do a study of the times that plants flower or when two or more species of birds nest.*

c. *Behavioral. Study the pollination mechanisms of some plants to see which are wind-pollinated and which are insect-pollinated. Place two species of crickets in a terrarium and observe their behavior toward each other. Examine the differences in the songs of similar bird species. If fireflies are found in your area, examine the patterns of flashes in two species.*

d. *Mechanical. Pollination mechanisms and flower shape can be examined again. Examine the reproductive organs of different insect species (note, for example, that the claspers of some male damselflies fit only on the females of the same species).*

e. *Genetic and Developmental. The eggs of starfish, sea urchins, or frogs can be artificially fertilized with the sperm of*

a different but similar species and egg development can be observed. Plants can be artificially pollinated with another species' pollen. Does fertilization occur? If so, how far does development proceed? [Several basic laboratory manuals in zoology provide the details for these exercises. See, for example: Laboratory Outlines in Biology III, *by P. Abramoff and R. Thompson (San Francisco: W. H. Freeman, 1982), and* Laboratory Studies in Integrated Zoology, *by F. Hickman (St. Louis: C. V. Mosby, 1979).]*

In spite of the occasional difficulty in the use of the term, species is a useful concept. It works much more often than it poses problems because it is defined by the biological process of reproductive capability. Higher categories such as orders and classes are artifacts and are open to more subjective interpretation

SUMMARY

Organisms have evolved into different types incapable of interbreeding; these types are called species. Individuals within a species vary, sometimes along a geographical gradient. New species arise by the geographic isolation of populations or by polyploidy. Individuals of different species do not interbreed because of isolating mechanisms.

STUDY QUESTIONS

1. Define species, isolating mechanism, polyploidy, panmixis, allopatric, sympatric, conspecific.

2. Describe and compare allopatric, sympatric, and parapatric speciation.

3. Explain how two populations of the same species acquire different gene pools via geographic isolation.

4. Describe and compare Bergmann's, Allen's, and Gloger's rules.

5. Does sharing the same gene pool allow for more or less variation among the members of that pool?

6. Why do you think that polyploidy has become so common among plants?

7. Species evolve and populations evolve. Does natural selection operate on individuals or populations? Can a population adapt or do only individuals adapt?

8. Give examples, other than those cited in the text, of each of the types of isolating mechanisms.

9. Is hybridization simply a failure of isolating mechanisms, or is it a way of introducing adaptive variability into a population? Explain.

10. Justify the stand of either the taxonomic "splitters" or the "lumpers."

Chapter 7

Dispersal

Previous chapters described the mechanisms by which new species arise. Different environments produce different selective pressures and, over time and generations, different species. We have discussed to some extent how environments change over time and will explore this further in Chapter 8. Here we will examine the movement of organisms into different environments, the causes of their movement, the methods of travel, and limitations to dispersal.

We are all aware of the vast differences, as well as the striking similarities, between the kinds of plants and animals in different parts of the world. Some of these similarities are due to **convergent evolution**, that is, the evolution of similar form in unrelated organisms. For example, some species of milkweeds in Africa resemble some cacti of North America because they are subject to similar environmental conditions (Figure 7.1a). In other cases, similarities in related organisms, such as the lungfish of Africa, South America, and Australia, are due to their evolution from a common ancestor; they later became geographically separated (Figure 7.1b). Our present discussion focuses mainly on the latter type of organisms.

The distribution of organisms raises interesting questions. Why have some organisms moved great distances from their center of origin and others very little? How did plants travel over thousands of

a

b

Figure 7.1 Why Similarities Occur. (a) Convergent Evolution. The euphorbs of Africa and the cacti of the United States are similar in appearance, not because they descended from a common ancestor, but because they evolved under similar environmental conditions, and were molded by natural selection. (b) Evolution from a Common Ancestor. The Australian, African, and South American lungfishes all resemble each other, and they all closely resemble the ancestral lungfish, which lived 200 million years ago. Lungfishes can survive in stagnant pools of water by breathing air with lungs. The African and South American lungfishes can burrow into the mud of a drying river bank and exist in a state of torpor until the next wet season.

miles of open ocean to colonize isolated islands? Why are some organisms so widespread and others so isolated?

Dispersal, the movement of organisms from their home site, occurs to some extent in virtually every species. It is particularly true in young organisms, seeds, spores, and other such **propagules**. Dispersal movements may be slow or cover only short distances, or they may be rapid and far-ranging.

MAJOR CAUSE OF DISPERSAL

The basic reason for dispersal is population pressure. Competition for resources in a group of individuals often causes some individuals to seek resources elsewhere or die. Young organisms in particular compete for needed requisites, and since it is extremely unlikely that all the offspring could survive to reproductive age in the same location, they must move to new areas. All the acorns of an oak tree cannot grow under the canopy of the parent tree. To survive, dispersal must occur, and it does—via birds, squirrels, and other animals. Sponges, **sessile** (attached) as adults, have ciliated larvae that swim to other areas. A mature juniper tree may produce 10,000 seeds; were it not for the dispersal of its seeds by birds, few, if any, seeds would become mature trees. All organisms produce more offspring than can possibly survive in the immediate area, compelling their dissemination to more suitable areas.

Population pressure can also force entire populations, adults as well as young, to spread over a wide area in search of resources. Ladybird beetles will leave an area when the population of their aphid prey drops. This movement may lead to a range expansion of the species. The cattle egret invaded South America via Africa within the last century and is now common in much of South and North America. The armadillo of the southwestern United States and Mexico has been expanding its range northward. Introduced species, such as the house sparrow, starling, water hyacinth, walking catfish, and the Japanese kudzu vine, have spread so rapidly and widely across some parts of North America that they are considered pests. The kudzu can grow 35 meters in one year—2 meters a week in favorable conditions. As long as organisms can expand into favorable habitats, their populations will continue to grow and force further expansion.

The spread of other organisms is extremely slow and limited. Observation of one species of coral, which occurs on rock reefs off the coast of western North America, has demonstrated that the ciliated larvae move only an average of half a meter from the parent coral before settling down.

OTHER DISPERSAL CAUSES

There are other reasons for the movement of organisms from their home site, in addition to population pressure. Natural phenomena, such as fires, floods, and volcanic activity, can make habitats unsuitable for organisms, which must then disperse or die. Residents of a California community were plagued by rodents and snakes when a nearby grassland was flooded, forcing these animals to seek another habitat. Weather can temporarily dislodge organisms. Exceptionally cold winters often drive organisms of arctic or alpine areas into more temperate ones; snowy owls have been seen a thousand kilometers south of their normal range during a severe winter. Mockingbirds have been expanding their home range northward across the United States, but contracting it during cold winters.

Also, slow, gradual changes in a habitat may occur that effect the movement of organisms both into and out of an area. As a sand dune becomes a forest or a pond becomes a terrestrial habitat, for example, both the plant and animal communities change. There could be no change if it were not for the continual dispersal of organisms into the area.

Further, organisms concentrated in an area are more likely to be preyed upon than those that are dispersed, as it is easier for a predator to feed on a clumped food source. Thus, the selective pressures of the evolutionary process again ensure some dispersal.

METHODS OF DISPERSAL

It is easy to picture a bird flying across a lake or caribou marching over the tundra, but how do spiders, clams, maple trees, and blackberry bushes disperse? How have fishes reached high mountain lakes? How have land tortoises reached the Galápagos Islands, 1000 kilometers from the Ecuador mainland?

Methods of dispersal are either active or passive. Active dispersion is simply directed movement under an organism's own power, such as walking, flying, hopping, or swimming (Figure 7.2). Some organisms are capable of more movement than others; for example, an earthworm travels shorter distances than a rabbit, but all active dispersers move by their own power (that is, they are **motile**). Passive dispersers, on the other hand, can be moved (that is, they are **mobile**), but by other forces, since they have no locomotory mechanism of their own. There are degrees of mobility, as some organisms move more readily than others. Major mechanisms of passive dispersal are wind, water, and other organisms.

Wind Dispersal

Organisms are moved about by air currents as mild as breezes or as strong as hurricane force winds. Many seeds have winglike appendages (like those of maple trees) or cottony tufts to catch the wind (Figure 7.3). Other seeds are simply light enough to be wafted away with the softest breeze. Some spiders spin a strand of silk and float through the air hanging onto that thread ("ballooning"). Light-bodied insects, such as ants and crickets, can be blown many kilometers;

Figure 7.2 Active Dispersion. Among organisms that actively disperse, birds are the most well known for their ability to move long distances. Many fly thousands of kilometers each year during migration. The Arctic tern probably flies farther than any other bird on its migratory route. From the far northern parts of the northern hemisphere to the extreme southern parts of the world, the Arctic tern flies about 42,000 kilometers each year.

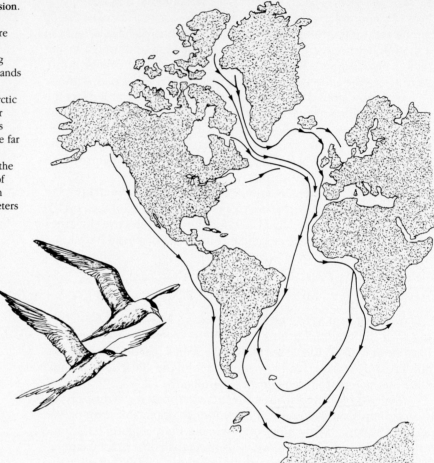

nets attached to airplane wings have collected insects at altitudes of several thousand meters. Tornadoes have been known to transport the contents of small ponds—plants, animals, and water—great distances. Occasionally, one reads in the popular press that a "rain of fish" has occurred. Although this may be a very rare event, it happens often enough to establish organisms in new areas. Fieldfares, summer resident birds of Scandinavia, established a permanent population in southern Greenland when blown off their migratory route by a storm. Many plants such as grasses depend on the wind for dispersal of their pollen. (Pollen, which is the male component of the flowering plant reproductive process, does not by itself result in plant dispersal, but the dispersal of pollen is necessary for reproduction to occur.)

Figure 7.3 Wind-Catching Seeds. Cottony tufts, filamentous strands, or other projections from seeds provide a sort of sail or parachute that catches the wind. Some seeds, such as those of some orchids, may be so light (2 micrograms or 500,000 seeds per gram) that they need no accessory mechanisms to be carried by the wind.

Water Dispersal

Trickles of water, rivers, and ocean waves all provide locomotion for organisms. Eggs, spores, larvae, and seeds are moved by water runoff after a rain, perhaps into streams and rivers, where they may be carried great distances before being deposited in an appropriate area. Logs floating down rivers may carry turtles, mice, and a myriad of small animals and plants. Floating islands of vegetation and soil may be quite large and contain trees and even deer. These may be short-lived phenomena or they may persist for considerable periods of time. The Galápagos Islands, 1000 kilometers west of Ecuador, arose from the ocean floor as a result of volcanic action; they were never connected to another body of land. Nevertheless,

large (600 kilograms) tortoises inhabit the islands; no suitable explanation other than "rafting" on mobile islands exists to account
for their presence there. Rafting probably also brought the land
snails, lizards, rats, and many of the plants.

A similar explanation holds true for the Hawaiian Islands. There
are 27 species of spectacularly colored land snails on the Hawaiian
Islands. Their progenitors are of unknown origin, but they probably
arrived on pieces of driftwood that floated across large expanses of
water before reaching Hawaii. (The Hawaiian Islands are the most
geographically isolated islands in the world. As a result of time and
isolation, 97% of their flora and fauna are found nowhere else in the
world.)

Animals as Dispersal Agents

Animals play a greater role in dispersing other organisms than
one might imagine. Pollen, spores, seeds, and small organisms can
be transported by birds, mammals, fish, and other animals.

Barnacles may attach to humpback whales, which annually migrate between the tropical Pacific and the cold Arctic oceans; barnacles that reproduce along this route scatter their offspring over a
large area. Some species of freshwater clams have larval forms called
glochidia, which attach to the gills of fishes where they begin to
develop; eventually they leave the fish to continue their development into adult clams. A fascinating example of an organism that
undergoes this type of development is the "pocketbook mussel";
the females of this species possess an extension of their body tissue
that resembles a small fish. When a large fish passes by, it is lured to
the mussel by this fishy-appearing lure. When the fish is close, the
mussel releases thousands of glochidia; some of these find their way
into the fish's mouth and then gills, where they attach and develop.
The fishlike flap of the clam serves to facilitate dispersal via another
animal. The fish host may travel many miles upstream or downstream during this larval development. Fleas, ticks, mites, lice, and
other arthropods that live on the skin of vertebrates move with their
hosts much farther than they could move alone.

Plants, and particularly pollen and seeds, take advantage of the
more motile animals for dispersal. Bees, wasps, beetles, flies, and
other insects that visit flowers for nectar or pollen, provide an efficient system of pollen delivery. Hummingbirds of North and South
America, honeycreepers of Hawaii, and the sunbirds of Africa, feed
on nectar. In the process, pollen grains stick to their bills and faces
and are deposited later on other plants. Hummingbirds are such
efficient pollinators that some flowers, called "hummingbird flowers" (Figure 7.4), have shapes especially adapted to exclude most
insects and admit only the hummingbirds' bills.

Figure 7.4 Hummingbird Flowers. Many flowers are trumpet-shaped to allow for the easy access of a hummingbird's bill and to restrict the entry of insects. These "hummingbird flowers" have evolved to attract hummingbirds because the birds are more efficient pollinators than insects. The flowers are typically red to attract the birds.

Many seeds are encased in a "sticker" or "burr" that attaches to an animal's fur or feathers when the animal brushes against the plant (Figure 7.5). Some seeds become trapped in mud and are carried about on an animal's foot. The fruits of many plants serve to attract animals, which then eat the fruit, pass the seeds through their digestive tract, and drop them elsewhere. Some seeds germinate faster after they pass through the gut of a vertebrate because the seed coat is scratched by the digestive processes and the waste products provide some fertilizer.

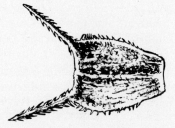

Figure 7.5 Seeds Dispersed by Animals' Fur. Many seeds have prongs, spines, or hooks, which enable them to attach to the fur of a mammal as it brushes against the plant, and to dislodge in another area.

Exercise 7.1 Dispersal Mechanisms of Seeds and Fruits.

Collect fruits and/or seeds from as many plants as possible. How do you think each one is dispersed? How many are specialized for one form of dispersal and how many may make use of several dispersal mechanisms? What characteristics are common to wind-dispersed seeds or fruits? Water-dispersed? Animal-dispersed?

Mistletoe, a parasitic plant, is distributed by birds in many areas of the world. The mistletoe seeds are coated with a chemical layer that protects them from digestion, and they exit the bird with a sticky coating. When the bird defecates in a tree, the seeds stick to a branch and penetrate the tree as they germinate. Toucans, large-billed tropical birds of Central and South America, regurgitate large seeds of several kinds of fruit after eating the surrounding fruit. Other seeds are eaten by animals for food. A macerated seed will not germinate, but in the foraging process many seeds are dispersed and not eaten. Squirrels bury acorns and may find fewer than half of them later; the others may become oaks. Pine seeds need to be buried beneath the forest litter to germinate; small rodents cache pine seeds, many of which are not found again and germinate.

Entire plants or parts may be transported by animals. Algae, for example, can be moved on the backs of turtles or in the hair of sloths. Moss can be carried from one tree to another by birds that use it as nest material. Duckweed, algae, moss, and fungi have been found in the feathers of waterfowl. Bedstraw, a plant native to Europe and introduced into North America, has small, stiff hairs on its leaves and stems that allow parts of the plant to attach to passing animals.

Other Dispersal Mechanisms

Organisms have evolved an infinite variety of means to expand their range; they are not limited to wind, water, or other animals. A few other mechanisms will be mentioned here.

The witch hazel produces nutlike fruits that, when dry, burst open to scatter the enclosed seeds. The touch-me-not produces capsules that burst open when touched, thus scattering seeds. The blackberry sends out shoots that root and form new plants. The cabbage butterfly tends to disperse its eggs over several cabbage plants rather than clumping them on one and forcing the larvae to compete. Chestnut blight, a fungus-caused disease, is often spread by birds building nests in healthy trees with twigs from trees killed by the fungus. Bark beetles have "mycangia" pockets that store fungal spores, which they then carry from tree to tree. The spores drop off in the burrows the beetles make and larval beetles feed on the growing fungus. The transmission of fungal spores by beetles is responsible for the spread of Dutch elm disease. Trains break off parts of, or pick up seeds from, plants that grow along railroad tracks and carry them long distances. In some cases, the wind created by a passing train is more effective in dispersing the seeds of plants than even strong natural winds.

BARRIERS TO DISPERSAL

If there are so many mechanisms for dispersal, then why are so many plants and animals restricted to certain areas? Why are there giraffes in Africa and penguins in Antarctica, but neither of these in Canada, for example? There are **dispersal barriers** that prevent the arrival or successful establishment of a dispersed individual. These barriers may be biotic or abiotic but often are a combination of the two.

Abiotic Dispersal Barriers: Physical and Climatic

One can easily imagine many impediments to the travel of plants and animals. Physical barriers, such as rivers, lakes, oceans, and canyons, cannot be traversed easily (if at all) by most terrestrial animals. A gopher, earthworm, or lizard has no mechanism to cross these areas. Fish and other aquatic organisms are unable to cross land barriers except under very unusual circumstances. Many organisms are where they are simply because they cannot reach other places.

Further, even if an organism is able to reach a particular area, that habitat may not be suitable because of an unfavorable climate. It may simply be too hot, too cold, too dry, or too wet for some organisms. As examples, the range of the black-bellied whistling duck is restricted to southern Texas and southward because it cannot withstand more severe winters, and the distributions of both mistletoe and mockingbirds are also restricted by the severity of the winter. Sea snakes, venomous reptiles adapted for a completely oceanic life, are restricted primarily to tropical latitudes because of their inability to survive at a temperature below 20°C. Temperature also restricts them to the Pacific, since the only route to the Atlantic would be through the cold southern oceans.

Biotic Dispersal Barriers: Resources, Competition, and Predation

Even if an organism is capable of coping with the physical elements of a habitat, it may not be successful in dealing with biological factors. Food, nest sites, burrows, and other components that allow the organism to survive and breed may not be present. The Everglades kite, for example, feeds only on one species of snail, which is found in southern Florida, and the bird apparently cannot survive elsewhere in the absence of that snail.

If an animal or plant reaches a suitable habitat with the proper climate and resources, it may still face problems—it may find itself

Exercise 7.2 Dispersal of Seeds by Animals: Field Observation.

A berry bush, an oak or maple tree, a thistle, or other plant that is bearing fruit or shedding seeds can be observed to determine its animal visitors. How many and what species of animals visit the plant? How long do they spend eating the fruits or seeds? How many fruits or seeds are dispersed? How many will probably be destroyed by eating and how many will survive to be dispersed? Compare different plants.

competing with organisms already in existence there. Very often, the resident or the invader will have to leave, or the species that come into contact with one another will be put into a predator–prey relationship that may end in the elimination of the invader.

INTRODUCED SPECIES

Although the present distribution of many organisms can be explained by physical barriers, the barriers have often been broken with the inadvertent or purposeful help of humans. The ability of humans to travel anywhere on earth, and their propensity to carry animals and plants with them has allowed organisms to inhabit areas they may never have reached otherwise. Often, with new suitable habitats opened to an exotic (nonnative), its populations rapidly expand. The Norwegian rat, house mouse, cabbage butterfly, chestnut blight fungus, starling, house sparrow, Russian thistle, tree of heaven, and peppermint have all been introduced into North America from Asia or Europe. The new environment may pose fewer restrictions to population growth than the old; fewer checks and balances result in a population "explosion," and the organisms become "pests." Rats and mice, introduced into the United States from Europe, are serious crop pests, as well as urban nuisances. A population of 17 mongooses, introduced into Hawaii from Asia via Jamaica to control the rat population, became serious pests themselves and have been a major factor in the extermination of many native bird species. The native Hawaiian land snails are being reduced in number by introduced rats and carnivorous Florida "cannibal" snails. European rabbits brought to Australia in the 1800s became severe pests until controlled themselves by an introduced viral disease called myxomatosis.

Introductions of nonnative species are not new. German merce naries who fought for the British in the Revolutionary War brought Hessian flies in their straw bedding. Along with wheat stem rust brought from Europe on barbery plants, the Hessian fly destroyed most of the United States wheat crops during World War I. The establishment of the Plant Quarantine Act in 1912 and subsequent laws authorized the government to establish inspection stations. Although many potential pests are stopped, our jet-age mobility facilitates the spread of exotic organisms. Large numbers of a scarab beetle pest arrived in the United States from France on 38 separate airplanes within a 2-week period; fortunately, the problem was discovered in time. The Mediterranean fruit fly (dubbed "medfly"), apparently a native of Africa, has spread to many areas of the world, where their larvae damage more than 250 kinds of fruit (Figure 7.6).

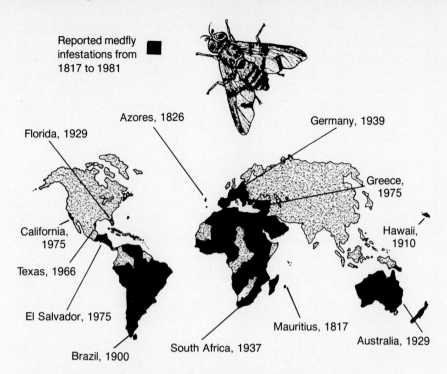

Reported medfly infestations from 1817 to 1981

Azores, 1826

Florida, 1929

Germany, 1939

Greece, 1975

California, 1975

Hawaii, 1910

Texas, 1966

El Salvador, 1975

Mauritius, 1817

Australia, 1929

South Africa, 1937

Brazil, 1900

Figure 7.6 Introduced Species. The Mediterranean Fruit Fly, better known as the medfly, probably originated in West Africa. First discovered in the early 1800s, it spread over many parts of Africa and north to Europe. In 1826, oranges in the Azores were damaged by the medfly, the beginning of a continuing series of attacks on crops by the insect. Since the medfly can only survive in warm climates, it is not found in the colder areas of the world. As the map shows, much of the world has been invaded by medflies, probably arriving in new habitats via fruit shipments.

Like many successful invaders, the medfly has excellent reproductive potential, perhaps having a dozen generations in a year, during which each female can lay up to 500 eggs.

The European carp, which was deliberately introduced into the United States in 1876 as a food fish and ultimately escaped, is now a widespread fish that often destroys the nests of other fish and up-roots aquatic plants as it forages.

If an organism is able to reach a new environment, its survival is not guaranteed. Many animals and plants have been introduced into new areas by humans only to die off. Numerous attempts have been made to establish populations outside their native habitats; many were failures. The chukar partridge was introduced into several areas of the United States but survives mainly in the Pacific Northwest.

Exercise 7.3 Introduced Species.

House sparrow

Gypsy moth

Green and black poison arrow frog

From whatever resources you can locate, try to determine how many species of plants and animals have been naturalized in the habitats of your area. The Hawaiian Islands are well known for their large number of nonnative species. Of 22 species of reptiles and amphibians in Hawaii, only 4 are native— three sea turtles and a sea snake. All the terrestrial forms are exotic, such as the poison-arrow frogs from Central and South America.

Several species of parrots and other exotic birds exist in small numbers in southern Florida and California (Exercise 7.3).

In summary, organisms, especially the young, disperse to find new habitats and to lessen competition from others of the same species. They may depend on their own locomotion or transporation by other biotic or abiotic elements. Their movement to, and survival in, other areas are dependent on a myriad of facilitating and inhibiting factors.

SUMMARY

Organisms spread to different geographic locations for many reasons and by many methods. The major cause of dispersal is population pressure. Methods of dispersal include wind, water, and animals. Barriers to dispersal are biotic and abiotic. Humans have broken down these barriers and introduced species into new areas, most often with unpleasant results.

STUDY QUESTIONS

1.   Define convergent evolution, dispersal, propagules, motile, mobile, dispersal barrier.

2.   Explain and give examples of convergent evolution.

3.   Discuss the major cause(s) of dispersal.

4.   Name the primary methods of dispersal and give an example of each.

5.   How have plants evolved to take advantage of animals as dispersal agents?

6.   What are the barriers to dispersal? Give examples of each.

7.   How have humans purposely or inadvertently introduced organisms into new areas?

8.   Purposeful introductions of exotic species have most often been detrimental. Have there been any successful introductions? What can be done to ensure that introductions will be beneficial— or should all introductions be avoided?

9.   What groups of organisms are or have been most widely dispersed? Weigh the relative importance of population pressure and the motility of organisms in dispersal.

10.   The chances of a dispersing organism finding a suitable habitat are low. Why would it not be beneficial for the organism to stay put and face the consequences of population pressure rather than disperse?

Chapter 8

The Development of Communities

Constraints exist on the ability of organisms to survive in various environments, as well as on their ability to disperse to those environments. If one randomly chose groups of organisms and placed them together on a bare plot of land, chaos would ensue. Many mobile organisms would leave and many immobile ones would perish. After a short time, few would be left to reproduce. But assemblages of plants and animals in forests, deserts, and coral reefs did not just happen; they evolved gradually into an interdependent, smoothly operating system. The creation of natural communities occurs through a process called **succession**.

ECOLOGICAL SUCCESSION DEFINED

If farmland is abandoned and fallow, it quickly becomes covered by what we commonly call "weeds." Gradually, plants of different species will arrive until the former farmland is fully covered. Over a span of several to many years, species of grasses, herbs, shrubs, and trees slowly appear and disappear. If this field is located in Ohio, a forest dominated by beech and maple trees may eventually form. Whatever community finally forms, however, it will be the native vegetation of that area, similar, if not identical, to the vegetation that existed there before it was cleared for farmland.

If we watch a pond in the midst of an Oklahoma grassland over a period of time, we will see it invaded by emergent vegetation (vegetation rooted under water and emerging above the surface) such as cattails, and floating plants such as duckweed. Year after year, as the aquatic and semiaquatic plants and animals reproduce and die, and nutrients accumulate in the pond from the water that runs into it, the pond will fill with organic matter. The floating vegetation will get thicker, the pond shallower, and the emergent vegetation will become denser and closer to the center. The pond will ultimately dry up and be covered by plants of the surrounding grassland (Figure 8.1).

These changes from one community to another, which continue until a state of relative stability results, demonstrate **ecological succession**, defined as the gradual and often predictable change in the composition of successive communities. Ecological succession involves a series of changes that, although continual, can be described in steps or **seres**. The dynamic process of succession stabilizes in the form of a mature community characteristic of the geographic area; this stabilized community is termed the **climax**. Climax communities do change, but very slowly and almost imperceptibly. Succession, then, is actually the process of ecosystem development.

CAUSES AND SEQUENCE OF SUCCESSION

Successional processes are observable everywhere: abandoned farmlands, ponds, sand dunes invaded by grasses, forests recovering from a fire or logging activities, newly formed volcanic islands, and weeds in sidewalk cracks. Diseases may kill the dominant trees in a forest, introduced species may displace native ones, or plagues of insects may destroy crops. Each of these factors will initiate some sort of successional process; both abiotic and biotic phenomena may stimulate succession.

In 1916 Frederick Clements divided the process of succession into six major phases:*

1. **Nudation:** the disturbance of the environment in such a way as to allow the invasion of organisms.

2. **Migration:** the invasion of the environment by seeds, spores, larvae, adults, or other **migrules** (organisms or their offspring capable of dispersal). Because **propagules** (organisms or groups of organisms capable of breeding and population increase) are perpetually dispersing, almost any area is open to constant invasion.

*Kenneth A. Kershaw, 1973. *Quantitative and Dynamic Ecology.* London: Edward Arnold.

Figure 8.1 Pond Succession. A pond, invaded by various kinds of emergent, floating, and underwater vegetation, is being encroached upon by grassland vegetation at the edges. As the aquatic plants die and the pond fills up with organic material, terrestrial plants will invade and the pond will ultimately disappear. In this case, the pond disappeared in two years.

Dispersal barriers will prevent the successful invasion of many habitats, but an area virtually devoid of organisms is ripe for colonization.

3. **Ecesis**: the germination, growth, and reproduction of the migrules. Often, the first organisms to occur are plants we often call "weeds." A better term for these is pioneer plants—hardy, resilient plants with a high reproductive capacity that are well suited for harsh habitats. Examples of these colonizers of new ecosystems are ragweed, lamb's quarters, thistle, dandelion, milkweed, and crabgrass.

4. **Competition**: the new colonizers vie for the same resources, which may be extremely scarce. Only the most fit survive. Although a forest may eventually develop on the site of a fallow field, only rarely would oak or maple seedlings be found as colonizers. As

seedlings, these and other plants cannot survive in the open sun-light and low moisture regimes found in the early stages of succes-sion. Their less efficient dispersal mechanisms and their attributes of growth and permanence rather than reproduction put them at a disadvantage to weeds, which may only live one or two years but grow quickly and produce prodigious amounts of seeds. Thus, the invaded area will be rapidly covered by pioneer plants, while oaks, for instance, will not be able to survive until the habitat has been colonized and modified by other plants over several years.

Pioneer plants have other adaptations for survival in the early stages of succession. They often have small leaves; thick, succulent (or dry and leathery) leaves and stems; and/or fuzzy leaves. All of these are adaptations to reduce water loss. To discourage herbi-vores, many are covered with spines (for example, thistle) or other protective mechanisms (nettle, for example, has fine hollow hairs that inject formic acid into animals that touch it), or contain distaste-ful substances (Figure 8.2). In early successional stages these plants are the only ones in an area and would be readily eaten if they were not protected. They also grow quickly, send down long roots to tap water far below the surface of the ground, and are resistant to severe climatic changes. Oaks and maples have no such adaptations and are very rarely colonizers.

5. **Reaction**: the change in the habitat due to the effects of the invader plants. Lichens, often pioneer plants, may colonize a bare granite rock, a harsh environment indeed (Figure 8.3). Lichens are very hardy and slowly decompose the rock surfaces by producing weak carbonic acid. The rock surface is etched by the acid, the lichens die, and organic and inorganic material accumulates on the rock, making it more hospitable to other organisms, such as mosses, which may now colonize the rock. The mosses retain moisture, and their matlike bodies eventually form a thin layer of soil. The soil now provides nutrients, water, and a substrate in which other plants may grow. Grasses, then shrubs, then trees invade over a number of years. Or trees may gain an early foothold in rock crevices in which organic matter has accumulated.

As the vegetation becomes denser and taller, the environment is substantially modified and earlier invaders disappear. Light penetra-tion and air movement are reduced, litter builds up, the soil pH changes, the soil is thicker, and moisture is retained longer. If the climax is a forest, the trees shade out many of the earlier invaders such as shrubs and grasses. Pioneers and many intermediates yield to the climax plants, whose characteristics of longevity and large size make them the fittest. This illustrates the fundamental mecha-nism of succession: the modification of the environment by one community to make it suitable for invasion by another (Figure 8.4).

Figure 8.2 Pioneer Plant. This bull thistle is one of the first invaders in an abandoned field. Its sharp spines and woody stems inhibit herbivores while its abundant seeds with cottony tufts ensure its reproduction and dispersal.

Figure 8.3 Lichen on Rock. Lichens, composites of a fungus and a green alga, are often the first colonists in a barren, rocky area. Resistant to desiccation, heat, and cold, they eventually modify the rock surface and deposit organic matter, making the rock suitable for colonization by other plants and animals. Two kinds of **crustose** (crust-forming) lichens are seen here.

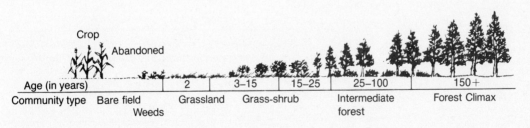

Figure 8.4 Example of Terrestrial Succession. It may take a century or more, but a climax vegetation will eventually develop as the preceding seres prepare the environment for the final, dominant stage.

6. **Stabilization**: the relatively unchanging climax community. The character of the climax reflects the influences of the climate, soil, physiognomy (physical lay of the land), and biotic factors of the area. Relative to the time taken by successional processes to reach climax, changes in the climax community occur very slowly, perhaps only over thousands of years. Climax vegetation is often used to describe the typical vegetation types of different geographic areas.

ANIMALS AND SUCCESSION

Successional processes are generally described in terms of plant groupings, as are climax types. But animal communities also change, along with the plant communities, influencing and being influenced by the successional process. Certain insects, perhaps analogous to weeds, are generally the first animal invaders.

Successional studies of habitats altered by fires or strip mining and of islands of sand created by dredging coastal areas show that as plant succession proceeds from simple to more complex communities, the variety of species and the number of individual birds also increase. In general, animal succession parallels plant succession because plants are not only the base of the food web but also provide shelter and nest sites for animals (Table 8.1).

COMMUNITY DEVELOPMENT TRENDS

As sere after sere heads toward culmination in a climax, certain tendencies can be observed. The number of species and their variety tends to increase, as does all other organic matter. Food chains become more complex food webs. Mineral cycles become more closed; fewer nutrients are lost. With greater numbers of plants and animals establishing themselves in a habitat, more efficient use can

Table 8.1*

Habitat:	Bare Ground	Early Shrub	Late Shrub	Bottomland Forest
Number of Bird Species	5	18	32	32
Number of Bird Species Pairs/10 Hectares	7	89	97	127

The figures in this table indicate the parallel trends of plant and animal succession on strip-mined land in central Illinois. There is good evidence to indicate that the diversity and number of birds are related to the density and complexity of the vegetation.
*Data from J. R. Karr, 1968. "Habitat and avian diversity on strip-mined land in east-central Illinois," *The Condor* 70:348–357.

be made of plant (primary) production. Rather than a few plants growing and dying, their bodies left to decompose down to chemicals that may wash away or leach through the ground before being used, there are herbivores that make use of the plant before and after death, carnivores to eat the herbivores, and other plants to reabsorb some of the minerals released as decomposition proceeds. Thus, the more complex climax communities are more efficient at recycling nutrients than the simpler, early stages of succession. Pine, fir, and hardwood trees have been shown to reabsorb from their leaves 20% to 60% of their annual nitrogen, phosphorus, and potassium requirements before their leaves fall. Thus, many of the minerals do not leave the tree and are not washed away. The role of decomposers becomes more important in the recycling of **detritus** (products of decomposition). Plant and animal species found in early successional stages tend to be generalists and are displaced by specialists. Organisms' life cycles tend to become longer and more complex. All these trends produce a relatively complex and stable system; it is more difficult to perturb than the simpler, early seres.

The final, or climax, stage is complex in terms of structure and variety of species, but it is not the most complex of all the stages. As the climax vegetation grows larger, it outcompetes some of the existing vegetation and the climax becomes somewhat less diverse than the immediately preceding sere, sometimes called a **subclimax**. But the climax is still a complex and stable system compared to most of the earlier stages (Exercise 8.1 and Exercise 8.2).

PRIMARY AND SECONDARY SUCCESSION

We are most familiar with the successional processes we observe along roadsides, in abandoned fields, and in our gardens and lawns; these have soil. Soil was formed from preexisting life, so succession occurred there once in the past. The invasion of land that once supported life is **secondary succession**. Secondary succession may occur after fires, logging, floods, avalanches, wind storms, and so on, as well as on unused roads, railroad beds, and flower gardens. The development of a community on areas devoid of soil is termed **primary succession**. The islands of Hawaii, the Galápagos, and Iceland were formed by volcanic eruptions from the floor of the sea. Never having been attached to a larger land mass, all of their flora and fauna had to arrive from other areas over vast expanses of ocean. Since no life existed on these newly formed land masses, the establishment of organisms there is primary succession. Landslides and volcanic eruptions that cover the land with ash, rock, or lava leave the environment devoid of life, but they are soon followed by primary succession.

Exercise 8.1 Succession.

Using the information in Exercises 8.2 and 8.4, compare the plant and animal species diversity in different successional stages. Also compare the changes in abiotic factors, such as soil type, soil depth, depth of litter layer, air temperature, and humidity, from stage to stage. Make a chart of some biotic and abiotic factors for each successional stage (sere) and compare them. What trends do you notice? What is characteristic of each stage? Which organisms persist through several stages? What special adaptations of organisms do you observe in each stage?

Exercise 8.2 Community and Habitat Analyses.

There are a multitude of ways in which habitats can be studied and analyzed. The thorough analysis of any one habitat is beyond the scope of this book, but you may wish to examine one or more terrestrial habitats with respect to the following features:

1. *Species diversity (see Exercise 8.4)*
2. *Vegetation analysis (see Exercise 8.3)*
3. *Dominant species*
4. *Stratification*
5. *Soil structure and temperature*
6. *Humidity*
7. *Air temperature*
8. *Wind*
9. *Topography*
10. *Zonation*
11. *Successional processes (see Exercise 8.1)*
12. *Light intensity*
13. *Precipitation*
14. *Chemical content of soil*

Analogous measurements can be made for aquatic communities. For specific exercises and methods you may wish to consult the following:

A. H. Benton, and W. E. Werner, Jr. 1972. Manual of Field Biology and Ecology. *Minneapolis, Minn: Burgess Publishing Co.*

J. E. Brower, and J. H. Zar. 1977. Field and Laboratory Methods for General Ecology. *Dubuque, Iowa: Wm. C. Brown Co. Publishers.*

G. W. Cox, 1980. Laboratory Manual of General Ecology. *Dubuque, Iowa: Wm. C. Brown Co. Publishers.*

R. M. Darnell, 1971. Organisms and Environment. *San Francisco: W. H. Freeman.*

R. G. Rolan, 1973. Laboratory and Field Investigation in General Ecology. *New York: The Macmillan Co.*

Primary and secondary succession differ only in how they begin. The processes of gradual and continual replacement of plants and animals by other plants and animals are very similar, if not indistinguishable, in primary and secondary succession, and the resulting climax communities are identical. But it is much easier for organisms to gain a foothold in soil than on rock, so secondary succession begins much more quickly than primary.

FIRE CLIMAX

Fires occur irregularly in many habitats, causing secondary succession to occur. If vegetation is burned regularly, succession is repeatedly set back. The fire-tolerant plants persist, but others must reinvade. Lodgepole pine is resistant to fire and its cones release seeds only when exposed to extreme heat. Thus, lodgepole pine stands are often maintained by fire. Grasslands of the midwestern United States would be invaded by herbs and shrubs were it not for fire; in the heavily farmed areas the only habitats that resemble native grasslands are those strips of land along railroad rights-of-way

Figure 8.5 Burned Area. This patch of burned pine forest is being invaded by low shrubs. Within 20 to 30 years it will look like the surrounding forest, and the only evidence of a burn will be the charred remains of trees.

where sparks from train wheels cause grass fires. Scrubby plants of the chaparral, the redwoods, the giant sequoia, and pines of the southeastern U.S. forests dominate their habitats because fire keeps out invaders and maintains an intermediate successional stage.

ECOSYSTEM RESILIENCY AND HUMAN PERTURBATION

When we dig weeds from sidewalk cracks, spray crops, fields, and orchards with herbicides, and clear brush from our backyards, we are fighting successional processes. Succession is a reflection of the strength and resiliency of an ecosystem to persist.

Most ecosystems can withstand a considerable amount of abuse. Avalanches, floods, hurricanes, and earthquakes are incredibly devastating. Yet, given sufficient time, the ecosystem reappears as if nothing had happened (Figure 8.5).

Effects of human activity can also be reclaimed by succession. Abandoned farms, roads, strip mines, dredger-tailings, clear-cut forests, and other denuded areas can and do recover if left undisturbed. As long as the appropriate plant and animal species exist, they can reinvade and repopulate a disturbed area. As optimistic as this sounds, we need to be cautious. Time will heal most injuries to the ecosystem, but the exceptions could be disastrous. Strip mines can leach acid for many years after abandonment, preventing the growth of vegetation. When a forest is clear-cut, erosion and leaching result in a loss of nutrients since tree roots are no longer able to absorb water and dissolved minerals. But studies have shown that if trees are immediately replanted, 5 years later any losses to the system will be about the same as they are in a 55-year-old forest. A coniferous forest may, however, take 50 years or more to return to climax.

AGRICULTURE AND SUCCESSION

Farming practices result in the expenditure of a great deal of energy, not to mention money, to control weeds. Over $1 billion was spent on herbicides in the United States in 1980 and untold more dollars were expended to apply them. Weeds are sprayed, burned, mowed, and plowed under to prevent their growth or spread. These are the costs that have to be paid in order to maintain a **monoculture**—a single species of crop plant. The forces of succession are always present, and the natural tendency of ecosystems to become more complex requires intensive and expensive farming practices. Additionally, the prevention of succession will slow or stop the rate of soil formation; farming practices in general cause a loss of soil and nutrients, requiring the use of artificial fertilizers. The maintenance of hedgerows between and around croplands will slow erosion and nutrient loss by holding some soil moisture and will reduce erosion by the wind.

ECOSYSTEM EVOLUTION

Just as life and organisms have evolved, so have ecosystems, through succession. Species have evolved mechanisms to cope with abiotic and biotic factors, and have produced an interdependent assemblage of biological organisms in an environment of physical forces. Large ecosystems that can be characterized by the similarity of their climax communities are collectively termed a **biome**—large ecosystems that are similar in form and function. To illustrate, there are deserts on every continent and each is unique; their physical environments and biological compositions are all somewhat different. But all deserts share enough characteristics that they can be described as one ecosystem, scattered over the earth; all deserts are similar enough to be collectively considered as the desert biome; the same applies to other large ecosystems. The biomes we will consider are:

- Tundra: arctic and alpine
- Coniferous Forest
- Deciduous Forest: temperate and tropical
- Sclerophyll Scrub

- Grassland
- Tropical Savanna and Scrublands
- Desert
- Tropical Rain Forest
- Fresh Water: lentic habitats and lotic habitats
- Marine: the ocean

Figure 8.6 shows the distribution of the major biomes of the world.

Ecotones

Not all of the biomes are clearly demarcated everywhere. There are many areas where one biome grades into another one. Coniferous forests become tundra at higher altitudes and latitudes. Scrublands become deserts in the direction of less rainfall. The faster the

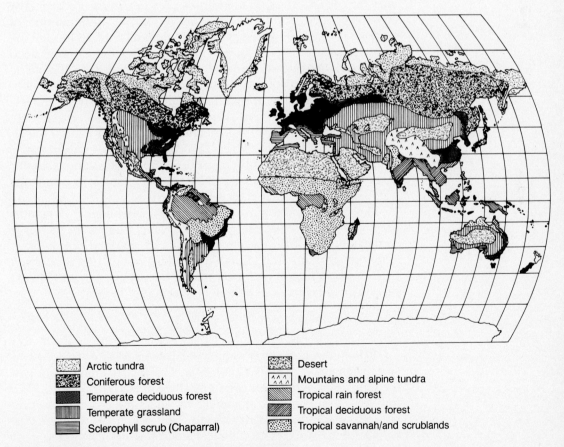

Arctic tundra		Desert	
Coniferous forest		Mountains and alpine tundra	
Temperate deciduous forest		Tropical rain forest	
Temperate grassland		Tropical deciduous forest	
Sclerophyll scrub (Chaparral)		Tropical savannah/and scrublands	

Figure 8.6 Schematic Map of the Major Biomes of the World.

environmental factors change, the smaller the area of integration is. These intermediate habitats containing a mixture of biomes are called **ecotones** and are often higher in species diversity than either of the bordering biomes, since species of both biomes are represented. This phenomenon is sometimes referred to as the **edge effect**. Keep in mind, then, that biome designations are handy for discussion but are much less clear in nature.

Tundra

There are two expressions of the tundra biome: arctic and alpine. The arctic tundra encircles the northern pole of the earth north of the coniferous forests. (Antarctic tundra is restricted to the southern tip of South America and a few islands off Antarctica.) Alpine tundra is found on high mountains all over the world (Figures 8.7a and b).

Tundra is characterized by low-growing plants such as lichens, mosses, rushes, grasses, and shrubs. Temperatures are generally cool to very cold, from a maximum of about 25°C to a minimum of −45°C. The arctic tundra is generally colder than the alpine, but the alpine areas are much windier, resulting in a more extreme wind chill factor. **Permafrost**, permanently frozen soil, is typical of arctic tundra; the soil may thaw a depth of 20–100 centimeters each summer before refreezing. Permafrost is rare in alpine tundra, and the soil thickness may be only 30 centimeters. Precipitation is generally much lower in the arctic (10–50 centimeters/year) than in the alpine (100–200 centimeters/year) tundra; in fact, the arctic tundra is rather arid, and many plants depend upon some permafrost thaw for water. Permafrost also acts as a barrier to percolation, resulting in a soggy marshlike habitat in the summer.

The arctic tundra growing season may only be 2 to 3 months long, with constant daylight during that time. The winter may have no daylight at all. In alpine environments, extremes in daily temperature occur because the 15-hour maximum photoperiod is followed by a rapid cooling of the environment due to the thin air (low air pressure) and clear skies at high elevations.

Organisms and Adaptations

Plants and animals obviously face considerable hardships in such harsh environments. Compared to a forest, the modification of the climate by vegetation is minor; physical factors dominate. The low vegetation, permafrost, and flat terrain of the arctic tundra provide few places for animals to seek shelter, so they are adapted to withstand the stress of the physical environment or avoid the most

a

b

Figure 8.7 Tundra. (a) Alpine tundra occurs above the treeline on high mountain tops. The vegetation is short and consists mainly of grasses and sedges, but flowering herbs may be very abundant. (b) Arctic tundra occurs above the treeline in high latitudes. Although it appears to be grassy, most of the vegetation consists of low-growing plants such as sedges, rushes, stunted trees and shrubs, and some grass. The lower latitude arctic areas tend to have thicker soil and denser vegetation than the high arctic areas.

severe seasons by migrating. The alpine tundra provides some shelter on rugged terrain, but some animals migrate to lower elevations for the winter.

The most obvious adaptations of plants are their short height and a tendency to be **herbaceous** (have a fleshy stem). Perennial plants can store carbohydrates in their underground roots and use this energy to produce new leaves and shoots the next year. Barely enough energy is left in tundra shrubs to produce a new crop of leaves each year, and there is little energy remaining for wood growth, so growth is slow. Most tundra plants, to take advantage of the short growing seasons, have high photosynthetic and growth rates; most are also perennials. Some species of alpine plants grow in a **rosette** (rose shape) close to the ground, minimizing the effects of the wind. They may also be woolly, reducing transpiration. Some African and South American species curve their leaves inward at night to protect the growing bud from frost.

Lichens and mosses occur at the highest elevations and latitudes of any plants. They can withstand lower temperatures than can the herbaceous or woody plants as long as there is surface water. They can photosynthesize at $-10°C$ though their optimum temperature is about $+5°C$. In the Antarctic there are 350 species of lichens and only two herbaceous plant species. The lack of roots in lichens and mosses allows them to grow on rocky substrates with little or no soil.

Seeds produced by herbaceous or woody plants such as heather are very hardy and may remain dormant for years. Seedling plants may take several years to become safely established, but once established, survival is very high.

Animals of the tundra are also adapted to extreme conditions (Figure 8.8). They tend to have a large body size and slow growth, and be well-insulated. (The arctic fox can withstand temperatures of $-80°C$ for short periods.) Compared to other biomes, species are few, and many are migratory. Invertebrates, amphibians, and reptiles are few in number and variety, but swarms of mosquitos, gnats, and other flies abound during the short summer.

White coloration is common among arctic animals such as the arctic fox, polar bear, ptarmigan, snowy owl, and arctic hare. Many of these animals have a dark color phase during the summer months. Their coloration (against a background of snow or bare terrain) conceals them from their prey or predators. Dark colors may help to absorb solar radiation, although recent evidence indicates that the polar bear, with no dark phase, has a coat of hollow white hairs that trap solar radiation.

Tolerating the cold tundra winters poses a major problem. Invertebrates such as insects and spiders winter as immature stages, which are particularly resistant to low temperatures. Many animals, such as the ptarmigan, burrow for some protection against the weather. Lemmings, common rodents of the Arctic, construct runways through the snow, a good insulator from the outside air. The only hibernator in the Arctic is the arctic ground squirrel.

A number of species of birds nest in the Arctic, but few are year-round residents. June and July are the peak nesting times for large numbers of shorebirds and waterfowl and fewer numbers of hawks, cranes, larks, pipits, and finches. In alpine tundra, ptarmigans, larks, swallows, finches, robins, choughs, and even hummingbirds can be found. During the winter almost all arctic and alpine birds migrate

Figure 8.8 Mountain or Bighorn Sheep. The typical habitat for the bighorn sheep is alpine tundra and timberline regions in summer, lower elevations in winter. Canadian Rocky Mountains.

to lower latitudes or altitudes. Ptarmigans, arctic hares, and snowy owls are among the few that remain during winter.

There are similarities between the arctic and alpine tundra flora and fauna; 40% of the species found in the alpine tundra are found in the arctic tundra. But the organisms of the Arctic are **circumpolar**— that is, they circle the globe, and thus are more similar to each other than alpine organisms, whose habitats are isolated on mountain tops around the world.

Human Use

Since the tundra offers few material resources to society, it has been all but ignored except by resident Indians and Laplanders. Recently, however, the discovery of fossil fuels has led to gas and oil exploration and to the construction of pipelines, roads, and other aspects of civilization in the tundra. We can only speculate as to the long-term effects, but the slow-growing tundra is one of the most sensitive and least resilient of all biomes because the extremes of environmental factors keep the biome simple in structure and number of species, characteristics that imply instability.

Figure 8.9 Coniferous Forest,
Wudle Mountains in the San
Juan National Forest.

Coniferous Forest

The coniferous forest biome, the northernmost part of which is called the **taiga**, is found chiefly in Canada, northern Europe, Siberia, and farther south on high mountains. Since high elevation environments are similar to high latitude ones, the coniferous forest biome "fingers" its way southward along the high mountain ranges. It is typically a forest of evergreen needle- or scale-leaved trees. The generally dense forest restricts light penetration to the forest floor (hence, the "Black Forest" of Germany), so the undergrowth is sparse. The evergreen spruces, firs, pines, and cedars photosynthesize whenever conditions are suitable. Most conifers are evergreen, but not all evergreens are conifers. "Evergreen" is a bit misleading, since the leaves do fall, but not all at once.

The structure of coniferous forests varies from open pine forests with grass as ground cover, to dense pine–fir forests with little undergrowth, to hemlock–cedar–fir forests of coastal areas with luxuriant undergrowth (Figure 8.9).

Figure 8.10 Quaking Bog in the Northern Appalachian Mountains. Note the beaver house in the background. Beaver activities keep some of the waterway clear of vegetation. The thick mat of vegetation over the water can be walked upon, giving one the impression of walking on a giant waterbed.

Temperatures during the winter may be as cold as those of the tundra, but the summer is warmer, averaging 15°–20°C, and provides a longer growing season of 3–4 months. The rainfall is moderate, and the lower temperatures reduce evaporation. Water may be tied up as snow and be unavailable to plants until spring, however. The soil is thin and acidic and forms slowly due to the low temperatures and the acid, waxy covering of the needles.

Bogs (marshy areas in northern regions) and lakes are also characteristic of coniferous forests. They were formed as glaciers receded and gouged out depressions in the earth during the Pleistocene epoch 1–2 million years ago. These lakes gradually become covered with a thick floating mat of vegetation that a person can walk on, giving one the impression of walking on a large waterbed (Figure 8.10).

Organisms and Adaptations

Although the coniferous forest is a cold biome, the climatic effects are ameliorated by the vegetative structure of the forest. The trees diminish the wind, filter or block the sunlight, and retain moisture. Coastal coniferous forests have a milder climate and are very wet due to moisture-laden air that is blown in from the ocean.

The plants are adapted to cold temperatures, wind, acidic and moderately fertile soils, snow, and a minimum of available water in the noncoastal areas (75–175 cm/year). Spruce trees can withstand temperatures as low as $-70°$C in the winter without damage. European cembra pine seedlings seem to be able to retain their photosynthetic ability even when buried by snow (light can penetrate lightly packed snow for a meter or more). Coniferous forest evergreen trees are typically thin, taper to a point on top, and have very flexible branches; these attributes allow the tree to shed or bear the weight of a heavy snowfall without damage. Even saplings bent over and covered by snow are elastic enough to straighten out when the snow melts. The small, waxy evergreen needles are adapted to withstand freezing and desiccation.

The typical soil of a coniferous forest has a thick litter layer because the cold temperatures slow down decomposition by bacteria and fungi. The decomposition of the waxy coniferous leaves produces organic acids that leach through the soil, taking minerals with them. As a result, the soil is thin and low in mineral content. The lowest soil layer is often hard clay, which restricts water percolation and root penetration. Trees grow slowly under these conditions.

Coniferous forest trees provide a habitat and food for animal inhabitants: shelter; nest sites; and food, such as seeds for rodents, sap for woodpeckers, and plant matter for porcupines, grouse, and insects. The undergrowth provides additional habitat and food such as berries. The more complex the habitat structure, the more kinds of animals are likely to inhabit it (Exercise 8.3).

Most animals in a coniferous forest are found in, on, or under the trees and forest undergrowth, and, like the plants, have evolved adaptations to the cold. Millipedes, centipedes, rodents, insects, shrews, moles, worms, some birds, and other organisms inhabit the undergrowth and the soil beneath it. In the upper story of vegetation, that is, in the trees and taller saplings, there are spiders, insects, birds, and rodents such as squirrels. They may overwinter in runways underneath the snow and brush (voles), hibernate (some squirrels), become dormant (bears, insects), or migrate to a warmer latitude (some birds and insects). Reptiles and amphibians can withstand the winter by burrowing into the mud of a lake or holing up in a burrow, but these animals are rare in coniferous forests because of their restricted ability to acclimate to cold. One lizard species of the high Andes, however, survives cold nightly temperatures by basking in the sun during the day to raise its body temperature. High metabolic rates, thick insulation, and large body size all contribute to the homeotherms' ability to survive in northern and high elevation coniferous forests.

On the higher mountain ranges of the world (Rockies, Sierra Nevada, Alps, Andes, and Himalayas) are communities that are sometimes called the montane biome, essentially a transition zone between the lower elevation forests and the higher elevation tundra.

Human Use

Coniferous forests serve many commercial, individual, private, and public uses, but the greatest visible use is probably timber production. The use of trees for wood and paper products has led to overexploitation of some forests, resulting in the virtual destruction of ecosystems through clear-cutting and subsequent erosion. More prudent management of forest resources—selective cutting and replanting—has diminished the impact of logging, but the ever-increasing demand for forest products is causing younger and younger forests to be cut. Mature coniferous forest ecosystems are becoming rarer.

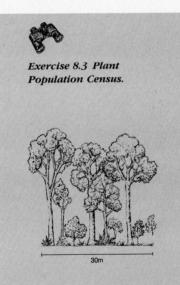

Exercise 8.3 Plant Population Census.

30m

The numbers and variety of plants in a community can *be measured in several ways. The two most common ways are by quadrats and transects. A **quadrat** is a square sample of the area to be measured. Its size depends on the plants to be counted; for trees it might be 25 meters on a side, for shrubs it might be 5 or 10 meters, and so on. Several quadrats should be established so as to be representative of the entire area. The **transect** is a narrow strip that runs through the entire length of the community. The width of the strip depends, again, upon the plants being measured—3 meters wide* *for trees, 2 for shrubs, and so on. The areas of the quadrats and/or transects are divided into the total area of the habitat to be measured and that figure, multiplied by the number of plants, gives an estimate of the total number of plants. This exercise can be used in conjunction with Exercise 8.4 to calculate plant species diversity. Relatively nonmotile animals such as insects, worms, millipedes, and so on can be measured by a similar method. Census techniques for motile animals are described on pages 203–205.*

Height in meters

Figure 8.11 Stratification. The arrangement of plants into recognizable vertical levels is called **stratification**. The various **strata**, or levels, provide homes and food for different types of organisms although many animals will use several strata. Generally, more complex stratification provides for a greater diversity of flora and fauna.

Deciduous Forest

Deciduous forests are dominated by trees that lose their leaves in the fall and produce new ones in the spring. The temperate deciduous forest covers most of the eastern United States, all but the northern and southernmost portions of Europe, and a portion of middle eastern Asia. The tropical deciduous forest is characteristic of parts of India and southeast Asia, a portion of northeastern Australia, and parts of central Africa and South America.

Leaf loss is an adaptation to the cold season in temperate climates; shedding leaves protects against freezing and water loss. In tropical areas, leaf drop during the dry season protects against water loss. Unlike the coniferous forest, the deciduous forest has a 5–6 month growing season in which to store the food reserves necessary to produce a new crop of leaves the next year. The lack of rainfall is the cue, but leaf loss may not occur if the dry season has above average rainfall. In temperate areas, shortening photoperiod initiates leaf loss.

A deciduous forest has several layers of plant communities, often two levels of trees, a layer of shrubs, and a layer of herbs (Figure 8.11). This forest is more open than the coniferous forest, allowing more light, and thus more growth, under the trees. A greater variety of plants provides a larger array of habitats, nest sites, and food, leading to a greater diversity of animals. The climate is considerably milder than in the tundra or coniferous forest, and the growing and breeding seasons are longer.

In the temperate forests, annual precipitation ranges from 60 to 200 cm/year; a similar amount falls in tropical deciduous forests but is restricted to a "wet season" (for example, May to October in Venezuela). Humidity is generally high, since the transpired moisture is trapped by the vegetation layers. In the temperate forests, temperatures may range from a high of $+30°C$ to a low of $-30°C$; in the tropical areas the extremes are $+20°C$ to $+38°C$ in the dry season with even less variability in the wet season. With milder temperatures in the rather **mesic** (wet) environment, growth and decomposition rates are fairly high and the soil humus thick.

Organisms and Adaptations

The structure and appearance of the forest changes considerably between the seasons because the trees become leafless in the fall and the herbs die back; many become dormant and grow from bulbs or roots the next year. Much of the deciduous forest becomes dormant in the winter or dry season, but sufficient energy is stored in underground parts or seeds to renew the forest the next year.

As a deciduous forest matures, the shade-tolerant trees become the dominant climax species. The first trees to grow in the area are sun-tolerant and produce a shaded environment under which shade-tolerant, but not sun-tolerant, plants grow. As shade-tolerant trees grow, they shade the sun-tolerant trees, which eventually disappear. Only shade-tolerant trees can grow under the dominant trees, which, at climax, are the same species. Beech seedlings, for instance, need very little light, but birch seedlings need 12%–15% of the incoming solar radiation. Thus, birch does well in an open forest sere, but beech trees eventually dominate.

The lower herb and shrub layer habitats are quite different from the tree layer. The lower part of the forest receives less wind, less sun, and has higher humidity. Many of the herbs do their growing and store food early in the year before the trees fully leaf out, because the forest floor plants receive 10 to 15 times more light at this time than they will later; they may even become dormant when the trees leaf out.

In the tropical deciduous forests, the dry season brings on leaf drop and dormancy, but dormancy is not as complete as in the temperate forest since rainfall can delay leaf drop and allow later flowering of herbaceous plants. The rise in temperature before the onset of the rainy season is the cue for the plants to break dormancy.

Animals are more abundant and diverse than in the tundra or coniferous forest because the deciduous forest has a less severe climate, a more complex vegetative structure, and greater species diversity, which provides a wide variety of habitat and food types.

There are many more species of both vertebrates and invertebrates. The milder temperatures allow the existence of many more poikilo-thermic animals. Most of the animals are concentrated in the soil and lower herbaceous layers. The seasonal dormancy of the forest requires that many animals also go dormant (insects, millipedes), hibernate (some rodents), or migrate (many birds), but a number of animals are able to lead an active life during the winter or dry season, living on the litter, nuts, and seeds produced during the growing season.

Human Use

Temperate deciduous forests have been logged, burned, or cleared to such an extent that little virgin temperate deciduous forest remains. Forests in the eastern United States are virtually all second growth. Indians burned the deciduous forests to open them up for easier hunting, and later forests were cleared for agriculture and logged for wood products. Recent increasing costs of fossil fuel have put pressure on forests for firewood.

The tropical deciduous forests have not been exploited to such an extent except in a few areas where the demand for firewood is so great that grassland exists where there once were forests.

Forest vegetation, deciduous and evergreen, contains approximately 90% of the carbon in the biomass of terrestrial ecosystems. Reducing forest land thus increases the amount of carbon dioxide in the atmosphere. In 50 years this could result in an increase of the earth's average temperature by 2°C, causing the polar icecaps to melt and crop productivity to change. There is considerable disagreement among scientists, however, on this speculation.

Sclerophyll Scrub

Sclerophyll is composed of dense, short, and shrubby vegetation with thick leathery leaves (**sclerophyll**). It is found mainly on the southwest coast of the United States and continues into northern Mexico, in southern Europe and northern Africa (the Mediterranean region), and in parts of the southern coast of Australia and Africa. In California this biome is called **chaparral**. It is dominated by short woody scrub, with grasses and herbs growing underneath (Figure 8.12a).

Organisms and Adaptations

Sclerophyll scrub grows in areas with a mild climate moderated by ocean breezes, with moderate rainfall during the mild winters; the summers are hot and dry. The shrub and herbaceous plants flower during the rainy season and the herbs dry up afterwards,

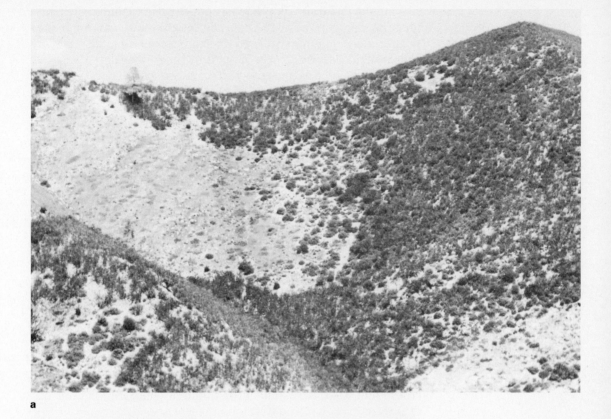

a

b

Figure 8.12 Sclerophyll Scrub. (a) Chaparral. Chaparral has few trees and the space between the shrubs is often barren, partly due to competition for water. A characteristic biome of California, chaparral often grows so densely that it is virtually impossible to walk through it. (b) Buckeye Butterfly. This common inhabitant of foothill chaparral uses its long tubular proboscis (tongue) to suck nectar from a variety of flowers. It lays two sets of eggs in a season, and the young caterpillar feeds upon a great number of chaparral plants.

making the environment very susceptible to fire. Often the chaparral is considered to be a fire climax because the shrubs sprout rapidly after a fire and trees are not able to become established. Many chaparral plants have seeds that open and germinate only after a fire. Some shrubs, such as buck brush, have an extensive root system that produces aboveground stems after a fire. Fire adds minerals to the soil and makes it more permeable to water. The well-adapted and fast-growing plants bring on a recovery of the habitat in only 10–15 years.

Animal inhabitants are not especially numerous because of the inhospitable dry season, but animals such as mule deer use the chaparral in the winter wet season and move to higher elevations and other habitats in the dry season. In the wet season it is also attractive to many insects and breeding birds, which leave in the hot summer (Figure 8.12b).

Human Use

The major effect on sclerophyll shrub habitat is its removal for other uses. Being in a mild climate, it is attractive for human residences and orchards. Ironically, its removal has often led to disaster for homeowners because the plants protect the soil from erosion; many towns built on chaparral communities in California have been inundated with floods or hit by landslides due to the removal of vegetation. Also, chaparral burns frequently and rapidly, and has destroyed many housing developments.

Grasslands

Grassland comprises one of the largest biomes, being distributed in the continental interiors across the central and western United States and Canada, much of Mexico, middle Asia, South Africa, Australia and southern South America. Its terrain is generally flat or slightly rolling (Figure 8.13a).

The major vegetation is grasses, but herbs are usually present, and there may be scattered shrubs and trees. There are tall, short, and mixed height grasslands, generally depending on the amount of rainfall. Most grass plants are **perennial**, dying each year and sprouting the next spring. The leaves and shoots become food for decomposers and raw material for soil. Stored materials in the roots provide energy for regrowth.

The climate ranges from cool to cold winters and mild to hot summers. Rainfall ranges from 25 to 75 centimeters and occurs primarily in the spring. Being exposed to the hot sun and drying wind, evapotranspiration is high. The late summer produces dry conditions, so the growing season is only about four months in spring and

a

Figure 8.13 Grassland.
(a) Grasslands cover more of North America than any other biome. Grasslands may have as their major components perennial **bunchgrasses** or **sod-forming grasses**. The bunchgrasses spread from buds near the base of the plant, resulting in a bushy clump of grass. The clumps are typically separated from each other, and the space in between may be filled by other plants or remain bare. The sod-forming grasses send out lateral runners (rhizomes), which penetrate the soil and form new plants in a continuous turf. (b) Grassland Savanna. As the environment becomes moister, trees may invade the grassland, forming a savanna habitat. Grasses are still the dominant plants, but trees, especially drought-resistant ones such as valley oaks, may be common.

b

early summer. The soil builds up rapidly and the humus content may be 10 to 12 times as much as that in a deciduous forest.

Organisms and Adaptations

Grassland vegetation structure varies. In the more mesic grasslands (the **prairies**), grasses form a thick sod by means of **rhizomes**, underground stems that produce aboveground stems and leaves. The roots may penetrate the soil to three or more meters. In the drier (**xeric**) areas, bunchgrasses grow in clumps, with areas of bare soil between; their roots are not as deep. Although trees may be scattered in the grassland, experimental evidence indicates that they lose when they compete with grass roots. In drier areas, lack of water alone prevents tree invasion. In addition to a hot and dry climate, fires and cattle grazing also prevent encroachment by trees. Fire favors grasses since perhaps 80% of the biomass of a grass plant is underground, protected from fire. Grasses are well adapted to a dry environment. They have an extremely branched and dense root system to absorb and store water. In dry periods they become dormant. Under short-term dry conditions such as a hot, windy day, the leaves of some grasses roll up edgewise to reduce the evaporative surface area. In wetter areas trees begin to invade to form a grassland savanna (Figure 8.13b).

Typical animals of grasslands are large herbivores (bison, antelope, and the many large grazers of Africa); mice, moles, worms, snakes, shunks, coyotes, many birds, and arthropods are also characteristic. Productivity is high in grasslands, so plant parts and seeds are abundant early in the year, supporting a high animal population. As the plants dry, smaller organisms, particularly insects, burrow into the ground or sod for the winter. Larger organisms may remain to feed on vegetation, or if they are **insectivorous** (feed on insects), migrate. The winter population of seed-eating birds may be even greater than the population during the breeding season, as birds from colder habitats may spend the winter in mild-climate grasslands.

Human Use

Most tall grass prairies, because of their thick, fertile soil, have been converted to agricultural fields. The short- and mixed-grass habitats are extensively utilized for cattle grazing. Although light grazing stimulates grass growth, heavy grazing results in the invasion of shrubs and trees. Similarly, the prevention of fire may allow woody plant incursion. In more populous areas, grasslands have been used for housing developments, since the thick soil is relatively easy to level with construction machinery.

Tropical Savanna and Scrublands

The tropical savanna is represented by extensive habitat in central South America, central and south Africa, more than half of Australia, and portions of southern Asia. Savanna is grassland with scattered individual trees or patches of trees, usually short and scrubby ones (Figure 8.14a). Sedges, grasslike plants, are also common in moist areas.

The savanna climate has three distinct seasons: a cool, dry season followed by a hot and dry period, and then a warm and rainy one. Rainfall varies from 50 to 150 centimeters. Fires are common as in the grassland and inhibit invasion by trees. Some savanna soils are rich in humus, but most are porous and well-leached and low in humus content.

Organisms and Adaptations

The plants of the savanna are adapted to dry, infertile conditions, and many are resistant to fire and grazing by wildlife. The grasses become dry and dormant after the rainy season and the trees lose their leaves. The trees are composed of hard, dry wood and store little water. Many of the trees and shrubs, such as acacias, have thorns or spines for protection against grazing by huge populations of wild **ungulates** (hoofed animals), as are found in Africa (Figure 8.14b).

Animal life is most abundant during the rainy season when water and green vegetation are present. During the dry season some animals become dormant and others migrate. Vast herds of zebras, for instance, migrate hundreds of kilometers to avoid the parched savanna. African aardvarks ("bush pigs" to the natives) dig large burrows for protection against the heat of the day. Giraffes stand in the shade while browsing on acacias. Many other animals are nocturnal and get water from plant matter, dew, or the moisture in their burrows.

Human Use

The African savanna is thought to be the home of the earliest humans. Mild climates have led to habitation by many primitive cultures in all savannas; the Bushmen of Botswana use the savanna as they did thousands of years ago.

Fire, agricultural practices, grazing, and hunting have all had an impact on the savanna. The confinement of large ungulate populations in Africa to restricted savanna preserves has resulted in deterioration of the habitat. But with the destruction of tropical forests for wood products, savanna land has actually been created, and there may be more now than there once was.

a

Figure 8.14 Tropical Savanna. (a) Typical tropical savanna characteristic of the Americas. Consists of a base of grasses and sedges, interspersed with clumps of trees that are a few centimeters to a few meters higher. During the rainy season, the grassy region may be flooded. (b) Ungulate Herds. Large numbers of zebra, wildebeest, antelope, elephants, giraffes, and other hoofed animals roam the savannas of Africa, eating myriad grasses, shrubby plants, and even trees. These many plants have evolved adaptations such as reduced leaf area, toxins, spines, or thorns to discourage heavy browsing.

b

Figure 8.15 Desert. This photo shows the alkali sink scrub community in northern Nevada, part of the Great Basin cold desert. The dominant plant here is greasewood. These low desert areas are often flooded during the rainy season, and a residual salty crust forms when the soil dries. Greasewood and other shrubs are very tolerant of a high salt content.

The seeds of greasewood and other associated plants are often buried by ants (in the process of storing them for food), and these have a better chance for germination than the unburied ones. During the growing season the seedlings must develop deep and extensive root systems to tap underground water.

Desert

Deserts represent one of the harshest environments of all biomes. Maximum temperatures are the highest in the world (49°–54°C). Although a desert environment is characterized by low precipitation (5–30 cm/year), and not by temperature, there are two types of desert: the hot and the cold. The higher latitude cold desert may actually receive snow and subfreezing temperatures in the winter. With little vegetative buffer against the wind and a usually cloudless sky, temperature differences in one day may be extreme (−5°C to +20°C). Growing seasons may be very short, as many plants are adapted to grow and flower only after a rain. The soil is thin and may consist only of sand (Figure 8.15).

The hot deserts are found in the southwestern United States, a narrow strip along the west coast of South America, much of north Africa (the Sahara), the Arabian desert that stretches from Iran to India, and the center of Australia. The cold desert is found west of the Rocky Mountains in the United States and Canada (the Great Basin), eastern Argentina, and much of Central Asia.

Organisms and Adaptations

The cycles of growth and reproduction in the desert are related to the water supply rather than to the temperature. In some plants, flowers, fruits, and seeds are produced only after a good rain, which may be only once every four years. The seeds may not germinate for years after that. Studies have also shown a direct relationship between the amount of rainfall and the production of plant mass. Some shrubs, for example, have only a few leaves, leaf out more fully after a rain, and lose their leaves as conditions become dry again. The Palo Verde tree has green photosynthetic stems and very small leaves to maximize photosynthesis and minimize water loss (Figure 8.16). Other plants have soft leaves that wilt under dry conditions, shut their stomates to prevent water loss, or are **succulents**—fleshy plants that store water in their leaves (aloe), stems (cacti), or roots (asparagus). The roots of desert plants tend to be large, widely spread, and shallow to absorb the maximum rainfall.

Nutrient and water resources needed by plants are scarce, so it is advantageous for individuals to be isolated from other plants. In some cases the plants produce a waste product that is released into the soil and chemically inhibits the growth of a plant nearby. This phenomenon is termed **allelopathy**. When depressions and hills are present, plants tend to be concentrated in the depressions and scarce or absent on the hills because the lower areas collect more rainfall and are somewhat more protected from the sun and wind.

Figure 8.16 Palo Verde. The Palo Verde is one of the desert shrubs that occurs near desert washes (arroyos). In times of stress brought on by drought conditions, the small leaves drop off, but since the stems are green, photosynthesis can still occur. The Palo Verde is a legume, able to live in the nitrogen-poor desert soil. It has large seeds and seed pods, as do many legumes. The seeds germinate only after being abraded by some mechanical action that exposes the seed to water and air. Most often this **scarification** process occurs as the seeds are washed down on arroyo plains after a storm, preparing them for germination as well as dispersing them.

The animals of the desert face the same water scarcity problems. Many live in burrows and restrict their activity to nights, when it is cooler and they need less water (Figure 8.17). Some, such as the reptiles and arthropods, have a thick outer covering that retards or prevents water loss. Birds and reptiles need very little water in order to excrete waste products. Some rodents, such as the kangaroo rat of North America and the jerboa of Asia, can metabolize water from dry seeds and do not need to drink. The camel of North Africa allows its body temperature to rise rather than cooling it by sweating; at night body heat is radiated into the cooler air with little water loss. Some birds and other animals are pale in color; experiments on caged birds demonstrate that such color allows the birds to reflect radiation and stay cooler. There are also many behavioral adaptations to avoid heat. An earless lizard of the Southwest hides in rock crevices or buries itself in the sand during the hottest part of the day. When it does venture out, it orients its body parallel to the sun's rays to reduce its exposure. In times of drought, many organisms become dormant.

Human Use

The desert, in spite of limited resources, has been the home of many unique peoples. The Piute, Hopi, and Navajos of the southwestern United States, the nomadic tribes of North Africa, the Bushmen and Hottentots of South Africa, and the aborigines of Australia are all natives of the desert. Their villages and/or migration are or

Figure 8.17 Spider Burrow. Like many animals of the desert, the predaceous spider is nocturnal and ventures out of its cool burrow only when the heat of day has dissipated.

were very much keyed to a water source, the most limiting factor in the desert.

Attempts to make the desert fertile via irrigation have had mixed results. The Aswan Dam of Egypt's Nile Valley was completed in 1967. It made irrigation of a half-million hectares of desert possible, but the resulting farmland is of poor quality due to lack of soil, and a decrease in the deposition of nutrients downstream has reduced the fertility of the Nile Delta. In North America the desert has been severely affected by off-road vehicle activity, the effects of which will remain for many years since vegetation grows so slowly. Poaching of cactus and desert tortoises has, unfortunately, become popular recently. Housing developments in the United States and Saudi Arabia have increased in deserts, requiring enormous amounts of water to be transported from other areas.

Tropical Rain Forest

Tropical rain forests occupy much of Central America, northern South America, a belt across equatorial Africa, and most of southeastern Asia, Indonesia, and New Guinea. They are characterized by heavy rainfall and luxurious evergreen vegetation in many strata. The forest canopy is dense, admitting little light to the interior. The undergrowth is not as thick as popularly imagined because of the lack of light. Only along the edge of the forest and where the canopy is open does the undergrowth become a thick jungle (Figure 8.18).

Rainfall, which exceeds 200 centimeters a year, falls throughout the year, except that some seasons are drier than others. Humidity is generally high, often reaching 100%. Temperatures are high and vary little; 20°–28°C is the average temperature, and the difference between the warmest and coldest season may be only 10°–15°C. The photoperiod is nearly constant throughout the year: 12 hours. Consequently, the growing season is continuous. Because of the warmth and moisture, decomposition is rapid and materials are quickly recycled into plant production, leaving little to be turned into soil; soils are thus thin.

Organisms and Adaptations

Partly because of the mild climate, abundant rainfall, and constant photoperiod, the diversity of plant and animal life is the highest of any biome (Exercise 8.4). Some grasslands or savannas may contain only 2 or 3 species of plants over a large area; a temperate deciduous forest may have 15 tree species per hectare; but a tropical rain forest may have 70 kinds of trees in the same area. A more or less solid mat of lower canopy trees is perforated by scattered taller trees. This tree-dominated forest structure provides varied habitats

Figure 8.18 Tropical Rain Forest. Island of Trinidad, West Indies.

for many organisms. A large number of trees have wide bases with projecting buttresses for support in the thin soil.

Many trees are covered with **epiphytes**, plants that use the trees for support and as a means to be closer to the canopy so that they receive more sunlight than they would on the forest floor. Some epiphytes are parasitic, but many, such as bromeliads, ferns, and lianas (thick, woody vines), derive their energy from photosynthesis. They receive water from the abundant rainfall typical of the rain forest (Figure 8.19).

In spite of the abundant rainfall, solar radiation is quite strong in the canopy and can dry out leaves exposed to sunlight. Thus, many plants have thick and leathery leaves; examples are the rubber tree and *Philodendron*. In the lower layer of the forest, the opposite situation exists: The air is saturated with water and water droplets condense on leaves. Many low-layer plants have "drip-tip" leaves to allow water to run off (Figure 8.20).

Seasonality is based mainly on the amount of rainfall, as the temperature and photoperiod vary little. Seasons are the wet season or dry season, although the dry season may not be dry at all but

Exercise 8.4 The Meaning and Calculation of Species Diversity.

You may have an intuitive feeling for the diversity, or variety, of species in a community. A tropical forest is generally assumed to be more diverse than a desert or grassland, for instance. But how do the diversities of a coniferous and a deciduous forest differ? Intuition fails here, so we need some sort of measure of diversity. Diversity of species in a community is measured basically by two parameters: number of species, and the distribution of the number of individuals among the species. The greater the number of species and the more equally distributed the individuals among the species, the higher the species diversity. Calculation of species diversity can be done with a number of formulae, but the most often used is the Shannon Index (see M. Lloyd, J. H. Zar, and J. R. Karr, 1968, "On The Calculation of Information—Theoretical Measures of Diversity," Am. Mid.Nat. 79:257–272). Simply stated, Diversity = $-p_i \ln p_i$, where p_i is the number of individuals in a species divided by the total number of individuals of all species.

You may wish to use the Lloyd et al. reference to calculate species diversity figures for a hypothetical population of, say, different colored beans or other objects in a classroom situation or for an actual population in the field, such as the tree community in a nearby woodland (see C. D. Monk, 1967, "Tree species diversity in the eastern deciduous forest with particular reference to North Central Florida," Am. Natur. 101:173–187).

Figure 8.19 Epiphyte. An epiphyte (literally: "on top of plant") may or may not harm the plant on which it grows. "Spanish moss" is neither Spanish nor a moss, and simply uses trees for support. It obtains its water from the air and thus can live only in very humid environments. It obtains nutrients from dust particles that land on its fine leaves. Some birds use the "moss" for nesting material, and it was also once used by people as packing material and in mattresses. Spanish moss is found in tropical areas from southern Brazil northward to the temperate deciduous forests of the southeastern United States.

Figure 8.20 Drip-Tip Leaves. Since plants exchange gases across their leaf surfaces during photosynthesis, water standing on a leaf would slow photosynthesis. Wet leaves are also prone to attack by fungi. Thus many plants in wet areas have leaf shapes which encourage water to run off.

simply less wet. Flowering occurs throughout the year. Many species of tropical trees bear flowers on lower, older branches or on the trunk, allowing animals such as bats and birds to pollinate them. Deciduous rain forest trees do not lose their leaves at any one time of the year. Different individuals of the same species may be bare or leafed out at the same time.

Rain forests are typified by **arboreal** (tree-dwelling) animals that utilize the tree canopies. Monkeys, birds, snakes, many insects, and even frogs find food and shelter many meters off the ground. Fruits and insects are the main staples of the larger animals. The epiphytes depend on animals for dispersal of their seeds, spores, and pollen. Birds tend to be brightly colored and have song frequencies that penetrate the vegetation so they can more easily find each other.

Birds and other animals have no particular breeding season as the photoperiod and temperature are so invariable. Breeding takes place throughout the year, and may be keyed to rainfall. Birds may have two or three clutches of eggs per year; clutch sizes are smaller than in the temperate zones because temperate species usually have only one chance to breed and put all their energy into that opportunity.

Poikilothermic animals such as reptiles, amphibians, and arthropods are well represented in the tropics and are very abundant. In one part of Ecuador, over 80 species of frogs can be found in a 3 square kilometer area. They also reach very large sizes. For example, some beetles are enormous: The Hercules beetle of South America is 16 centimeters in length, and the goliath beetle of West Africa reaches 12 centimeters. One species of New Guinea spider is large enough (9 centimeters) to capture small birds and mammals. A South American frog eats small rodents.

Human Use

Tropical forests are disappearing at a dismaying rate. Every year an area the size of Great Britain is removed. Trees are felled for lumber, and forests are cleared for agriculture or mining. Although a productive ecosystem, the tropical forest does not provide fertile land for agriculture due to the thin mineral soil. An area of cleared forest may support crops for a few years, but the heavy rains quickly leach or erode the soil and create an infertile patch of land that takes many years to recover. Even artificial fertilizers applied to the soil are of no help since the soil is unable to retain nutrients. South America is known for the "slash and burn" technique used by natives who cut down forests, burn them, farm the land, and move on after a few years.

Being primarily in third world countries, tropical forests have little protection against indiscriminate exploitation by a rapidly expanding human population. If an entire biome can be considered endangered, the tropical rain forest is a prime candidate.

Fresh Water

Aquatic communities composed of fresh water contain perhaps 1% of all the water on the earth. The ocean contains 97%, and the remainder is tied up in ice. There are **lentic** habitats—still water, such as lakes, ponds, bogs, and swamps—and **lotic** habitats—moving water, such as rivers, streams, and creeks. The physical characteristics of each of these environments around the earth are similar, even though the surrounding terrestrial biomes may be quite different.

Fresh Water: Lentic Habitats

Lakes are depressions in the earth ranging in depth from a few meters to 1500 meters or more. They are formed from glaciers, natural or artificial damming of streams, or geologic activity. In many lakes there are distinct temperature differences from the sur-

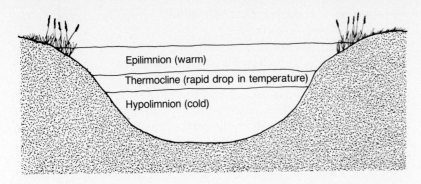

Epilimnion (warm)

Thermocline (rapid drop in temperature)

Hypolimnion (cold)

Figure 8.21 Temperature Layers of Lake. The warmer epilimnion lies over the colder hypolimnion, with the two layers separated by a zone of rapid temperature change, the thermocline. Water is densest at 4°C, so as the lake's surface cools in the fall the surface waters sink and mix with the lower layers. As the entire lake reaches the same temperature the waters mix easily. This stratification and mixing alternate in the summer and fall with winter and spring, resulting in fall and spring "overturn" (see text).

face to the bottom; this is **thermal stratification**. Top and bottom layers tend not to mix because of these temperature differences. The layer of water between the surface water (**epilimnion**) and the bottom water (**hypolimnion**) is called the thermocline—it is a layer in which rapid temperature change occurs (Figure 8.21). In the summer the surface is warmer and the bottom colder. As colder fall weather sets in, the surface waters cool and sink toward the bottom; as they do, the entire lake mixes, and the thermocline and the hypo- and epilimnion disappear. This is the **fall overturn**. As the surface waters cool in the winter to 4°C or less, the lake stratifies again, this time with cooler water near the top and warmer water toward the bottom. In the spring, as the surface water warms (or thaws if frozen), the entire lake mixes again; this is the **spring overturn**. Overturn is a significant phenomenon because it distributes nutrients to the upper layers and oxygen to the lower layers. In the polar regions overturns are not as pronounced.

There are other physical features that characterize lakes. Currents influence the circulation of oxygen, nutrients, and heat. Surface waves may have an effect on floating and emergent vegetation and on the shoreline. The **turbidity** (clarity of water dependent on suspended particles), nutrients, acidity or alkalinity, and solar radiation can each create quite different aquatic environments.

Lakes that are rich in organic matter from water inflow, runoff, erosion, pollution, or other sources are called **eutrophic**. **Eutrophication** is simply the accumulation of nutrients and is part of the successional process. Eutrophic lakes are usually shallow and low in oxygen due to decomposition. **Oligotrophic** lakes are poor in organic matter but usually deep and high in oxygen content.

Bogs are lakes that have no outflow and accumulate organic matter rapidly (Figure 8.10). A low oxygen content allows only partial decomposition, which results in the formation of **peat**—partially decomposed organic matter that releases acids into the water and tinges the water brown. Species diversity is low, but unusual species are found, such as the pitcher plant, whose members digest insects that fall into them (Figure 8.22).

Marshes are wet areas, generally shallow, whose dominant species include cattails, rushes, bulrushes, and sedges (Figure 8.23). **Swamps** are wooded wetlands whose dominant vegetation consists of trees such as willows, alders, and bald cypress (Figure 8.24).

Lakes, bogs, marshes, and swamps contain a variety of plants and animals in a number of different communities within the aquatic biome. In larger bodies of water, especially lakes, there are three major habitats: the **littoral zone** at the edge of lakes and ponds, extending inward to the inner portion of emergent and floating vegetation; the **limnetic zone**, the open water through which light penetrates and photosynthesis exceeds respiration; and the **profundal zone**, found below the limnetic zone, where light does not effectively penetrate.

Organisms and Adaptations

Plankton are organisms that are immotile or effectively so and are moved by currents or waves. These are generally very small organisms and are most abundant in the littoral zone since it is highest in nutrients. **Phytoplankton** are small plants such as diatoms and algae. **Zooplankton** are the animal plankton, including protozoa and crustaceans. **Nekton** are motile organisms and are typically larger than plankton. Nekton are characteristic of the limnetic zone and are less common in the profundal zones; some examples are fish, tadpoles, crayfish, and some large insects. **Neuston** are organisms that restrict themselves or are restricted to the surface—some algae, and insects like water striders. **Benthos** are organisms that inhabit the bottom where they are motile or attached; examples are snails, clams, worms, and catfish (Figure 8.25).

Figure 8.22 Pitcher Plant. The pitcher plant is one of several insectivorous plants that are characteristic of bogs. Insects fall down the sides of the pitcher, which is lined with downward-facing hairs, and are digested by enzymes in the liquid at the base of the pitcher. Insects are a source of mineral nutrients that are often lacking in bogs. In spite of the enzymes some species of mosquito larvae are able to live in the pitcher's fluid "stomach" and hatch into adults. Perhaps the wriggling larvae serve as "bait" for larger insects.

Figure 8.23 Marsh. A marsh typically contains emergent vegetation such as bulrushes, reeds, sedges, and cattails, and usually floating and submerged plants as well. This marsh is part of the Okavango Delta in northern Botswana. Drainage from this delta sinks ultimately into the sands of the Kalahari Desert. Note hippos.

Figure 8.24 Swamp. Cypress swamp near Columbia, South Carolina. Bald cypress dominates the lower coastal plain areas, which are flooded for most of the year. The southeastern evergreen forest can be seen in the background.

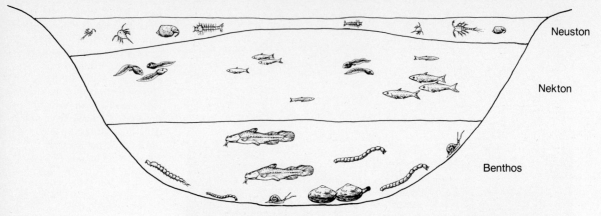

Figure 8.25 Plankton, Nekton, Neuston, Benthos. The nekton, neuston, and benthos are those aquatic animals who move under their own power. The plankton are those that are at the mercy of the current, and may be found at any depth. Plankton is composed of both plants and animals.

Fresh Water: Lotic Habitats

Streams and **rivers** are systems of continually flowing water. The speed of the flow depends upon the size, shape, and depth of the channel, its steepness, configuration of bottom, and rainfall. These ecosystems should be considered as part of a larger ecosystem, the **floodplain**, which is the area adjacent to the flowing water and subject to flooding. Rivers and streams flood the banks and often deposit **silt**, fine-particled organic matter, which forms a fertile soil. Streamside (**riparian**) vegetation is usually different from the surrounding countryside, since the plants must be tolerant of water and occasional flooding (Figure 8.26).

Streams develop from a water source such as springs, a lake overflow, or rainfall runoff. The path of the stream is determined by the topography; the continual movement of water cuts a permanent channel. Recently formed streams are narrow, have a rock bottom, move rapidly, and are high in oxygen. Older streams are wider and slower, have a muddy bottom, and are low in oxygen.

Organisms and Adaptations

Plants are not numerous in a stream or river since anchorage is difficult. Filamentous algae covering the stream bedrocks and planktonic algae are often the major producers. Floating or emergent vegetation may be found along the edge in very slow portions of the stream. Animals are characterized by their ability to swim in the current or hold onto the substrate to avoid being washed downstream. The larvae of many aquatic insects, such as stoneflies and

Figure 8.26 Riparian Habitat. Riparian or streamside vegetation must be able to withstand flooding. Not only must the vegetation be tolerant of inundation, but it must be anchored firmly to avoid being washed away. The extensive and strong root systems of riparian plants such as willows and alders are thus important in minimizing erosion and maintaining the stability of the stream bank.

dragonflies, have long appendages or claws on their legs by which to hold on to rocks. Darters, slim-bodied fish, have long fins to wedge against and between rocks. Some minnows, trout, and other fish species are streamlined and sufficiently muscular to swim in the current, although in fast water they spend much of their time in slower areas behind rocks. Snails and flatworms tend to inhabit the bottoms of rocks.

Streams find their way to larger rivers, and most rivers ultimately empty into the ocean. (One notable exception is the Okavango delta of Botswana, where rivers meet to sink into the sands of the Kalahari desert.) Where fresh water meets the salt water of the ocean, the generally slow moving, marshy area formed is called an **estuary**. Much organic matter is deposited in an estuary, so it is a very productive area and serves as a breeding area for many fish, shrimp, oysters, and other organisms. Organisms living in an estuary have to be adapted to moving water (as fresh water comes in and the tide goes in and out), so they are strong swimmers, attach to the bottom, or burrow in the mud. Cordgrass, an emergent plant, is adapted to varying water levels by having air spaces in its leaves and roots so that gas exchange can continue to occur, even when underwater. Organisms are also faced with fluctuating levels of salinity; some are tolerant of salinity changes, others are not. Thus, species composition on the river side of an estuary may be very different from that on the ocean side. A few organisms are **anadromous**—they move from salt to fresh water (and in some cases back again) to breed. Salmon, steelhead trout, and shad are a few well-known examples. The

American and European eels are **catadromous**, living in fresh water streams and moving to the ocean to spawn.

Human Use

Humans have settled on the banks of streams, rivers, lakes, and estuaries for eons because of their need for water and the natural fertility of floodplains. Water sources became places for hunting, fishing, washing, irrigation, transportation, and waste disposal. Floodplains are often called "cradles of civilization." As long as populations were low, people had an indiscernible effect on these systems. But as the need for water increased, water shortages and water pollution also increased.

The Mississippi River, which begins in northern Minnesota and empties into the Gulf of Mexico at New Orleans, drains many smaller river systems. Many communities and industries have arisen along these rivers. The water receives incredible amounts of organic and inorganic matter that the rivers' biological action cannot cope with: This is **pollution**. Pollution can make the water unusable by people and unlivable for organisms. There may even be a link between the pollution of the Mississippi and the extremely high rate of stomach cancer in the area where it empties into the ocean.

The Colorado River has been and is being used by Arizona and California for irrigation of farmland and domestic uses. One and a half million hectares of desert are made productive by the river. Although the cities of Phoenix, Denver, and Los Angeles could not exist without it, farmers and ranchers use 90% of the Colorado's water. By the time the Colorado reaches Mexico, where only a trickle drains into the Gulf of California, it is extremely salty. Increasing demands for water indicate that needs will exceed supply by the year 2000, with no workable solution in sight.

Lakes do not have the input or outflow of water that rivers do, so what solid matter does flow into the lake tends to settle in the lake basin. This natural characteristic of lakes makes them temporary phenomena, as they will eventually fill in, successional processes turning them into terrestrial habitats. Unfortunately, they will also accumulate substances caused by human activity.

In some cases, eutrophication is accelerated; in other cases, unnatural materials poison the lake. Constructing buildings on lake basins, logging, and other activities, for example, cause erosion of the soil, which runs into the lake. Disturbance of a stream bed, even far upstream from the lake to which it leads, can add considerable amounts of nutrients to the lake. Sewage, industrial wastes, oil spills, acid rain, pesticides, and many other sources of human pollution have turned many lakes into essentially sterile bodies of water.

Although the lakes can be cleaned up, it takes many years; but unless their sources of pollution are stopped, there is little hope for recovery.

Marine Habitats

Nearly three-fourths of the world is covered by salt water, and its evaporation provides most of the world's rainfall. Photosynthesis of marine algae may supply 70% of the world's oxygen. In addition, the temperature of the oceans affects the world's climate and wind circulation. Since all the oceans are connected, organisms tend to be widespread, limited only by temperature and differences in salinity. Life evolved in an ocean environment, and some of the oldest and some of the most unique organisms live there.

Fresh water has a salt content of about 0.5%, while the ocean's average salinity is around 3.5%, most of it due to sodium chloride (table salt). The salinity of the ocean results from the deposition of salts from freshwater systems, and it is constantly increasing, since the ocean is the ultimate dump for river contents.

Oceanography, the study of ocean environments, is very complex because of the variety of ocean habitats. We will considerably simplify the discussion of these habitats by reducing them to three: seashore, **pelagic** (open ocean), and benthic.

The seashore may consist of a rocky shore, tide pools, or sandy flats. The rocky shore is usually beaten by waves, so organisms consist of plankton and those species that are able to cling tightly to rocks, such as limpets, barnacles, chitons, mussels, and algae (Figure 8.27). Some may move up and down with the tides. Tide pools are depressions along a flat rocky shore that are filled with water with the incoming (**flood**) tide and emptied with the outgoing (**ebb**) tide. Many organisms are active with high tide and become inactive when the water is low. Typical inhabitants of tide pools are sea anemones, snails, sea urchins, and some fish such as gobies, which have large fins to keep themselves positioned in the moving water. Sandy beaches or flats tend to be hit with less force than rocky shores or tide pools, but provide no protection for organisms unless they burrow into the sand. Small snails, crabs, sea urchins, clams, and worms are typical inhabitants. Some fish, such as the grunion, bury their eggs in the sand at the ocean's edge. Farther into the ocean, where the wave action is not so severe, starfish, sea cucumbers, and sea urchins forage across the sandy bottom.

The pelagic environment of the ocean is by far the largest of all the world's habitats. The upper portion is the **photic zone**, where light penetrates and photosynthesis occurs. The lower portion is the **aphotic zone**, where no light exists. Unlike deep lakes, there are

Figure 8.27 Tide Pools. The surging surface and the alternation of inundation by water followed by exposure to air makes tide pools a harsh environment. Pictured are sea anemones and barnacles.

usually no clear temperature zones; temperature slowly and steadily decreases toward the bottom. Producers are algae; consumers range from small zooplankton to larger plankton to larger nekton, such as shrimp and fish. The nutrients of the aphotic zone originate in the photic zone, as few producers are present where light is absent.

The open ocean is not as productive as is sometimes thought; since matter—organic and inorganic—eventually falls to the bottom, it is concentrated in a relatively inaccessible area. Much of the open ocean, in fact, is a biological wasteland. However, in areas where **upwelling** (winds producing currents that bring nutrients from the bottom to the pelagic zone) occurs, such as off the coast of Peru, the open ocean teems with life.

The benthic environment, devoid of light, is capable of life only because of the rain of nutrients from the pelagic zone. Most benthic animals are sessile for most of their life cycles. Corals, tube worms, and sponges are examples. Because of the darkness, some fish have light-producing organs that enable them to find each other. The darkness has eliminated the need for color, so many deep sea organisms are black or some other dull color. In addition to the lack of light, water pressure is also very different in the deep benthic environment than it is in a pelagic one—up to 2000 kilograms per square centimeter. The food web is simpler than in the pelagic zone; benthic organisms are primarily decomposers. Generally, the variety and number of organisms decreases with depth; there may be 200 times the number of organisms per area in a pelagic zone than there are on the bottom.

Human Use

The ocean has been historically used for transportation and food. In more recent times, technology has allowed its exploration for minerals and oil. It has also been used as a dumping ground for garbage (by New Jersey and France, for example) and nuclear waste. Although waste dumping and oil spills receive much attention, the two greatest problems that face the ocean are perhaps overfishing and pollution from terrestrial inflows. Fish and shellfish populations are getting smaller and younger, and competition among fishermen of all nations is increasing. Whales have been hunted to the extent that many species are in imminent danger of extinction. Rivers polluted with human and industrial waste and eroded soil deposit their contents into the ocean. Although the enormous size of the ocean can dilute great amounts of matter and degrade much of it, the seas are not infinite and will ultimately be polluted, diminishing a tremendous food and oxygen source.

CONCLUSION

All biomes of the world are being altered by human activity. Whether they are built upon, mined or harvested, or polluted, we are seeing the disappearance of large portions of some biomes and the extinction of some species in all.

Since we are dependent upon our environment for survival, it behooves us to learn as much as we can about the composition and workings of natural ecosystems. Yet they are being altered or destroyed faster than they can be studied. It is estimated that some 10,000 to 15,000 plants in Latin America have yet to be scientifically described and named; many may turn out to be economically or medically valuable. (Approximately 45% of the medical prescriptions written in the United States contain chemicals of natural origin.) Fish populations in the Amazon Basin are an important source of food for the people there, but 40% of the fish species have not been described and named. Important crop plants such as corn were derived from wild plants, and new potential crops continue to be discovered (the oil palm and the jojoba, for example, which produce commercially useful oil). Mesquite bushes of the southern California deserts, eradicated as noxious weeds, have only recently been shown to add a considerable amount of nitrogen to the nitrogen-poor desert soils. In addition, the seed pods can be fermented for the production of alcohol, the wood has a high heat value and is made into charcoal, and the gum has many industrial uses.

So the preservation of biomes in their natural state is important, not so much for the esthetic appeal, but for the fact that they are living laboratories in which studies crucial to human survival can and must occur.

SUMMARY

The ability of organisms to survive under different environmental conditions, and their ability to disperse to different geographical locations, explain the present-day distribution of plants and animals and the composition of ecoystems. Communities develop by the process of succession, either primary or secondary.

Ecosystems characterized by the similarity of their climax communities are termed biomes. The biomes considered in this chapter are:

- Tundra
- Coniferous Forest
- Deciduous Forest
- Sclerophyll Scrub
- Grassland
- Tropical Savanna and Scrublands
- Desert
- Tropical Rain Forest
- Fresh Water
- Marine

STUDY QUESTIONS

1. Define succession, ecosystem, biome, permafrost, pollution, epilimnion, eutrophication, allelopathy, succulent, deciduous.

2. Name the major biomes of the world; list their major environmental features and their characteristic organisms.

3. Discuss the causes and sequence of the successional process. Compare primary and secondary succession.

4. How do the various trophic levels change in relative importance as the process of succession proceeds?

5. Why do climax communities remain relatively unchanged for long periods of time; that is, why does succession stop? Or does it?

6. In which biomes is hibernation most common? Estivation? Migration?

7. Compare and contrast the arctic and alpine tundras.

8. What biomes have been most affected by human activity and why?

9. Explain the concept of overturn in a lake. How does it affect productivity?

10. What arguments can be made for preserving large areas of biomes in a state of wilderness?

Chapter 9

Population Dynamics

Organisms carry on with their necessary activities constrained by the environments in their biome. The success of an organism in an environment is often measured by the number of offspring produced that in turn go on to reproduce. The frequency of reproduction, the number of offspring produced, their survival rates, and the ability of the environment to support them are the topics this chapter will address.

One often sees tidbits of information that report astonishing biological facts. We read, for instance, that if a bacterium reproduced unchecked, its descendents would cover the earth to a depth of 30 centimeters in a day and a half. Or, a single juniper tree may produce 1 million offspring in its lifetime. And, in 900 years, the human race will cover the earth to a density of 120 per square meter. Obviously, these are exaggerated statements based on potential reproduction, not realistic projections. With the rare exception of enormous population outbreaks of locusts, cicadas, and other insects, the numbers of most animals and plants in a stable environment fluctuate mildly about an average (Figure 9.1).

Birth and death rates and longevity vary considerably among living things. Since explaining this great variance has filled many books and journals, we will restrict ourselves to an overview. Keep in mind that the biological determinants of these life processes (genetic constitution, metabolism) are not easily separated from environmental determinants (weather, resources).

Figure 9.1 Population Fluctuations in Stable Environment. Over a period of 19 years the populations of five species of raptors have neither decreased nor increased, although there were obvious fluctuations. The populations of each of the five species showed similar changes. (The sixth species, the Bald Eagle, probably declined due to DDT usage.)

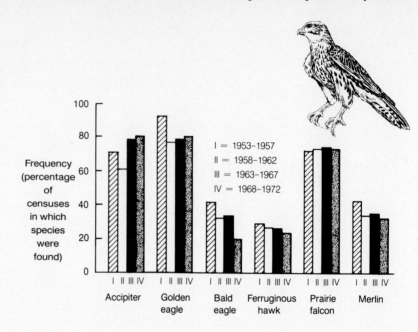

POPULATION AND SPECIES

We have defined communities as all the living organisms in an area. A **population** is a subset of the community; it is a group of organisms of the same species. All the indigo buntings living in a deciduous forest in Illinois, for example, constitute a population, as do all the coho salmon in Lake Michigan, the mongooses on the island of Hawaii, and the mangrove trees in the southern part of Florida. A population encompasses only a portion of the individuals of a species, except when the number of individuals is low, as is the case with the endangered California condor or Everglade kite; in their cases the population comprises the entire species.

DENSITY

We can measure the numbers of plants and animals to get an estimate of the size of their population. Population sizes are usually expressed in terms of **density**, the number of individuals per unit area. There may be 20 spiders per square meter, 500 pine trees per hectare, or 500 Cuban crocodiles on their namesake island. This density measure is crude and can be misleading because, for instance, the island of Cuba is thousands of hectares in size, but the crocodile is confined to about one hectare in the Zapata Swamp.

A better measure of density is **ecological density**—the number of organisms per area of appropriate habitat. Thus, we talk about the number of white-tailed deer per area of deciduous forest, the number of goldenrod plants in a hectare of abandoned farmland, and the number of worms in the humus layer of a deciduous forest. Determining just what area is suitable and how many organisms of what kind can survive in it is a challenge to ecologists and wildlife biologists.

What Determines Density?

The density of any population of organisms varies with time. It may change in a few hours—thousands of migratory blackbirds may substantially, although temporarily, cause an increase in a population's size, for instance. Or it may change only over many years—the number of pine trees in a successional sere of a coniferous forest increases until climax, for example. Animal populations are more subject to fluctuations because of their generally greater motility.

Considering only the relatively simple circumstances of nonmigratory organisms in stable (climax) communities, we typically find seasonal or annual changes in their numbers (Exercise 9.1). Numerous factors affect or cause population changes: weather, food supply, predator populations, nest sites, water, competitors, and so on (Figure 9.2). In a species of ciliated protozoan, the reproductive rate is directly proportional to the food supply; as food increases so does reproduction, and vice versa. Even the size of the prey item determines the rate of the protozoan's reproduction. The most drastically fluctuating cycles are found in organisms whose populations are most easily affected by external factors and who have short generation times. Thus, insect populations oscillate considerably, while whale and maple tree populations change almost imperceptibly.

In unstable situations such as during succession or continual disruption, there may be a constant increase or decrease. The number of deer increases in an oak–hickory forest succession as the seres proceed from abandoned farmland to forest. The number of sardines in the Pacific Ocean along the South American coast has steadily decreased due to overfishing.

DETERMINING POPULATION SIZES

One of the most important pieces of information to collect about an organism's population is its size. There are several methods used by ecologists and wildlife biologists, and volumes have been written on the topic. The reliability of many of these techniques is hotly contested, however, especially when political decisions must be made on the basis of a wildlife census. For example,

Exercise 9.1 Phenology.

Phenology is a branch of ecology that deals with seasonal events. Phenological studies can easily be done on a habitat or on one or more groups of organisms in a habitat. Some examples:

a. Flowers. Which species bloom first? How long does each persist?

b. Seeds. Determine the sequence of seed production in a number of plant species. Which go to seed first? Which species have the longest period of seed production?

c. Insects. Which species emerge first in the spring? Which disappear last in the fall?

A chart or graph of various seasonal events for a habitat can easily be constructed over the period of a semester or year. Comparing two or more habitats can be informative. For example, how do similar habitats at different altitudes compare?

Figure 9.2 Density. (a) Population Sizes of Yellow-headed Blackbirds at Eagle Lake, CA. Changes in the breeding population sizes of yellow-headed blackbirds were due to changes in the water level of the marsh they nested in and the density of the vegetation. (b) A classic example of population cycles is that of the horseshoe hare and the lynx. The lynx feeds extensively on the hares, so their population fluctuations are closely related. (Modified from E. P. Odun, 1971. *Fundamentals of Ecology.* W. P. Saunders Co., Philadelphia, Pa.)

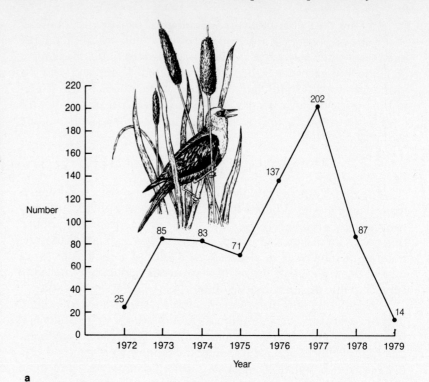

a

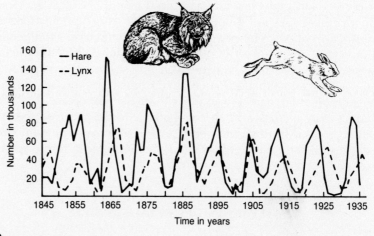

b

the vicuña, a relative of the camel and similar to a llama, was near extinction 20 years ago. Today it is protected on a national reserve in Peru. Estimates of its present numbers range between 15,000 and 49,000. The higher figure indicates that its population should be thinned so it does not overgraze the habitat; the lower figure indicates a reasonable population. A debate continues in Peru over its management.

If the population of 2-meter tall vicuñas in the short grassland of Peru cannot be reliably estimated, then one can imagine the difficulty in censusing whales, seabirds, aphids, thistles, worms, or mushrooms, due to their size, habitat, mobility, abundance, or ubiquitous distribution. Some methods lend themselves nicely to certain groups of organisms and not at all to others. We will briefly examine some major census techniques.

Total Population Census

This method involves directly counting all of the individual organisms in a given area. For certain organisms in small areas this works well. Some examples: ash trees in a 60-hectare forest, bison on 100 square kilometers of grassland, sea anemones on 500 square meters of coral reef, and seals on an offshore island. Large, slow-moving, or immobile organisms can be counted this way with very accurate results. This is a **true census**.

Sample Census

Since it is only rarely that an entire population can be counted or an entire habitat censused, a sample is usually taken and the total population size projected from that. Square, circular, or rectangular **plots** (quadrats) of a sample area are marked off and the plot censused. The plot must be representative of the entire area if it is to be used to derive an estimate of the whole area; that is, for example, it should not contain a pond if there are no other ponds nearby. Plots work best when in **homogenous** (uniform in composition throughout) habitats. In **heterogenous** (having dissimilar parts) habitats, several plots in different areas or a **transect** (straight narrow strip) census through the entire area provides a better sample (Figure 9.3). Obviously, the larger the sample area the better, since a larger size makes it more representative of the entire area. Plots and transects can be used for censusing almost any animal or plant population; the exact methods are numerous, and they vary with the organism, the time of year, the size of the area, and other factors (Exercise 9.2).

Exercise 9.2 Sampling.

Referring to Exercise 8.3 again, do a plant population census of a selected habitat and compare census methods. Do you get the same results with rectangular and circular plots as you do with transects?
Do a total count of all organisms of one species in a habitat (trees would work well for this) and compare the actual figure derived from that with the number calculated from a sample plot. How close does the sample approximate the actual figure? How many samples should be taken to approach the actual figure with some degree of accuracy?

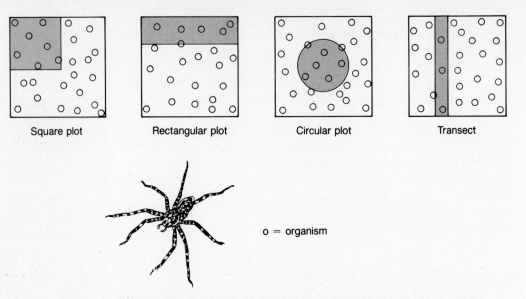

| Square plot | Rectangular plot | Circular plot | Transect |

o = organism

Figure 9.3 Samples of Entire Area. There are four common ways to sample a larger area. Which method is used depends on the structure of the habitat and the organisms to be sampled.

Most animals are more difficult to census than most plants because animals move and are harder to detect. A technique that is often used by wildlife biologists to census larger animals is the **mark-recapture method**. A number of animals are trapped and marked; for example, rabbits can be captured and then marked with a spot of dye on their fur. They are then released and allowed to wander through the habitat as they did before they were captured. Two weeks later another group of rabbits is captured. Some will be marked with dye (recaptures) and others not (new captures). From the ratio of recaptures to new captures we can estimate, with the simple formula

$$N = \frac{S(T)}{M},$$

the population of the entire area in which traps were set. In this formula:

T = number of animals captured and marked
S = total number of animals captured with second trapping
M = number of marked animals that were recaptured in second trapping
N = population estimate

Suppose a biologist captures, marks, and releases 39 rabbits. Two weeks later 34 rabbits are captured; of these, 15 are marked and 19 unmarked. So $T = 39$, $M = 15$, and $S = 34$:

$$N = \frac{S(T)}{M} = \frac{34(39)}{15} = 88.4$$

Thus, the estimate for the total rabbit population of the area is 88.

There are some problems with this and similar techniques, however. It assumes that there is no movement of animals into or out of the area and that a significant number are neither born nor die in the period between the first and second trapping. Marked animals may have a different mortality than unmarked ones; perhaps predators can see marked animals better, for example. And, once captured, the animals may become "trap-shy" and avoid traps, or "trap-happy" and seek them out for food and shelter. The accuracy of this method is variable, therefore; few studies have compared the results of this method with the total population census technique (Exercise 9.3).

There are several other direct sample techniques which will not be discussed here. The *Wildlife Management Techniques Manual* published by the Wildlife Society provides considerable information on such techniques.

Indirect Census

Another roundabout method used to census animals is the indirect census. Instead of trying to locate or capture the animals, signs of their presence are used as indicators of abundance. Tracks, nests, fecal pellets, regurgitated remains of prey, songs, dens, burrows, dead individuals, and shed antlers have all been used. Some are much more reliable than others; some studies have shown deer fecal pellet counts to be so variable, for example, as to be useless.

As with many aspects of scientific work that require observation, there can be discrepancies in the results of different observers. One observer may be more alert and experienced than another and record more organisms. The weather, time of day, and many other factors may not only affect the actual population of organisms present, but the investigator's ability to measure it, and different census techniques often yield different results. In some instances, the differences are not meaningful; it may not matter much whether indirect techniques showed that 398,000 starlings comprised a group but that the figure was 306,000 as determined by direct count (Table 9.1). There is nearly a 25% difference in the two census methods. Although this is a considerable difference, the goal of the study determines its significance. If the purpose of this study is to deter-

Exercise 9.3 Laboratory Demonstration of the Mark-Recapture Method.

Obtain a colony of mealworms (larvae or adults), crickets, or other suitable organism. There should be at least 200 individuals in a fairly large container. A 40 liter fish tank works well.

Using a net, a can, or your hand, capture a sample of individuals, mark them with a dot of nail polish, and release. A few hours or days later, take a second sample (recapture). What is your estimate of the total population? How close is it to the actual population? How large do you think a sample must be to give a good estimate of the actual population size?

Table 9.1. Numbers of Birds in a Roost*

Species	Indirect Method	Direct Method
Starling	398,719	306,707
Red-winged Blackbird	2,103	1,753
Common Grackle	2,533,576	1,444,153
TOTAL	2,934,398	1,752,613

*Modified from Paul A. Stewart, 1973, "Estimating numbers in a roosting congregation of black-birds and starling," *The Auk* 90:353–358.

mine whether or not this group of birds will have an impact on local agriculture, the exact numbers are not needed. If the goal is to determine how much grain can be eaten by this flock, then accurate numbers are more essential. But when a species population is changing drastically and may even be on the verge of extinction, accurate figures are much more important.

POPULATION INCREASES AND DECREASES

We have just considered census methods under the assumption that the populations are stable; they rarely are. Changes in population size are generally attributed to external factors such as weather and food supply, but there are characteristics of the population itself that cause changes in its size and density. There are births, deaths, immigration, and emigration (Figure 9.4).

The greatest influence on population increase is usually the birth rate (**natality**). Maximum natality is the greatest number of offspring that can possibly be produced under ideal conditions (Table 9.2). These numbers of offspring are not produced each breeding season, year, or generation because rarely are conditions ideal. If the parents are small, young, or physiologically weak, their potential to produce young is minimal. Also, food shortages often result in fewer eggs being laid by birds or in the resorption of embryos by lizards. High population densities can cause competition for resources, aggression, stress, and a lower frequency of mating, resulting in a lowered natality. See Table 9.3 and Figure 9.5 for examples.

The death rate, **mortality**, also affects population size, often substantially. Shortages of resources such as food, water, nutrients, and sunlight, as well as predation, cannibalism, failure of the embryo to develop or the egg to hatch, parasitism, and weather are major causes of death. In the Sierra Nevada and Rocky Mountains, the mortality rate of pikas is close to 40% per year. Only 1 of 300 teasel seeds becomes a reproductive plant. In some species of corals, mor-

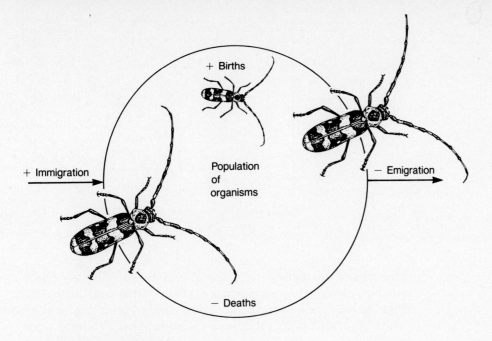

+ = Increases population size
− = Decreases population size

Figure 9.4 Population Fluctuations. These can be explained via four internal population parameters. When the plusses exceed the minuses, the population grows; when the opposite occurs, it shrinks; when they are equal, the population is stable. In an undisturbed climax habitat, populations should fluctuate around an equilibrium.

tality of the larval stage exceeds 99%. About 340,000 seeds need to be produced by a species of a tropical palm tree to yield a reproductive tree (which may take 50 years). Generally, the younger stages are subject to the highest mortality rates.

It is often stated that organisms that have the highest mortality among their offspring (most plants and invertebrates) produce the greatest number of eggs (or spores, young, and so on) to ensure that

Table 9.2. Maximum Natality

Tarantula Spider	600 eggs
Bluegill	400–61,000 eggs
Codfish	61 million eggs
Quail	15 eggs
Deer	1–2 fawns
Giant Sequoia	200–300 seeds per cone

Figure 9.5 Natality Rate. In an experimental study of a clam, the number of offspring produced per litter dropped as the population of adults increased.

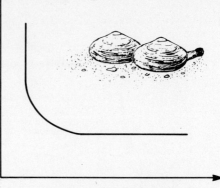

Litter size
per
reproductive
adult

Density of parental population

some survive. An alternative explanation is that the mortality rate is high because so many offspring are, in fact, produced, since each offspring tends to be smaller, contains less energy, and must compete with many others for resources. Both arguments have some validity.

Immigration and emigration also affect a population's size. Often, the rate of movement both ways is nearly the same, and the population size is not much affected. Changes in the abundance of resources or predators or climatic changes can affect movement in and out of an area. Temporary changes in populations of migratory animals can be considerable. During the dry season, African ungulates such as zebras, wildebeests, and many antelopes move hundreds of kilometers in search of water. Populations of annual or

Table 9.3. Reproduction in House Mice as Food Supply Dwindles.[*]
As the food supply (a stack of oat sheaves) was eaten, the percentage of house mice capable of producing offspring diminished and so did the total population.

Time (months)	Food Supply	Percentage of Mice in Breeding Condition	Total Number of Mice
0	100%	100%	6
2.5	diminished	100%	68
6.1	considerably diminished	100%	966
10.5	14%	49%	448
15.2	4%	27%	163

[*]Modified from A. E. Newsome, 1971, "The ecology of house-mice in cereal haystacks," *J. Anim. Ecol.* 40:1–15.

perennial plants appear from seeds or bulbs in the wet or warm seasons and disappear in the dry or cold ones as if they migrated into or out of the area. Immigration and emigration of plants is most easily demonstrated during the process of succession, but since individual plants do not generally move, we really are considering plant migration in terms of seed and spore movement.

To obtain a clear and systematic picture of natality and mortality, we can construct a **life table** such as the one shown in Table 9.4. A life table is a summary of some vital statistics of a population. Similar versions are used by life insurance companies to calculate insurance risks and rates for humans.

Life tables and variations on them are very useful to natural resource managers. Whether harvesting animals commercially or for sport, culling confined animals to prevent overgrazing, logging trees, introducing species into a new habitat, stocking a stream with fish, or replanting a forest after a fire, life tables provide information on which to base intelligent decisions. With our intense concern for

Table 9.4. Life Table of a Coral.*

The table tells us that: (1) these corals have a generally decreasing mortality as they get older, (2) their life expectancy increases as they get older (until about age 50 when it decreases), (3) the older the individual, the more offspring it produces and (4) corals do not breed until 10 years of age.

Age (years)	Mortality Rate (%)	Life Expectancy (years)	Number of Offspring Produced in Age Interval
0–1	25	10.68	0
1	5.8	13.05	0
2	8.3	12.85	0
3	9.2	13.00	0
4	0	16.5	0
5	0	15.5	0
6	14.5	14.50	0
7	0	15.90	0
8	6.1	14.90	0
9	6.6	14.80	0
10	2.1	14.80	29,177
11	1.7	17.70	36,798
•	•	•	•
•	•	•	•
•	•	•	•
68	100	0	1,077,606

*Modified from Richard W. Grigg, 1977, "Population dynamics of two gorgonian corals," *Ecology* 58:278–290.

rare and endangered species, we need to know both the minimum environmental needs of the species in question as well as the minimum population size needed to maintain a stable or growing population. Below that minimum size, known as the **extinction level**, the population would shrink irreversibly.

SURVIVORSHIP CURVES

From life tables, survivorship curves can be created. **Survivorship curves** represent the mortality rate of a population of individuals. There are three types of idealized survivorship curves, as shown in Figure 9.6. The survivorship (or, conversely, mortality rates) of almost all organisms fits on or between one of these curves. A large population of newborn individuals is assumed at the start; the curves track the decline in population size as individuals die. Curve A represents organisms whose mortality is low throughout their life, then increases drastically at an advanced age. Curve B represents those organisms whose mortality is constant throughout life; at no

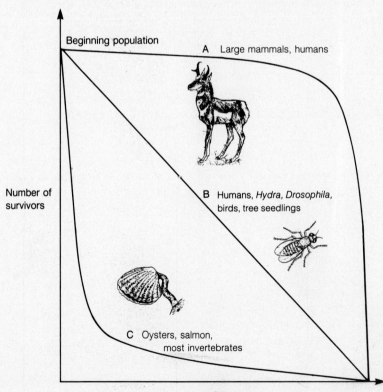

Figure 9.6 Three Idealized Survivorship Curves. See text for explanation.

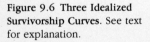

time during their lives are they more likely to die than another. Arguments can be made to put humans on the A or B curve; most likely they would fit somewhere in between. Curve C is the most common curve and probably applies to the majority of organisms; mortality rate is very high in the early stages of life but low for individuals who survive past those stages.

THE GROWTH CURVE

The survivorship curves can be called death curves because they measure only the mortality of a declining population. **Growth curves**, on the other hand, represent the net changes in population size resulting from both natality and mortality. Figure 9.7 shows a typical growth curve of a population of organisms under ideal conditions. Sometimes called a sigmoid curve because of its S shape, it is composed of three parts. To explain each of these parts, or phases, let us consider an example of a population of grasshoppers restricted to a patch of grass; no immigration or emigration occurs.

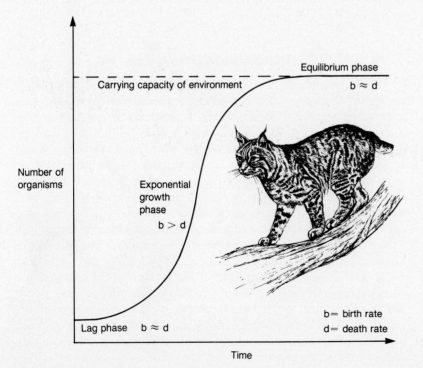

Number of organisms

Carrying capacity of environment

Equilibrium phase
$b \approx d$

Exponential growth phase
$b > d$

Lag phase $b \approx d$

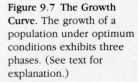

b = birth rate
d = death rate

Time

Figure 9.7 The Growth Curve. The growth of a population under optimum conditions exhibits three phases. (See text for explanation.)

Lag Phase

A small population, perhaps a pair, of grasshoppers is introduced into an area devoid of grasshoppers; food is abundant, and we
assume that the abiotic factors are ideal, and competitors and predators are scarce or absent. Under these conditions, maximum natality
and minimum mortality are expected. When first introduced into
the habitat, the grasshoppers must acclimate to the conditions, find
each other, and begin the cycle of reproduction. During this period
no offspring are produced, so the population does not change. This
is the **lag phase**, where birth rates equal death rates (both equal or
close to zero).

Exponential Growth Phase

Eventually reproduction begins. Eggs are laid, and the larval
nymphs grow to reproductive adults. If the original pair of grasshoppers produces two pairs of offspring, the population doubles. If
those two pairs each produce two pairs, the population doubles
again. This doubling (or nearly so) of the population each generation is represented on the growth curve by the **exponential growth
phase**. The reproductive rate continually increases, and although
there are deaths, the new organisms added to the population far
outnumber the number dying so that the population increases rapidly. Natality exceeds mortality.

Equilibrium and Carrying Capacity

Since grasshoppers do not blanket the earth, their populations
must not increase endlessly. Eventually the rate of reproduction is
approximately equal to the mortality so that the population size
fluctuates only slightly, as an earlier example (Figure 9.1) demonstrated. This **equilibrium phase** occurs when natality decreases and
mortality increases, and the curve flattens out. There are numerous
reasons for birth and death rates approaching equality. Food becomes scarcer (grass can grow only so fast), egg-laying sites become
rarer, and waste products accumulate. Crowding interferes with
movement and behavior and makes hiding places difficult to find.
Predators increase as their food (the grasshoppers) becomes more
abundant and easier to find. Thus, there is a limit to the number of
individuals a habitat can support. This maximum limit, called the
carrying capacity, is the point at which equilibrium occurs. A population grows until it reaches the carrying capacity, when mortality
equals natality. Under stable conditions as found in climax communities, all populations are at or near carrying capacity. The carrying
capacity may change with the weather, food supply, or other factors.

Under some circumstances the growth curve does not level off at carrying capacity indefinitely. In successional processes, for example, the numbers of organisms change from sere to sere (Figure 9.8); or animals may increase rapidly, overexploit their food supply, and exhibit a population "crash" (Figure 9.9).

DENSITY-DEPENDENT AND DENSITY-INDEPENDENT FACTORS

The changes in, or stability of, population numbers are due to many environmental components. Almost all of these components can be categorized as either density-dependent or density-independent factors.

Density-dependent factors are environmental factors whose effect on the population is proportional to the density of the population; that is, the denser the population, the greater the effect. Food and other resources (nest sites, shelter, and so on) become scarcer as the population grows; there are fewer resources per individual. Predators are often attracted to prey species that have high densities. Competition with other individuals and species becomes more severe. Communicable diseases spread through denser populations more rapidly. Reduced resources, increased predation, and increased disease potential all increase mortality, and the population becomes smaller. In a controlled experiment, toad tadpoles were found to grow at a slower rate under high densities, and the proportion of tadpoles that survived to become toads decreased with increasing population size. In general, as population size is reduced, resources become relatively more abundant. Predation, competition, and disease are reduced, and the population increases. Density-dependent factors, then, are a feedback mechanism that helps keep the population in equilibrium with its environment.

Density-independent factors affect a population with the same severity regardless of the population size. Wind, cold, heat, flooding, and fire are all density-independent factors. Example: A population of salamanders in the Blue Ridge Mountains of Virginia suffered a 99% mortality during a month-long drought; population size was immaterial. Independent factors are generally capricious, so their effects on populations are unpredictable.

Since overcrowding is detrimental to organisms, it is logical that self-regulating feedback mechanisms evolved to maintain populations at carrying capacity: These are the density-dependent factors. It is much more difficult, if not impossible, for organisms to evolve adaptations to deal with density-independent factors because their occurrence and frequency are so erratic. It is difficult to imagine

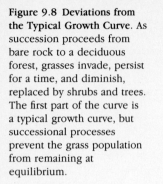

Figure 9.8 Deviations from the Typical Growth Curve. As succession proceeds from bare rock to a deciduous forest, grasses invade, persist for a time, and diminish, replaced by shrubs and trees. The first part of the curve is a typical growth curve, but successional processes prevent the grass population from remaining at equilibrium.

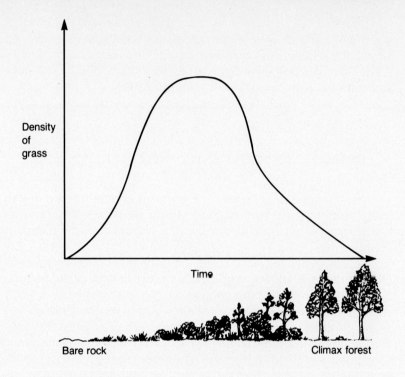

earthworms evolving adaptations to deal with floods that occur per-haps once a century. Consequently, density-independent factors can prove disastrous to organisms.

Like many other designations in ecology, there is no clear divid-ing line between density-dependent and density-independent fac-tors. A light snowfall may make some seeds inaccessible to birds, thus reducing the food supply; in this case snow would have a densi-ty-dependent effect. A heavy snowfall, however, would cover all the seeds, and all the seed-eating birds would starve, irrespective of population size: a density-independent effect.

HUMAN INFLUENCES

Humans have always had some effect on plant and animal popu-lations. There is a lot of speculation but little information that attri-butes the decline of a plant or animal species directly to human intervention. Egg collectors, meat hunters, fur trappers, and the like no doubt took their toll in the past. Sport hunting, more efficient commercial hunting, pest control, exotic species introduction, and pollution have had a considerable impact in recent times.

The greatest threat to an organism's population, though, is loss

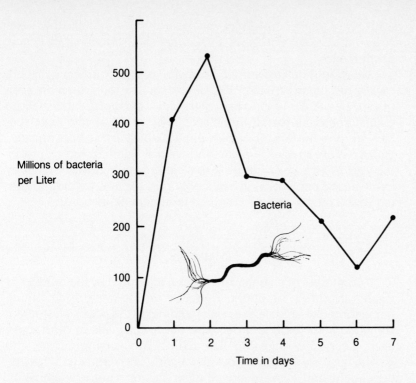

500

400

Millions of bacteria
per Liter   300

Bacteria

200

100

0

0   1   2   3   4   5   6   7

Time in days

Figure 9.9 Under very favorable environmental circumstances, animal populations may increase faster than their food supply, exceed the carrying capacity, and eventually deplete their food. The population then drops precipitously. This often happens when the animals' predators are reduced. The growth curve shown here is of bacteria in a laboratory culture with a high level of food. No predators (protozoa) are present.

of habitat. Logging, strip mining, road construction, agriculture, and housing all usurp space that was once a habitat for a community of organisms. And the smaller a habitat becomes, the fewer the number of individuals able to survive there and the more likely a population decrease will lead to extinction. Wildlife biologists, foresters, and other professionals who are charged with the management of natural resources are learning more and more about ways to increase the carrying capacity of habitats. Grains are planted for waterfowl, forests are burned to provide deer browse, nest boxes are provided for birds, forests are selectively cut (rather than clear-cut), and hunting and fishing regulations are designed to have minimal impact on animal populations. All of these and many other such practices must continue if we are to reduce or compensate for habitat disappearance.

EXTINCTION

Extinction is a natural process, a result of organisms not being able to evolve adaptations suitable to the changing environment. Most scientists believe that over 95% of all species that have ever existed are now extinct. Estimates are that the average species evolves, survives, and disappears over a period of about 25,000 years—the life span of a species, so to speak. Extinction is not a new phenomenon.

We should be concerned, however, by the extinctions that are caused by human activities, since these shorten the natural life span of a species. Since 1600, at least 300 species of animals have become extinct due to human disturbance, and up to 4% of vertebrate species are now in danger of becoming extinct. By some estimates, 500,000 more species of plants and animals will become extinct by the year 2000. Some of these losses are probably inevitable, but by knowing the population dynamics of other species, we may be able to prevent or at least forestall these imminent extinctions.

SUMMARY

The success of an organism is often measured by the number of viable offspring it produces. There are various ways to determine population sizes and densities, depending on the organism studied. Populations increase and decrease due to changes in birth rates, death rates, and migration. Life tables, survivorship curves, and growth curves measure various population characteristics. Density-dependent and density-independent factors control population sizes.

STUDY QUESTIONS

1.   Define population, density, exponential growth phase, density-dependent factors, heterogenous.

2.   List as many density-dependent and density-independent factors as you can. Can density-independent factors have an effect on density-dependent factors? How?

3.   Compare a plot sample census to a transect. Under what circumstances is each the best?

4.   Using the mark-recapture method, 85 alligators were marked and, 2 months later, 102 were captured, 65 of which were marked. What is your estimate of the alligator population size?

5.   What are the advantages and disadvantages of the mark-recapture method?

6.   Create a hypothetical life table for any animal. How could having such information be useful in analyzing that species?

7.   What is a survivorship curve? Draw the main types. What kinds of organisms tend to exhibit each type? Give specific examples.

8. What is carrying capacity? What factors affect the carrying capacity of a habitat?

9. What is meant by extinction level? Can human intervention change that level?

10. Are birth and death rates related in such a way that when birth rates are high, death rates are also high, and that when one is low, the other is also? Or are they independent of each other?

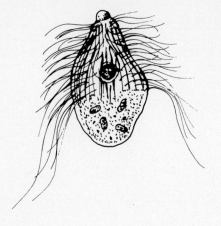

Chapter 10

The Interaction
of Organisms

Populations reproduce and grow within environmental limitations. The interaction of organisms, whether of the same or different species, often has considerable effect on population regulation. In addition, the ways in which organisms relate to one another determines to some extent their roles in the community. This chapter examines the major ways in which organisms interact.

In any community there are a few to a thousand or more populations of organisms. A functioning ecosystem requires that many of these species interact with each other. The types and intensities of these relationships vary immensely—from a total separation of life histories to such a close affinity that one species cannot exist without direct contact with another. Energy flow, nutrient cycling, and population dynamics are all linked to the interaction of species with each other.

A species normally interacts with several other species, but its role may change with each interaction. A ladybird beetle is a predator as it feeds on aphids but becomes prey to a trout when it falls into a stream. We will now consider the major types of relationships two species may have.

TYPES OF INTERACTIONS

No Interaction

In some ways all organisms within an ecosystem have an effect on all the others, but it may be only in the broadest terms. For example, dead leaves of an elm tree are recycled into the soil; some of the soil nutrients may be incorporated into a flower; the flower may provide nectar for bees, which make honey that may be eaten by a bear. So there is a connection between the elm and the bear, but it is so diffuse that we have to say that in this instance bears and elms do not interact. The same is true for hawks and moose, mosquitoes and worms, and so on.

Predation

We most often consider a predator to be an animal and prey to be an animal that the predator kills for food, but in the broadest sense a **predator** is any organism that eats any other organism; a **prey** organism is any one that is killed and eaten. With few exceptions, predators are animals and prey are both plants and animals. A predator–prey pair that may come to mind is, say, the cheetah and antelope of Africa, or coyote and deer. But insect-eating birds, worm-eating moles, and the starfish that eat oysters are just as certainly predators.

The relationship between a predator and its prey may be very close, or in other cases very loose. Some predators are generalists and feed on many types of prey. Crows and foxes will eat eggs, insects, fruits, worms, and most anything else that they encounter. Some are more specialized; swallows and nighthawks eat only insects, but they eat a variety of them. Others are more restricted to a particular type of food, such as the fish-eating osprey (Figure 10.1). Still others are extreme specialists. The Everglade kite, for example, feeds only on one kind of snail in its Florida Everglades habitat. Sometimes this specialization by a predator can be an advantage to humans. The Klamath weed (or St. John's wort) is a native of Europe that became a serious pest in the United States in the 1940s, usurping large areas of rangeland and making them unsuitable for cattle grazing. The U.S. Department of Agriculture imported a leaf-eating beetle that was known to attack the weed in Europe. The beetle flourished and brought the weed under control. The beetle is an extreme specialist and is no threat to other plants since its life cycle is closely attuned to that of the Klamath weed and it cannot reproduce without it.

Even the most generalist of predators are restricted to certain categories of foods. Specialist predators may be very effective at

Figure 10.1 Osprey. The osprey is a somewhat specialized predator. It eats fish exclusively and is well adapted for catching them. Like most birds, it has four toes, three in front and one in the rear. When it swoops from the air to catch a fish, however, it swings one forward toe backward to make a two front–two back toe arrangement that, along with sharp claws and prickly soles on the toes, is very effective at snatching and holding onto fish.

capturing certain kinds of prey. The oystercatcher birds, for example, can open up a mussel or clam by cutting the mollusc's muscles. To become proficient takes them two to three years, but few other animals are capable of preying on the oystercatchers' food source so effectively. Conversely, some prey have been able to evolve defenses against specific predators. Some sea squirts contain acid fluids and a high concentration of vanadium, which deter some fish predators. Any one chemical defense is not effective against all predators, but predation by a few specialist predators probably has assisted in the evolution of the chemical defenses of sea squirts and other sessile aquatic animals such as sponges, sea anemones, and marine worms.

The population cycles of a predator and prey may parallel each other if the predator depends heavily or exclusively upon one or two prey species. As the number of lemmings in the Alaskan tundra increases, so do the numbers of snowy owls and Pomarine jaegers, which are predators on lemmings. In a Canadian pine plantation, the numbers of small mammals preying on cocoons of the pine sawfly increase proportionately to the numbers of cocoons, but to different degrees, reflecting their amount of dependence on the cocoon as a food source.

Predators play an important role in the ecological balance of a community because they are often one of the major density-dependent factors that keeps a prey population at its carrying capacity. Prey populations often increase drastically when their predators are removed, demonstrating the importance of the predator's role, but the prey's increase may only be temporary. If the prey population grows unchecked by predator pressure, the prey may overwhelm its food sources and undergo a population crash.

The balance between predator and prey may be very fragile. For example, ladybird beetles feed on aphids. It has been shown that under stable environmental circumstances a certain number of ladybird beetles will maintain a certain number of aphids at that level indefinitely. But an increase in temperature causes the beetles to overeat and ultimately wipe out the aphid population. A decrease in temperature causes the aphid population to increase beyond the capacity of the beetles to control it. Although this was a laboratory study, it serves to indicate how complex the natural situation is.

Biological control is the term applied to methods of pest control that use other organisms rather than pesticides or other artificial means. The Klamath weed-eating beetle is an example of biological control, as is the introduction of myxomatosis virus to control rabbits in Australia. There are a number of success stories like these but probably an equal number of spectacular failures. The mongoose, imported into Hawaii to control rats, became a pest itself. Biological control has enormous potential for being inexpensive, permanent, and very effective. But a thorough knowledge of the life cycles of both predator and prey is essential.

Competition

Competition is the striving for resources that are in short supply. Whenever two or more individuals of the same or different species eat the same food, need the same nest sites, or try to sink their roots into adjacent places in the soil in search of water or nutrients, there is likely to be competition for limited resources. The more similar the organisms' requirements, the more intense the competition. **Intraspecific**, or within-species, **competition**, is most severe since the requirements of individuals of the same species are nearly identical. Competition is rarely strong between individuals of different species, but it does occur and is termed **interspecific competition**.

Competition is sometimes mistakenly interpreted as a "tooth and claw" struggle for survival. There is some direct contact between males of some vertebrate species for territories or mates, and fighting, injury, or even death may ensue (Figure 10.2). Elephant

Figure 10.2 Animal Aggression. The caribou, or reindeer, of the far north is polygamous and males will fight other males in their quest to mate with 10–15 females.

seal bulls fight other bulls to maintain their harem. The African kob antelope aggressively defends his mating site. Robins may physically defend their territories against other robins. But it is not adaptive for an organism to brawl with another, because the chances of injury are too great and the consequences severe. The best competitors are not fighters, but those organisms that are the most efficient at obtaining needed resources (or at avoiding predators, germinating the fastest, withstanding extremes of the environment, or producing the most offspring). "Survival of the fittest," another well-used term, refers to the survival and reproduction of the best competitors.

Competitors commonly coexist, but rarely are their competitive abilities equal. As a result, the population of one is usually much lower than the other and is sometimes relegated to a poorer habitat. Figure 10.3 illustrates a classic experiment in competition. Two species of single-celled protozoa, *Paramecium caudatum* and *P. aurelia,* were placed in jars with a culture of bacteria for food. When the species of protozoa were raised separately, their population

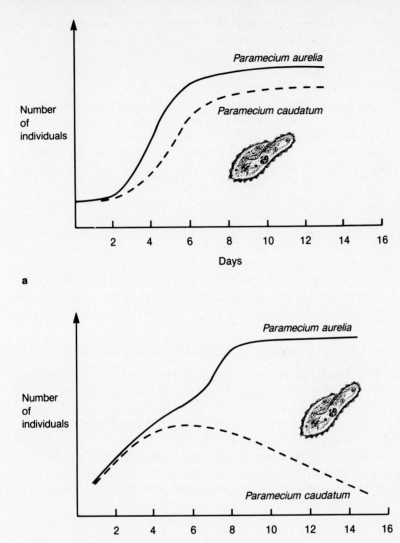

Figure 10.3 Growth of Protozoa. (a) In Separate-Species Cultures and (b) In Two-Species Cultures. Experimental evidence of competition in two species of *Paramecium*. See text for explanation. (Modified from G. F. Gauge, 1934. *The Struggle for Existence.* Baltimore, Williams & Wilkins Co.)

growth curves were typical, reaching equilibrium in about two weeks (Figure 10.3a). The carrying capacity of the environment was slightly higher for *P. aurelia,* but both species survived well in single-species environments. When raised together in the same jar, however, they competed for the food supply (Figure 10.3b). At first, when food was relatively abundant, competition was low, and both populations grew quickly. But as the protozoan populations grew, food became scarcer. *P. aurelia* was a better competitor (that is, was more efficient at feeding on bacteria) than *P. caudatum,* so its population increased to an equilibrium level similar to the one it had

This experiment, or some version of it, can be conducted in the laboratory using Paramecium, crickets, Tribolium (flour) beetles, Drosophila flies, or other easily cultured animals. Collect data from which to draw population growth curves under varying conditions. Competition for resources can also be demonstrated by germinating grass, beans, or radish seeds in small containers. The growth rate of individuals under varying degrees of crowding can be measured. Will additions of resources (such as fertilizer) reduce competition? How important is space as a resource? How important are minerals?

reached in a noncompetitive environment. *P. caudatum,* however, was a poor competitor. When competition became severe, after about 5 days, its population declined; after 16 days it became extinct. Many organisms do not coexist (or exist) today because they are (were) inferior competitors (Exercise 10.1).

But it is not simply a case of a species either existing by itself or going extinct; there are numerous instances of two or more species coexisting in the same habitat. They do so by using slightly different resources, or by using them at different times or in different ways, to reduce competition. Figure 10.4 shows how two species of starfish are able to coexist. A large starfish and a smaller one compete for food and space in the intertidal zone of the Pacific Ocean. They both feed on a variety of foods—barnacles, limpets, mussels, and snails. But by feeding at slightly different times during the year and mainly by feeding on different size prey, they are able to coexist. The size of the smaller species is, however, a result of its poorer competitive ability, as experimental removal of the larger species resulted in an increase in size in the small species since it then had more food for growth.

Competitive Exclusion Principle

The protozoa and starfish in Figures 10.3 and 10.4 are examples of the **competitive exclusion principle**: No two species can coexist indefinitely on the same resource. Since many organisms share similar requirements, there are a number of ways in which they avoid or lessen competition; for example,

1. Live in different geographic areas. Geographic separation reduces or eliminates competitive interaction. The bison of North America and the kangaroos of Australia are both grazing herbivores of grasslands, but they feed on separate food supplies.

2. Live in the same geographic area but in different habitats. One species of goldenrod may grow in a wet meadow adjacent to a forest containing a different goldenrod species.

3. Live in the same habitat but use it differently. The starfish example applies here; similar examples among birds are also common. When red-winged blackbirds and yellow-headed blackbirds nest in the same marsh, the red-wings nest on the outer edges and the yellow-heads use the center. Species can be separated vertically, horizontally, or in their use of different sizes or types of food and other resources.

4. Live in the same habitat but use it at different times. Many animals are **nocturnal**, using resources at night that are used during the day by **diurnal** animals. The resources are often different; bats and birds both eat flying insects, but the insects eaten by diurnal birds are different from those eaten by nocturnal bats. Some trout, living in the same stream, lay eggs in the spring; other species breed

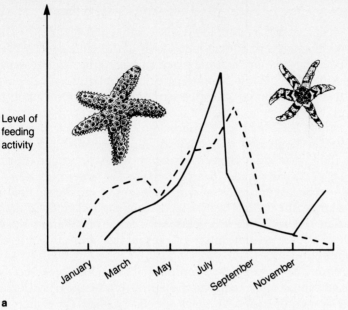

a

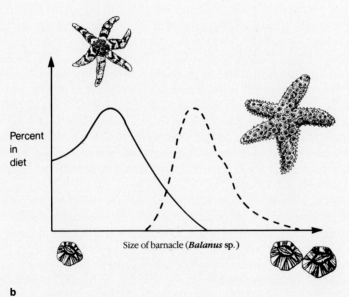

b

Figure 10.4 Coexistence. (a)
Feeding at Different Times
and (b) Feeding on Different
Size Prey. See text for
explanation. (Modified from
B. A. Menge, 1972.
Competition for food
between two intertidal
starfish species and its effect
on body size and feeding.
Ecology 53:635-644.)
Leptasterias indicated by
solid line and *Pisaster* by
dotted line.

in the fall. Also, there is good evidence to indicate that the breeding
seasons of all species of birds in a habitat are shifted in time by about
one-third. For example, if the time it takes to lay eggs, incubate
them, and raise the young to **fledging** (leaving the nest) is four
weeks, one species of birds may breed from mid-March to mid-
April, and another species from the beginning to the end of April.
The breeding season puts a maximum demand on resources, so
competition is lessened by offsetting the breeding times.

INTRASPECIFIC COMPETITION

It is somewhat easier to imagine individuals of different species avoiding competition than individuals of the same species because conspecifics live in the same place and need the same resources at the same time. But just as interspecific competition is reduced by partitioning resources, so is intraspecific competition, but on a finer level.

In the Yarrow's spring lizard, which lives in rocky areas at high elevations in Mexico, individuals divide the resources of their habitat, thus allowing a larger population to exist. Larger lizards eat larger prey than smaller lizards; females eat larger prey items than males. The larger lizards feed in the morning and the smaller in the afternoon; the larger lizards feed from higher perches than the smaller. These differences between lizards of different sizes and sexes allows partitioning of resources and thus lessened competition.

Numerous examples of morphological, behavioral, ecological, and/or temporal mechanisms are utilized by conspecific organisms to lessen competition. In many species of birds, for example, one sex has a larger bill than the other, resulting in the sexes feeding on different size foods. In the extinct huia of New Zealand, the male had a short, straight bill and the female a long, curved one. The male mosquito sucks plant juices while the female feeds on the blood of vertebrates. Immature and adult animals of the same species may be so different as to totally avoid competition between the two forms. Tadpoles are herbivores and frogs are carnivores; dragonfly larvae are aquatic and the adults terrestrial; and some voracious plant-eating caterpillars become nectar-feeding butterflies. A common method to avoid competition is partitioning resources by keeping competitors away. This is called territoriality and will be discussed in more detail in Chapter 11.

Aside from the obvious differences in size, behavior, and so on, there is variability in all populations. Not only is this variability the same material for evolution, but it provides a mechanism by which organisms reduce competition because, in one or several ways, even organisms of the same species differ slightly in their requirements for survival (Exercise 10.2).

Symbiosis

Symbiosis (literally "living together") refers to a close, long-term relationship of two organisms of different species. One or both species are often dependent on the other for survival. There are three types of symbiosis: commensalism, mutualism, and parasitism.

Exercise 10.2 The Measurement of Competition: Field Exercises.

There are a number of ways to observe and measure the means by which organisms reduce competitive interactions. As an example, observe the ways in which birds use a particular species of tree; a few, or even one, tree will suffice. When a bird lands in the tree, note the following information: species of bird, height in the tree, position in the tree (near trunk, on top, etc.), behavior (feeding, singing, etc.), and the time it spends in each part of the tree and/or devotes to each behavior. Put in table form with headings such as bird species, height in tree, position in tree, behavior, and time.

After you have a sufficient number (30 or more) of observations for each of several bird species, determine the percentage of time each species spends in the various parts of the tree (by height— 1,2,3 meters— and by position—on trunk, in middle, on periphery.) Also determine the percentage of time it devotes to each behavior, such as pecking, probing, flying, and calling. Then determine the amount of spatial and behavioral overlap between two bird species by comparing the percentages for each

category; the lowest percentage is the amount of overlap. For example, if species A spends 27% of its time, and species B 11%, at a height of 1 meter in the tree, the two species overlap 11% of the time at that height. Do the same for other positions in the tree and for behavior. The overlap figures must be interpreted with the bird's ecology in mind. An owl may sit in the same place in a tree as a pigeon, and the two birds may overlap a good deal in space, but they eat totally different foods so their behavioral overlap would be very small as would competition between them. Two insect-eating warblers, however, would probably show a good deal of overlap in behavior, but little spatial overlap. This same method can be used for insects in vegetation, the distribution of flowering plants in a field, arthropods in a rotting log, tide pool animals, and so on.

Commensalism is the symbiotic relationship in which one organism profits by the association and the other is neither harmed nor helped. The shark sucker, or remora, has a modified fin on its head with which it attaches to sharks for a free ride; it may also pick up leftovers from the shark's meals (Figure 10.5). The shark apparently derives no benefit nor is hindered in any way. Epiphytes, such as lichens, mosses, and bromeliads, grow on the trunks and branches of trees, where they receive more sunlight, water, and nutrients than if they grew alone; the trees are not affected. Humans provide homes for ferocious-looking but harmless microscopic mites in their eyebrows. Looser commensal relationships are those in which animals take advantage of human habitation for food and shelter; the house sparrow, house mouse, rat, cockroach, and termite are good examples.

**Figure 10.5
Commensalism.**

A *Remora,* or shark sucker, adheres to a shark by means of its modified dorsal fin. Note that the *Remora* is actually upside down. The *Remora* gets a free ride from the shark and picks up scraps of food from the shark's meal. The shark is apparently unaffected.

Mutualism is a type of symbiosis in which both organisms benefit by their relationship (Exercise 10.3). Termites are able to digest wood because of protozoa in their gut; the protozoa have a home and food. The sloths of Central and South American rain forests are sluggish, herbivorous mammals that spend most of their time hanging from tree limbs in an upside-down posture. Because blue-green algae grow in grooves in the sloths' hairs, the sloths blend in quite well with the foliage, thereby escaping easy detection by predators. Lichens are a "superorganism" formed by the mutualistic association of an alga and a fungus, neither of which could live alone. The alga produces food for the fungus and the fungus provides some protection against adverse conditions such as drought. Several species of "cleaning" shrimp rid fish of external parasites, which the shrimp eat. The shrimp advertises its cleaning services by waving its antennae and the fish signal their wish to be cleaned by remaining motionless. Tarantulas and narrow-mouthed toads share an underground burrow. The tarantula defends the burrows, and thus the smaller toad, against predators. The toads have an appetite for ants, which are predators of spider eggs and spiderlings.

A widespread form of mutualism is the pollination of flowering plants by insects, birds, or bats. The flowers have evolved to attract the animals visually or via an odor and to reward the animal with pollen grains and/or nectar. The animal receives nutritious food, and the plant benefits by having its pollen transferred to other flowers by the animal (Figure 10.6). Some species of bees visit orchids to collect a fragrance from the flower that helps the bees find each other; the bees are thus able to form a mating group and the orchids are pollinated in the process. Nectar-feeding bats of the South American tropics are nocturnal, have a good sense of sight and smell, weak teeth, a long snout, and a long, prehensile tongue. The flowers upon which they feed open at night, are white in color, and emit an odor that attracts bats. The flowers are located away from the foliage, providing easy access for the bats. The bats lap up nectar, eat pollen, and in the process coat their face with hundreds of pollen grains, some of which are transferred to the next flower the bat visits.

There are also many examples of birds, mammals, and insects dispersing the seeds or spores of plants. Squirrels and jays store acorns in the ground, some of which are never retrieved and become oaks. Berry seeds pass through the guts of birds and pass out away from the parent plant. The animals obtain food, and the plants have their seeds dispersed. Several kinds of insects live in close association with fungi. The fungi may provide food, digestive enzymes, or protection for beetles, midges, ants, termites, or wasps. Some termites actively cultivate a fungus garden. The fungi benefit by being provided food (the rotting wood in which these insects live) and by having their spores dispersed.

One of the most fascinating cases of mutualism is the interaction between ants and acacia trees in Central America. The acacias grow as shrubs or small trees and have many small leaflets and large swollen thorns. Ants live inside the swollen thorns. They forage on nectar produced by **nectaries** (nectar-producing organs) on the twigs of the plant and on solid food on the tips of the leaves packaged in a special form called **Beltian bodies**. The ants thus are provided food and shelter by the acacia. In return, the ants "patrol" the acacia and attack any other insects or larger animals that attempt to feed on the acacia or any plant that begins to grow under or near the foliage of the acacia. Acacias that do not have ant colonies associated with them are destroyed by rodents and insects within 6 to 12

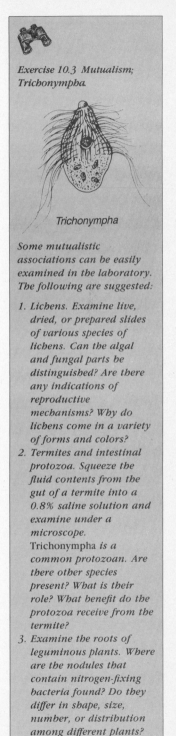

Exercise 10.3 Mutualism; Trichonympha.

Trichonympha

Some mutualistic associations can be easily examined in the laboratory. The following are suggested:

1. *Lichens. Examine live, dried, or prepared slides of various species of lichens. Can the algal and fungal parts be distinguished? Are there any indications of reproductive mechanisms? Why do lichens come in a variety of forms and colors?*

2. *Termites and intestinal protozoa. Squeeze the fluid contents from the gut of a termite into a 0.8% saline solution and examine under a microscope.* Trichonympha *is a common protozoan. Are there other species present? What is their role? What benefit do the protozoa receive from the termite?*

3. *Examine the roots of leguminous plants. Where are the nodules that contain nitrogen-fixing bacteria found? Do they differ in shape, size, number, or distribution among different plants?*

Figure 10.6 Mutualism. The insect in the center of the flower is a soldier beetle, which feeds on flower pollen. In the process of feeding, the beetle carries pollen to other flowers. The flower, then, provides food for the beetle and the beetle disperses the flower's pollen. (The cucumber beetle in the upper right is a predator and feeds on the plant's body, causing injury.)

months or are shaded by competing plants and grow very slowly. This development of interdependent features by two species is **co-evolution**. Coevolution is not limited to mutualism, however; virtually every interaction between two species shows some degree of coevolution. Parasitism is also a form of coevolution.

Parasitism is a form of symbiosis in which one organism (the **parasite**) lives on or in and at the expense of another (the **host**), which is harmed to some degree. There are **ectoparasites** that live on the outside of a host's body; these are such things as mites, ticks, fleas, lice, leeches, and some fungi. Mistletoe sends its roots into the circulatory system of a tree branch to tap nutrients. Fungi cause plant diseases such as potato blight, rust, and mildew. Ringworm and athlete's foot are fungal infections common to people. Ectoparasitic infections can be serious, but **endoparasites**, those living inside of the host, are usually more detrimental. Viruses, bacteria, fungi, flukes, tapeworms, protozoa, and roundworms are all well known for their potential effects. All manner of plant and animal diseases are caused by endoparasites: tobacco mosaic (virus), pneumonia (bacterium), malaria (protozoan), schistosomiasis (blood fluke), hookworm (roundworm), and so on.

Many parasites have a complex life cycle, requiring more than one host. Most endoparasites are incapable of dispersing very far under their own power; they depend on a series of hosts for dispers-

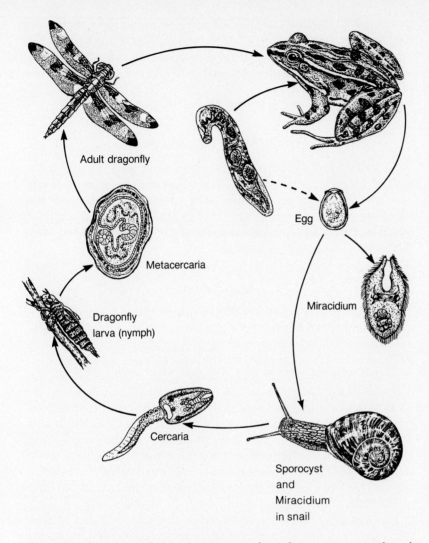

Adult dragonfly

Metacercaria

Dragonfly larva (nymph)

Cercaria

Egg

Miracidium

Sporocyst and Miracidium in snail

Figure 10.7 Life Cycle of Frog Lung Fluke, *Hematoloechus*. There are various stages in the life cycle of the frog lung fluke involving hosts other than the frog. The eggs of the fluke are produced by the adult fluke in the frog. They are passed to the outside where they are eaten by a snail. The egg hatches in the snail, releasing a ciliated **miracidium** larva, which moves to the digestive gland of the snail and becomes a **sporocyst**. The sporocyst produces many tailed larvae, called **cercariae**, which leave the snail, swim through the water, and burrow into the tissues of a larval dragonfly. They encyst in the dragonfly and become **metacercariae**. When a frog eats the immature or adult dragonfly, the metacercariae leave the cyst, travel to the lungs and develop into adult flukes ready to lay eggs and begin the cycle again.

Cercariae sometimes burrow into the wrong intermediate host and die. If they bore into a human's skin, they produce sores and rashes, an irritating but not serious ailment called "swimmer's itch."

al. In the frog lung fluke, the intermediate hosts are a snail and dragonfly (Figure 10.7). Some asexual reproduction may occur in these hosts. The frog is the **definitive host**, in which the adult parasite reproduces sexually.

The close relationship between internal parasites and their host has allowed the parasite to become extremely dependent on the host and unable to live alone. Many parasites are simply reproductive machines full of gonads but lacking a digestive system, a circulatory or respiratory system, or any means of locomotion. In most cases it would be nonadaptive for a parasite to kill its host. It is also nonadaptive for a host organism not to be able to tolerate some parasitic infection. Thus, there has been coevolution of host and parasite for coexistence.

Parasitic infections and disease are quite similar and sometimes synonymous. Fungi, bacteria, and viruses that cause diseases such as blight, botulism, and rabies are also parasites. Parasites and disease organisms can be considered specialized types of predators, in the broadest sense of the term.

GROUPS OF ORGANISMS

Often we find a group of individual organisms clumped together. The individuals may be grouped because of their attraction to one another or simply because of a common requirement that brings them to one place. A cluster of organisms is frequently composed of individuals of the same species (conspecifics), but often mixed-species clusters are found.

Aggregation

An **aggregation** is a group of individuals clustered together because of nonsocial factors; for example, they are clumped because of environmental conditions. Thus, in the Sahara Desert clumps of palm trees are found in oases where there is water. Aggregations of isopods ("pill bugs") are found under rocks and logs where it is dark and moist. Hundreds of cliff swallows build mud nests on the underside of a bridge because the bridge is an appropriate habitat, and mud for nest construction is available nearby. Cows often huddle together to reduce the effects of stormy weather; some birds spend the night in a tight cluster to reduce loss of body heat; garter snakes winter en masse in limestone caves in Manitoba. These organisms are not in social groups, because they are reacting to environmental conditions and not to other individuals (Figure 10.8). Organisms are generally distributed in the same manner as their resources, and since resources are usually concentrated, so are organisms.

Social Groups

Organisms also gather with other individuals if that interaction is beneficial to the individual. It is a social group, or society, if the organisms respond to each other as individuals. Typically, the members of a society have different roles that contribute to the adaptiveness of the society. Although there are differing opinions, most scientists agree that social groups exist because each individual benefits by its association in a society; societies as a whole do not benefit and do not evolve except as their component individuals evolve (Figure 10.9).

Figure 10.8 Ladybird Beetles. Ladybird beetles (lady bugs) gather in large clumps under leaves or twigs, where they spend the winter protected from extremes of weather.

Many species of predators hunt in a group. White pelicans can scoop up more small fish if they first head them into the shallows, a task that is more easily done by a group. Killer whales have been known to frighten seals off an ice floe into the jaws of other whales. Coyotes, wolves, and hyenas can stalk and bring down large prey only if they hunt as a team.

In many species, reproductive processes require the aggregation of many individuals to ensure that reproductive individuals

Figure 10.9 Termites. Termites are among the most social of all animals, living in groups with a well-developed caste system. Kings, queens, workers, and soldiers have specialized duties, and all are dependent on the others. The kings and queens are the reproductive individuals, while the sterile workers collect food and the sterile soldiers guard the nest.

meet each other and/or to provide the stimulation to breed. The palolo worms swarm to the surface of the ocean from their burrows in the sand in enormous numbers on two days during each October and November. The bodies of the 40-centimeter long worms burst open and release great quantities of eggs and sperm. Were it not for this behavior, the otherwise solitary worms would have little chance for their gametes to meet. (This habit also makes it easy for the islanders of Samoa to gather the worms, which they consider a delicacy, live or cooked.) The passenger pigeon, now extinct, once nested in flocks numbering over a billion individuals. Apparently they needed the stimulation of immense numbers of conspecifics to reproduce successfully. As their populations declined through overhunting, their demise was probably hastened by the lack of sufficient social stimulation. Seals, penguins, mayflies, and numerous other organisms flock together for all or part of the year, primarily for reproductive purposes. Flocks, herds, and other gatherings provide the opportunity for sexes to meet and choose mates and provide some protection against predators (Figure 10.10).

Traveling in a school must have an advantage for some fish, as nearly 20% of all fish species do school. There is good evidence to indicate that large numbers of prey fish confuse predators. Schooling also seems to facilitate the finding of food and social stimulation to feed may even promote faster growth. Both fish schools and flocks of birds may provide hydrodynamic and aerodynamic advantages, respectively, to the individual members in the group (Figure 10.11). Baboons travel in troops, antelope in herds, and some birds

Figure 10.10 A Flock of Gulls Nesting on an Island. The noise and commotion made by the disturbed birds is often enough to scare away potential predators.

in flocks. In some cases protection derives simply from the fact that there are so many pairs of eyes, predators can be easily spotted. For example, a flock of small birds cannot defend itself against a hawk or owl, but it can avoid or evade these predators because it has a greater chance of spotting them. A group of baboons, however, can physically drive away a cheetah or lion.

In the next chapter we will discuss some important outcomes of the interactions of organisms.

SUMMARY

Organisms interact with one another, thereby defining their role in the community. We can define major types of relationships that organisms have with one another. A particularly significant interaction is competition, sometimes leading to the exclusion of one organism or species from an area. Some organisms live in a close, long-term relationship called a symbiosis. Organisms also group together in various social relationships.

Figure 10.11   Flight Flocks of Birds. Some typical patterns of the coordinated flight flocks of some bird species such as waterfowl or pelicans. Individual birds seem to profit from flying in this formation because they gain aerodynamic lift from an adjacent bird's wingtips or reduce friction by flying behind another bird. The V formation may also provide for easy communication between individuals. Unorganized flocks, such as seen in migratory songbirds, seem to provide no aerodynamic advantage, but the benefits of increased communication, accurate navigation, predator protection, and food location make flocking adaptive.

STUDY QUESTIONS

1.   Define symbiosis, predation, competition, diurnal, commensalism, intraspecific.

2.   Explain and give an example of the competitive exclusion principle.

3.   Differentiate between parasitism and predation.

4.   Define and give an example of each of the three types of symbiosis.

5.   Which do you think is stronger, inter- or intraspecific competition? Why?

6.   What are ways that organisms have evolved to reduce or avoid competition?

7.   Why is it necessary for both parasite and host to evolve adaptations to the relationship? In other words, why cannot the parasite simply evolve adaptations to parasitize the host?

8.   Design an experiment to determine whether protozoa in the gut of a termite are mutualistic, commensalistic, or parasitic.

9.   Differentiate between an aggregation and a social group.

10.   For what reasons are social groups adaptive? Under what circumstances could they be nonadaptive? Give examples.

Chapter 11

Territoriality, Habitat, and Niches

Organisms can interact in a wide variety of ways. The ways in which they are able to interact with each other and the environment determines where they are able to live and what roles they fill in the ecosystem. Often, to be able to live in their "homes" and be successful at their "jobs," organisms must also defend necessary resources.

For many organisms to function effectively and reduce or avoid competition, they must occupy some sort of semiexclusive space. Animals wander, some widely and others only a short distance, in an area termed a **home range**, which is simply defined by any place they go. A **territory** is a portion of the home range that is defended against intruders of either the same or different species. Territoriality is a common phenomenon in higher vertebrates, and is especially well documented in birds, but it also occurs in lower animals and plants. Among animals, territoriality is behavioral, but in plants it is chemical (alleopathic). The concept is most often used with reference to animals, however.

HOME RANGE

Although a species may populate a wide geographic area, an individual of that species will move over only a small portion of the area, even though it may be able to reach and survive in other parts

Figure 11.1 Home Range. Although the range of the Western fence lizard covers thousands of square kilometers, an individual may have a home range of only 0.1 hectare.

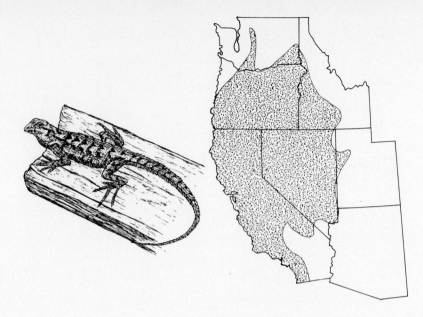

(Figure 11.1). One reason an animal restricts its movements to a home range is simply because it has become familiar with its surroundings and is better able to find food, water, shelter, and other resources than if it continued to wander to new areas. Also, the size of the organism greatly affects the area it can cover. Larger organisms need more food and must travel greater distances to find it. Recalling the ever-decreasing amount of energy as one goes up through higher trophic levels, we know that organisms higher on the trophic levels have scarcer food resources and thus must travel more (Figure 11.2).

INDIVIDUAL DISTANCE

If you have ever observed birds perched on a wire or cows in a pasture, you have noticed that the animals maintain space between themselves and other individuals. This personal space around an individual is known as **individual distance**. Intruders may be pecked, butted, pushed, or verbally abused. Individual distance may be smaller between parents and young or between mates than between unrelated individuals. Also, at the approach of a predator, individuals may gather together more closely than usual for protection. They may huddle in tight groups for protection against the weather. Typically, however, they maintain a personal space that they hesitate to let others invade (Figure 11.3).

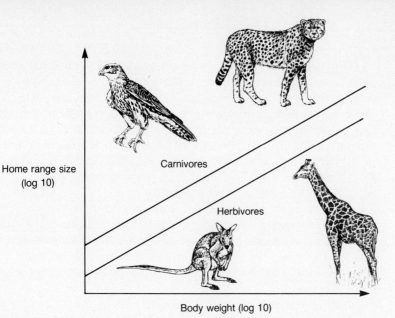

Figure 11.2 **Relationship of Home Range Size to Body Weight for a Variety of Mammals.** Note that (1) home range size increases with body weight, and (2) carnivores have larger home ranges than herbivores of the same weight.

Home range size (log 10)

Carnivores

Herbivores

Body weight (log 10)

Figure 11.3 **Individual distance.** These cliff swallows sitting on a wire are good examples of animals maintaining their individual distances. The birds appear to space themselves rather regularly. The greater the number of individuals, the closer they perch to each other, but there is always a minimum distance separating them. In the case of some colonial birds, such as murres or cormorants, nesting space is at such a premium that their territory consists of the distance the birds can reach while sitting on their nest. In that case, the territory and individual distance are the same. Humans also maintain an individual distance; think how you would react if someone conversed with you with your noses nearly touching!

TERRITORY

A territory is an exclusive or semiexclusive space defended by an individual. It is essentially an extension of the personal space. Since time and energy are required to defend a space physically, territoriality must be advantageous to the defender. Territoriality is most well known, and apparently most common in, the higher groups of animals such as mammals and birds. It is also found among numerous invertebrates, however, such as spiders, lobsters, dragonflies, limpets, some wasps and bees, ants, termites, and some flies. Territoriality serves a number of possible functions:

1. Provides familiarity with the area
2. Provides for distribution of resources
3. Reduces aggression by spacing out competitors
4. Reduces disease by decreasing the density of individuals
5. Attracts one or more females to a territory-holding male
6. Provides one, some, or all of the resources necessary for raising young
7. Provides an exclusive food source

Many other possible functions of a territory have been postulated, but basically territories disperse individuals so that the territorial organisms have sufficient resources. Nonterritorial individuals of that species are at a disadvantage since they are excluded from these territories and have to compete continually for resources with many others. Thus, territoriality is advantageous and is selected through evolution (Exercise 11.1).

Sizes of Territories

Territories can be neither infinitely large nor infinitely small. The maximum size is limited by the ability of the animal to defend it; if too large, the organism is unable to defend it properly. To be of value, a territory must contain enough resources or be large enough to serve its intended purpose (Table 11.1).

Territory sizes vary among individuals, and from season to season and year to year in the same individuals. The function of a territory partly determines its size; for example, a territory used for feeding, mating, and raising young is generally larger than a winter feeding territory. But even the same types of territories vary in size. There are at least three major reasons for this variability: resource abundance, competition, and energy needs, all of which are closely associated.

Consider a winter feeding territory, as is held by many birds. The main function of the territory is to protect a food source during a harsh part of the year. If the food is scarce, territories will be large to encompass sufficient foodstuffs; if abundant, territories will be smaller (Figure 11.4).

Exercise 11.1 Territoriality.
The territory of an animal can be determined by watching its defensive or aggressive behavior and movement and plotting them on a map. It has been done for many species, but it is most easily done for birds in the field during the breeding season. Marsh or grassland dwelling birds are easiest to work with since they are more easily observed than forest species.

Choose a study area several hectares in size and prepare a map of it. For one or two hours each day over a period of two weeks during the beginning of the breeding season, walk through or around the study area. Locations of individual birds (of one or more species) should be plotted on the map. Aggressive encounters between birds should be noted as well as singing and nest sites. At the end of two weeks there will be many clusters of points on the map. The clusters can be enclosed by a line, which then will loosely indicate territorial boundaries. Singing and nesting often take place near the center of the territory and aggressive encounters at the edges; these observations should help you determine the boundaries and the size of

the territory. See Figure 11.4 for examples of territory maps.

If competitors are few, territories will be larger, because the amount of defense required will not be as great as it is with more competition. It has been shown in mating territories of a fish, the blue wrasse, that mating success declines with time needed for defense. If a male fish has to devote time and energy to driving out intruders, he naturally has less time to court and mate with females.

If the weather is cold, windy, and rainy, more energy and thus more food and a larger territory are required than they are during milder weather. Thus, the size of a territory is a function of the interaction of several variables.

Table 11.1 Some Typical Territory Sizes.

Animal	Size of Territory (in hectares)
Canada Goose	0.5
Sagebrush Lizard	0.00004
American Coot	0.4
Common Snipe	8
Red-breasted Nuthatch	1
Yellow-bellied Marmot	1
Beaver	25
Golden Eagle	25,000

Figure 11.4 **Relationship of Food Supply to Territory Size.** (a) A number of territories of the Townsend's solitaire on 50 hectares of juniper woodland. (b) The territories of these birds one year later. In the second year the territories overlap and are smaller due to an increase in the birds' population and their food supply (juniper berries). Shaded areas indicate places where territories overlap.

Recent evidence indicates that normally territorial animals may become nonterritorial under some circumstances. If food is unusually abundant and/or competitors rare, there may be no need to defend an area in order to protect a resource. In fact, it may be disadvantageous to do so. On the other hand, food may be so scarce and/or competitors so abundant that defending a territory may be more energy costly than simply wandering through the home range and feeding opportunistically.

Defense of Territories

During the establishment of territories, antelope, deer, and sheep butt heads, male fish flare their gill covers and engage in locked-jaw tussles, and birds chase and peck intruders. Physical skirmishes are not uncommon at the outset, although injuries are generally rare and minor. Once the territory holders have staked out their respective boundaries, however, territory maintenance is accomplished through visual, olfactory, or auditory signals rather than physical contact. The bright colors or songs of male birds continually advertise the territory; only an occasional confrontation occurs. The dik-dik, a small African antelope, has large circular scent glands just below its eye. The antelope impales twigs into this gland, leaving a gob of sticky substance whose odor identifies the territorial boundaries. The females of some weevils and fruit flies mark their food sources with a chemical that keeps other females from laying eggs on that food source. The bitterling fish of the eastern United States lays its eggs in the gills of a mussel and defends the area around the mussel. Since the mussel moves, so does the territory!

Territoriality in Plants

Plants, having few behavioral defenses, can still maintain a territory of sorts through allelopathy. Chamise, a common shrub of California chaparral, emits a chemical into the soil that prevents the germination of other shrubs nearby (Figure 11.5). This inhibition ensures that the roots of chamise do not have much competition for water in the soil. The chemical defenses of plants may have originally evolved in response to predator pressure, as predators will avoid foul-tasting plants. The existence of allelopathy may be a secondary effect, since the advantage of allelopathy in some plants is not clear.

Nonterritorial Organisms

If territoriality is so advantageous and common in some groups of animals, why are not all animals territorial? There are two possible explanations. One, territoriality requires an active defense and expenditure of energy to protect resources or young or to attract a mate. The social systems of lower animals are often so simple that mating occurs very opportunistically (two earthworms cross paths in the soil) and rarely is parental care given (many fish and invertebrates simply scatter their eggs into the surrounding water). Further, food may be impossible to defend. How, for instance, can a clam or sponge, which filters lake water or seawater for fine organic particles, defend its food source? Territoriality, then, is not advantageous for all organisms.

The second explanation is that some sort of territoriality does exist in these organisms, but we have not discovered it yet. For example, rabbits have long been considered nonterritorial animals, but evidence now exists to show that at least some species are territorial.

Figure 11.5 Allelopathy— Territoriality Among Plants. Allelopathy is apparently common in areas where water and nutrients are scarce, such as in the desert pictured here. Some plants may simply be better competitors than others, thus excluding other plants without actually "defending" the resource. Obtaining resources by being a better competitor is sometimes known as "scramble" competition; defending resources by territoriality or other means may be termed "contest" competition.

Interspecific and Intraspecific Territoriality

Intraspecific territoriality is that which occurs between individuals of the same species; one blue jay excludes another blue jay from its territory. Since competition is greatest between individuals of the same species, it follows that territoriality should exist primarily to exclude conspecifics.

Interspecific territoriality occurs between individuals of different species and is much rarer, since different species have different requirements, but it does occur between some closely related species (Figure 11.6).

HABITAT

The home range and territory of an organism, or simply the place it is rooted or attached, are within its habitat. We have used the word throughout the text without really defining it. To give it a specific definition would limit its usefulness, so we will simply use **habitat** to mean a set of conditions suitable for an organism's survival, or more simply, anywhere an organism lives. We can speak of different levels of habitat, such as **macrohabitat**, which refers to larger entities such as forests, lakes, and prairies. The concept of

Figure 11.6 Interspecific Territoriality. Yellow-headed and red-winged blackbirds are similar in their habits and requirements and often exclude from a territory members of the other as well as their own species. The yellow-heads occupy the center of the marsh and the red-wings the fringes.

microhabitat, such as a fallen log, the litter layer, or the tree canopy, narrows the concept to a more specific physical place.

Habitat Selection

Every organism is limited in its adaptations to the environment and is thus restricted to one or a few habitats. In the case of organisms with limited mobility, their arrival in a habitat is determined by factors outside of their control. If they land in a suitable habitat, they have the chance to survive. Active, mobile organisms, on the other hand, have the capability of choosing an appropriate habitat.

There are clues that enable these active organisms to make a proper choice. A simple experiment can be done with terrestrial isopods (sowbugs, pillbugs). If put in a container with a light source at one end and a darkened opposite end, they will move around quickly until they reach the darkened end where they will remain. The same can be done with a moisture gradient. This is not to imply a conscious or learned choice (although that may be the case in higher animals). Isopods live under rocks, decaying logs, and in other such habitats that provide a dark, moist atmosphere. Any isopod that had the tendency to choose a warm and dry habitat would desiccate and die, so only the ones whose genes "tell them" to go to the proper habitat survive. They instinctively move from low to high humidity and light to dark.

An experiment with two species of deer mice, one that normally inhabits grassy areas and one that inhabits woodlands, demonstrated that habitat selection is genetic, but that after being raised in the laboratory for 12 to 20 generations, this genetic preference was weakened somewhat. Thus, some learning must occur in order for them to choose the proper environment.

Whether genetic or learned, there are factors in the environment that allow or force an organism to make certain choices. These choices may be modified by various conditions. For example, four species of flycatchers spend time feeding on insects at certain levels in forest vegetation in the spring. In the summer, during the breeding season, only two species are present, but they show a change in their preference for feeding heights, indicating that the presence of the third species caused a difference in habitat selection (Figure 11.7). So competition, as well as the physical environment, apparently affects habitat selection.

Habitat Structure and Diversity

For various reasons, some ecosystems provide many more habitats than others. The structure of the habitat is dependent on the topography of the land and the physical form of the vegetation. The flat terrain of an alkaline desert contrasts starkly with the habitat of a coniferous forest in a mountainous habitat (Figure 11.8). In general, the more diverse the environment, the more habitats, and, thus, the more kinds of organisms present.

NICHES

Niches and the Competitive Exclusion Principle

A **niche** (pronounced "nitch") can be defined as the role an organism plays in a community—how it interacts with the environment and other organisms. Is it a predator or prey, does it burrow or fly, and does it photosynthesize or feed on decaying matter? Any description of an organism's function is a part of its niche. In a perhaps oversimplified statement, we can say that an organism's habitat is its home and its niche is its occupation.

We stated in Chapter 10 that competition occurs whenever two or more organisms strive for the same limited resource. The closer the organisms are in their needs, the greater the competition. In situations of potentially severe competition, one species usually excludes another. In two species of cattails, one is restricted to

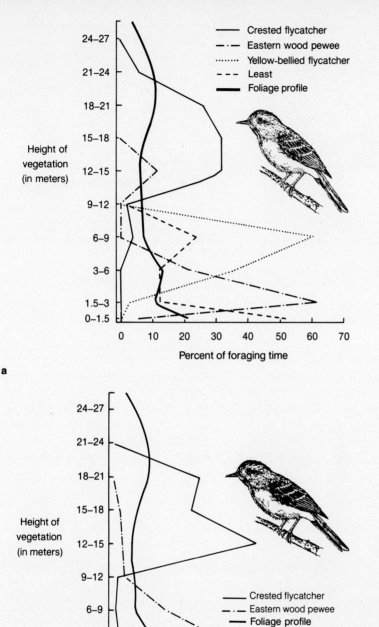

Figure 11.7 Competition and Habitat Selection. Flycatchers feed primarily by perching on a tree branch and flying out to capture insects. Different species of flycatchers prefer a different range of heights at which to feed, perhaps because their preferred prey occurs most often at those levels. Figures 11.7a and b show the preferred foraging heights of some flycatcher species. (a) Four species that occur in the same habitat together in the spring. Note how they prefer different heights and thus avoid competition. (b) Two of those species in the same habitat later in the year after one of the species has left (migrated northward). Note that the two species generally prefer the same heights as they did previously, except that they have expanded the heights over which they feed to fill the void left by the now-absent species. This indicates that habitat selection is a function of competition as well as other factors.

Figure 11.8 Habitat Structure. (a) An alkaline desert in central Nevada is a harsh, relatively barren ecosystem. The hard ground, lack of water, and scattered shrubs of greasewood provide few and poor habitats for both plants and animals. (b) The varying topography, increased precipitation, and tall stands of coniferous trees in the Idaho mountains provide much more food and space (more habitats) than the desert.

a

b

deeper water and the other to shallower because of competitive interaction. This, as you recall, is known as the "competitive exclusion principle." Stated another way: "No two species can occupy the same niche."

The **fundamental niche** is the niche that can be fully filled by an organism in the absence of competition. An organism can eat all the food appropriate to it or sink its roots into any suitable soil, for instance, because competition does not restrict its activities (its niche). Rarely, if ever, does this occur, however. Rather, the niche of a creature is restricted by competition; it cannot do everything it is capable of doing or utilize all suitable resources in the habitat. It has to restrict its niche to avoid competition. This restricted niche is the **realized niche** (Figures 11.9 and 11.10).

There are a number of dimensions to a niche. A niche includes physical factors such as soil pH, weather, nest sites, and shelter, as well as biotic factors such as mates, food supply, competitors, predators, and parasites. Any attempt to measure an organism's niche must necessarily be incomplete since there is an infinite number of factors to consider. The niche is sometimes considered an **n-dimensional hypervolume**, that is, a space with an infinite number of measurements (Exercise 11.2).

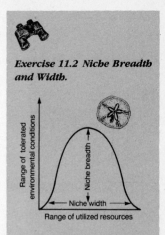

Exercise 11.2 Niche Breadth and Width.

Niche breadth is the range of environmental conditions used by a species. Niche width is the range of resources that a population utilizes. Mathematical formulae exist to quantify these niche measurements, but their sophistication and the fact that there is much disagreement about their validity, make their discussion inappropriate here. However, you may be able to create a method to measure these characteristics. We discussed variability in Chapter 9 and said that morphology, physiology, and behavior among individuals in a population vary in such a way to produce a bell-shaped curve. If you can determine, or search the literature for, the niche width of several populations (using one or more niche dimensions—see text), you can then determine the niche breadth of a species.

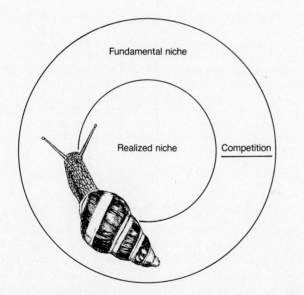

Figure 11.9 Fundamental and Realized Niches. The difference between the potential, or fundamental, niche and the actual, or realized, niche is caused by competition. As in Figures 11.7a and b, competition causes a species to retract and concentrate on the resources it is best able to compete for. In a situation of lessened competition, the realized niche expands.

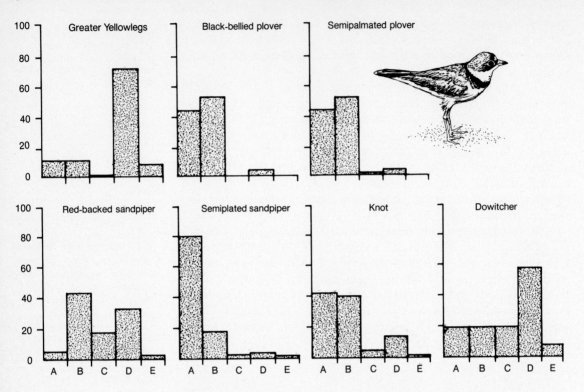

Figure 11.10 Shorebird Niches. The niches of many shorebirds are similar. During migration they walk along the shores of marine habitats, picking up or probing for invertebrate animals in the sand. However, although their niches are quite alike, they avoid competition by feeding in different areas along the shore. Those species that are most alike in their requirements migrate at somewhat different times so their peaks of population abundance do not overlap. In addition, their bill lengths are different, so they probe to different depths in the sand. The vertical axis represents the percentage occurrence of the species on the New Jersey shore. The letters A through E represent the shoreline zones from the beach without a film of water to more than 0.3 meter in the water. Zone C is the water's edge.

The Niche Redefined

It is not quite accurate to say that the niche is simply the organism's role. It is much more than that: It is the way in which the organism reacts to its environment. The niche is a dynamic concept, changing from moment to moment. It encompasses the organism's behavior, physiology, competitors, and the physical habitat. It is the way in which the organism fits in its environment and relates to other organisms.

If two organisms cannot occupy the same niche and they have developed ways to reduce competition, then it should also be possi-

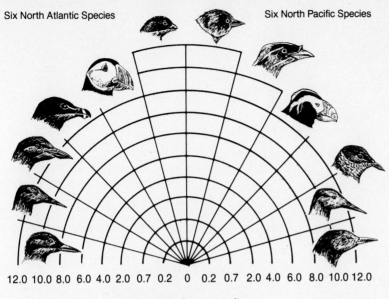

Six North Atlantic Species   Six North Pacific Species

12.0 10.0 8.0 6.0 4.0 2.0 0.7 0.2 0 0.2 0.7 2.0 4.0 6.0 8.0 10.0 12.0

Kilometers from nest site

Figure 11.11 Niche Segregation in Seabird Communities. See text for explanation.

ble for several organisms with similar niches to subdivide the resources. Figure 11.11 demonstrates how six species of cliff-nesting ocean birds divide the fish resource upon which they all depend. The zonation of the feeding area seems to be the major mechanism allowing their coexistence. This is **niche segregation**. The term **guild** is sometimes applied to a group of organisms that have similar niches. A group of animals that feed on the leaves of trees may be put into a "leaf-eating guild" and include birds, insects, primates, and others, regardless of their evolutionary relationships. Or we may find organisms separated along a gradient of environmental changes. The Himalayas stretch over 2500 kilometers, with many mountaintops exceeding 30,000 meters. Topographical features and elevational changes cause considerable changes in the vegetation. The varied environment provides niches for several species of hoofed animals (Figure 11.12 and Exercises 11.3 and 11.4).

In theory there is an infinite number of niches. Organisms evolve to fill slightly different roles, thus evolving new niches. Sometimes biologists speak of "empty" niches, meaning that a habitat appears to be suitable for a particular species, which does not exist there. The Hawaiian Islands, for example, have no native amphibians, reptiles, or mammals other than one bat species. Eighteen species of amphibians and reptiles and several species of birds and

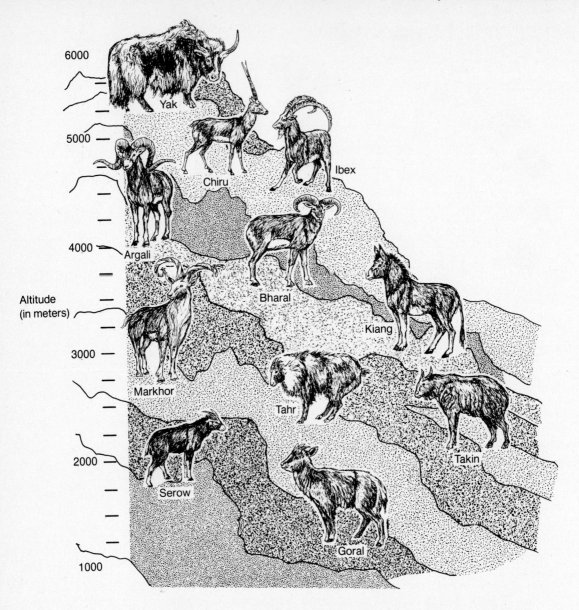

6000

5000

4000

Altitude
(in meters)

3000

2000

1000

Yak

Chiru

Ibex

Argali

Bharal

Kiang

Markhor

Tahr

Takin

Serow

Goral

Figure 11.12 Each Has Its Niche. The Himalayan mountain range is 2500 kilometers long, with many peaks above 6000 meters. Climate and vegetation change drastically with elevation, creating diverse living areas for the specialized needs of these animals that live there. This figure shows the hoofed animals that inhabit the Himalayas.

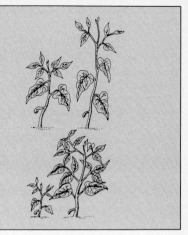

Exercise 11.3
Environmental Gradients:
Laboratory Study.
Bean, tomato, or radish
seeds can be easily
germinated in small
containers of potting soil.
Environmental factors such
as sunlight, water,
temperature, and fertilizer
can all be varied to see what
effect they have on growth.
For example, 15 plants may
be used to examine the effect
of temperature on growth. A
set of three plants can each
be grown under 16°C, 18°C,
20°C, 22°C, and 24°C. Does
temperature have an effect
on the rate of growth, total
growth, number of leaves,
and so on? Would similar
effects be seen in the field
with wild plants? Do the
same for the other factors
and graph the data.

Exercise 11.4
Environmental Gradients:
Field Study.
Choose a study plot several
hectares in size. In this plot
you may measure several
gradients of environmental
factors. Some suggestions:

1. Measure the density of
   vegetation at different
   heights. The greatest
   amount of vegetation may
   occur at the greatest
   height, or in the middle of
   the range of heights. This
   works best in a forest
   habitat.
2. Measure the density (or
   height) of vegetation from
   one side of the plot to
another. Is there a
gradient? Why?
3. Collect animal species
   along a transect across the
   plot. Worms, isopods,
   spiders, insects, or other
   kinds of species are
   suitable. Do their numbers
   differ? Why?

If you do find differences in
the numbers or kinds of
organisms, horizontally or
vertically, try to determine
why the differences exist by
taking environmental
measurements along the
gradient. Soil depth, pH, or
temperature; air
temperature; humidity;
wind; or sunlight
penetration can all be
measured. Draw a graph, if
you can, depicting the
relationships of a species' (or
group) distribution to changes
in environmental factors. Are
gradients you observe due to
acclimation (nongenetic) or
adaptation (genetic)?

mammals have been successfully introduced into the island, indicating that there were niches to be filled. One can argue in some cases, though, that introduced species usurped the niches of some native species. Competition from introduced birds caused the extinction of 22 native Hawaiian bird species over the past 200 years.

The idea of the niche has and is much debated, but the general concept is understandable: Organisms share resources by doing things somewhat differently; if their roles are not sufficiently different, the least fit will be excluded. As Dr. Seuss once put it:

> And NUH is the letter I use to spell Nutches
> Who live in small caves, known as Nitches, for hutches.
> These Nitches have troubles, the biggest of which is
> The fact there are many more Nutches than Nitches.
> Each Nutch in a Nitch knows that some other Nutch
> Would like to move into his Nitch very much,
> So each Nutch in a Nitch has to watch that small Nitch
> Or Nutches who haven't got Nitches will snitch.*

SUMMARY

For many organisms to function effectively, they must occupy some sort of semiexclusive space. A home range allows an organism to become familiar with an area. A territory is a space defended against invasion by other organisms. Territories serve many purposes. The habitat is anywhere an organism lives, and its niche is its job. There are an infinite number of niches and habitats and no two organisms can occupy the same niche.

STUDY QUESTIONS

1. Define territory, home range, habitat, niche, niche segregation, microhabitat.
2. Compare territory and home range.
3. What is meant by individual distance? Is it related to territory in any way?
4. What are the purposes of territories?
5. What factors affect territory size?
6. What environmental conditions would be favorable to territoriality and under what circumstances would territoriality be disadvantageous?

* *From:* Dr. Seuss, 1955. *On Beyond Zebra.* New York: Random House.

7. What factors affect the selection of a habitat by organisms? In other words, how does an organism "choose" an appropriate habitat?

8. Discuss the competitive exclusion principle.

9. Discuss the concept of niche, including the realized and fundamental niches and n-dimensional hypervolume.

10. Why is the "guild" a useful way of grouping organisms?

Part Two

Natural History

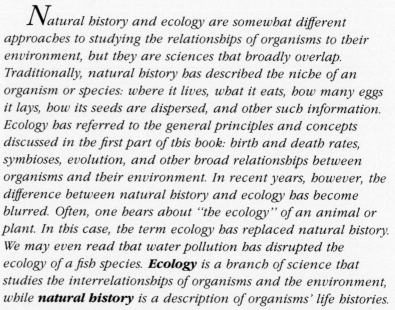

*Natural history and ecology are somewhat different approaches to studying the relationships of organisms to their environment, but they are sciences that broadly overlap. Traditionally, natural history has described the niche of an organism or species: where it lives, what it eats, how many eggs it lays, how its seeds are dispersed, and other such information. Ecology has referred to the general principles and concepts discussed in the first part of this book: birth and death rates, symbioses, evolution, and other broad relationships between organisms and their environment. In recent years, however, the difference between natural history and ecology has become blurred. Often, one hears about "the ecology" of an animal or plant. In this case, the term ecology has replaced natural history. We may even read that water pollution has disrupted the ecology of a fish species. **Ecology** is a branch of science that studies the interrelationships of organisms and the environment, while **natural history** is a description of organisms' life histories.*

It is only a mental exercise to try to separate the two terms in a meaningful way. The science of ecology depends upon natural history. Natural historians were the first to recognize that all organisms depend upon their biotic and abiotic

environment for survival; the more recent science of ecology organized natural history and developed general principles from that information. Ecology is often quantitative while natural history is descriptive; natural history can perhaps be described as ecology without numbers.

The beginning of all natural sciences consisted of casual observations. People saw that some trees lost all of their leaves in the autumn, some birds disappeared and reappeared with the seasons, and worms came out of the ground after a rainfall. More meticulous observations demonstrated such things as the fact that spiders inject their prey with a digestive chemical and suck the digested innards out; that many plants are dependent upon animals for pollination; and that rings on fish scales indicate the fish's age. Experimentation has revealed even more. Taking organisms from their habitat and placing them under controlled conditions in the laboratory has shown, for example, that a species of ragweed from Montana has a different tolerance to cold than the same species found in Texas. The defensive mechanism of Paramecium, its discharging of trichocysts, was found to be stimulated by mechanical or chemical means. It was learned that newly hatched geese

imprint *(instantaneously learn) that the first moving thing they see is their mother. These experimental observations could have been made under natural conditions only with great difficulty, if at all.*

Casual observation can uncover information, but the inherent lack of rigor in such a process may result in the inclusion of errors and the exclusion of important facts. Pliny (A.D. 23–79) wrote the 37-volume Natural History, *which included both fact and fiction. Myths about plants and animals are often based on a core of fact embellished to fill in the "facts" that observation could not provide. Manatees and dugongs (sea cows) became mermaids; the cotton plant was regarded as a half-plant, half-animal organism that gave rise to sheep; and the barnacle goose was originally presumed to arise from barnacles. Unicorns, bigfoot, the hoop snake (which supposedly put its tail in its mouth and rolled downhill to escape), and the transformation of horsehairs into worms are all myths created by incomplete observation. Closer, more methodical examination over longer periods of time eventually dispelled these myths (except bigfoot, which some people are convinced exists, in spite of a notable lack of evidence).*

On the basis of observations, scientists construct a **hypothesis,** *an idea about the cause of the observations, and test the idea with experiments under controlled environmental conditions. From many experiments a* **theory** *(a tentative explanation of the observed phenomena) is derived. The science of ecology is based on observations of natural history, so we might say that ecology can be defined as theories of natural history.*

This part of the book will briefly examine the natural history aspects of the major groups of organisms, with two goals in mind. The first is to show how organisms fit into the ecological framework; that is, how are they examples of ecological principles? The second is to demonstrate the incredible variety of organisms and point out both their similarities and differences. It is not intended to be a comprehensive survey of living organisms, but a potpourri of information. A more comprehensive examination of their structure, taxonomy, life cycles, and physiology of organisms can be found in most basic biology texts.

"What is man without the beasts? If all the beasts were gone, men would die from a great loneliness of the spirit. For whatever happens to the beasts soon happens to man."

—Chief Seattle, 1854

THE KINGDOMS OF ORGANISMS

*Most organismic surveys are based on evolutionary relationships. Organisms are grouped **phylogenetically**—by their presumed evolutionary history. This gives us a "family tree" of life, with the earliest evolved organisms lowest down on the tree. The lowest group, the kingdom Monera, consists of single-celled organisms that lack nuclear membranes and cellular organelles. Called **prokaryotes,** they may have evolved as early as 4 billion years ago. All higher organisms possess a nuclear membrane and organelles and are termed **eukaryotes**. Kingdom Protista consists of eukaryotic, single-celled organisms with varied forms and life histories, which first appeared on the earth about 1½ billion years ago. The eukaryotic and mostly multicellular organisms that comprise the kingdoms Plantae, Fungi, and Animalia began with a burst of rapid evolution in the Cambrian period. This "Cambrian explosion" started nearly 600 million years ago and lasted for about 10 million years, during which time most of the major groups of animals apparently evolved. By 130 million years ago, all of the life forms familiar to us had evolved.*

If we look at the representative forms of the kingdoms, we see that they can also be categorized into an ecological scheme. The kingdom Plantae consists mainly of the autotrophic organisms, and the kingdoms Fungi and Animalia of the heterotrophic ones. The heterotrophic fungi feed by absorption, and the heterotrophic animals feed by ingestion. Plantae generally contains the producers; Fungi, the decomposers; and Animalia, consumers and decomposers. The Protista and Monera cannot be so easily categorized, as they contain a heterogenous mixture of organisms. It is not necessary or desirable here to try to describe each group of organisms from a taxonomic or evolutionary standpoint, as this is done quite well in most elementary biology texts. Rather, we will look at the natural history of organisms in their roles as producers, consumers, and decomposers.

Classification of organismic groups is derived from several sources but is based on Whittaker's five-kingdom model. No attempt is made to include all groups, so the classification scheme used here is considerably simplified.

Chapter 12

Kingdom Monera

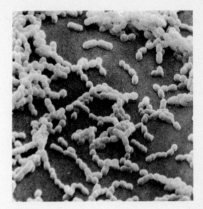

The kingdom Monera is composed of the prokaryotic organisms. In addition to the characteristics mentioned in the introduction, prokaryotes differ from eukaryotes in that the DNA of prokaryotes is in the form of a single DNA molecule, whereas in the eukaryotes the DNA is associated with proteins in complex structures called chromosomes. There are three major groups of Monera:

Division Schizophyta: bacteria
Division Cyanophyta: blue-green algae
Division Prochlorophyta

As a group, Monerans are of enormous significance. They are producers, consumers, decomposers, parasites, mutualistic symbionts, and organisms of commercial as well as ecological importance.

THE BACTERIA

Bacteria play a much larger role in the environment than their small size may indicate. These single-celled organisms decompose organic remains, are involved in soil production, incorporate free nitrogen from the air into nitrogen compounds, produce organic matter, cause diseases, and help to break down food in the digestive tracts of animals. In any environment where life is present, bacteria will also be found (Figure 12.1 and Exercise 12.1).

As Decomposers

Bacteria are enormously abundant in the air, water, soil, and in or on other organisms. Most are beneficial, and perhaps their most important function is that of reducing dead organisms and their waste products from complex organic compounds to simpler chemical substances that can be recycled into other organisms. Because bacteria are so plentiful, decomposition begins immediately after the death of an organism or the fall of a leaf, if not before. A body ripe for decomposition is rapidly invaded by bacteria (and other microorganisms) and because the nutrient level is so high in this fresh habitat, bacteria multiply rapidly.

Some bacteria are most suited to early stages of succession and others to later stages. We can measure their importance by their abundance. It has been determined that if 10^7 bacteria per milliliter are present, they are probably playing a significant role in that habitat. As decomposition progresses, the types of bacteria present change. Chemical changes, differences in humidity, and changes in temperature caused by microbial action result in successional changes in the populations of bacteria, just as higher organisms modify the environment to produce seral stages. For example, *Bacillus mesentericus* increases the water content as it degrades a

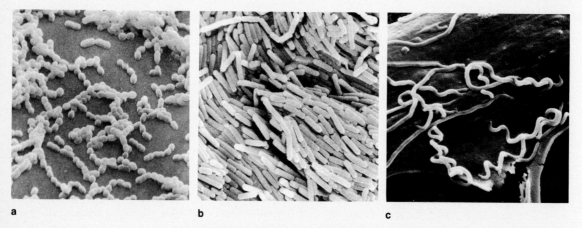

a   b   c

Figure 12.1 Typical Shapes of Bacteria. The three main shapes of bacteria are spheres, rods, and spirals. The spheres are known as cocci and may be found singly or in pairs, chains, or clusters. *Streptococcus*, seen in (a), is in a chain configuration. Rods (bacilli) (b) and the spiral **spirilla** bacteria (c) occur singly or in chains also.

piece of dry bread; as it does, it allows the growth of *Clostridium botulinium,** the bacterium that produces botulism toxin.

Since animals and plants live in environments rich in bacteria, they must possess mechanisms of defense against bacterial action.

---

* Scientific names are used here rather than common names, as has been the practice to this point, because these organisms, like many small or obscure creatures, have no common names.

Exercise 12.1 Detection of Microscopic Decomposers. *The ubiquitous presence of bacterial and fungal decomposers can be easily demonstrated by inoculating (see discussion that follows) nutrient agar plates with different sources of decomposers.*
Procedure: Group I. Collect about two tablespoonfuls of soil and place them in a beaker. Add just enough water to cover the soil. Mix thoroughly with a glass rod that has been sterilized by heating. Inoculate the petri plates with the mixture. Do not put soil directly on plate. Label as soil.

Group II. Collect a handful of fallen leaves that are partly decayed and place them in a large beaker. Cover them with a little tap water. Mix thoroughly with a sterile glass rod and inoculate the petri plates with the mixture. Label dead leaves.

Group III. Obtain several live leaves from trees and shrubs and place them in a beaker of water. Mix thoroughly and inoculate the petri plates with the mixture. Label live leaves.

Group IV. Take the petri plates outside, take the covers off, and expose them to the air for 15–20 minutes. Then cover and tape them shut. Label as environment.
Inoculation: To inoculate a petri plate, take a wire loop, heat it in a flame, let it cool, and then dip it in the mixture of water and organic material. The loop will pick up a small drop of the mixture. Run the loop lightly over the surface of the nutrient agar mixture in a zigzag line. Tape the top of the plate to the bottom of the plate and label. Incubate the plates at 30°C until the next class period (2–3 days). Examine the growth on the petri plates with a dissecting microscope. You will find two types of growth. The shiny, generally circular growths, usually white but very often cream, gray, pink, yellow, or red, are bacteria of various types. The white or gray fuzzy growths are fungi. Were the petri plates that were inoculated with your source of decomposers more infected than others? Where do you find decomposers most commonly?
To examine the bacteria

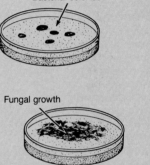

Bacterial colonies

Fungal growth

more closely, take a small bit of bacterial colony (using an inoculating loop) and place it in a drop of water on a microscope slide. Add a small drop of methylene blue to the slide and mix. Dry the slide by passing it through the flame of a bunsen burner (do not boil). Examine the slide under high power. The small rod or dotlike forms are various types of bacteria. The petri plates can be incubated under various conditions of temperature and moisture. Compare petri plates placed in a refrigerator with those incubated at room temperature.
Inoculate petri plates in which the gel-like agar has dried. What growth occurs on them?

In living organisms, bacteria usually occur only on the outside of the organism, where they soon die. Those that do invade internal parts are suppressed or killed by the immune systems of the organism. (Symbiotic bacteria, which live in the digestive tract of higher animals, are nonpathogenic. These will be discussed later.) But as soon as an organism dies, these defenses are no longer present, and decomposition begins quickly. A weak or injured organism is also susceptible to invasion, and the multiplication of bacteria already present begins. At this point, the bacteria assume the role of **pathogens** (disease-causing organisms).

As bacteria (and fungi) decompose dead plant or animal matter, chemicals are produced during this **putrefaction** that render the decomposing matter unappealing or toxic to larger decomposers such as insects or vertebrates. Hydrogen sulfide (a rotten-egg smell) is often produced, for instance. One scientist has suggested that the unappetizing process of decomposition, with its unappealing smells, is the way that microbes compete with larger organisms for food. Bacteria and fungi growing on a fallen peach may discourage larger animals from eating it.

As Pathogens

As pathogenic organisms, bacteria are filling roles as parasites. Bacteria cause crown gall on fruit trees, fire blight on fruit and ornamental trees, and tobacco angular leaf spot, as well as a variety of other diseases (Figure 12.2). In animals, various kinds of infections may cause effects from simple rashes to death. Diseases such as pneumonia, rheumatic fever, scarlet fever, food poisoning, gonorrhea, syphilis, and meningitis are caused by bacteria.

Bacterial diseases may serve as density-dependent population controls. Diseases that are communicable and spread from one individual to another either directly or by means of a vector such as a mosquito are spread more rapidly if the host population is dense. For example, tularemia, or "rabbit fever," infects rabbits, muskrats, and other small mammals, and may be epidemic in times of dense populations.

As Producers

Bacteria are often present near the base of a food web, and are eaten by protozoa, snails, arthropods, and other small creatures. Bacteria can ingest and incorporate small organic particles into their bodies, which can then be consumed by other animals. In this way, they are simply passing energy from one level to another as a con-

Figure 12.2 Crown Gall. Crown gall is a bacterial disease that often causes economic damage to pecan trees. Wartlike growths from a few inches to a foot or more in diameter form on the base of the tree trunk and larger roots. Here the disease is shown on nursery stock. Such trees should be burned to prevent spread of the disease.

sumer. There are some bacteria, however, that produce or fix energy in a manner similar to that of green plants.

Some bacteria are **photosynthetic** and are found in marine waters, sulfur springs, and on the surface of mud, where little or no oxygen is present. They produce organic compounds, but they do not contain the same type of chlorophyll or produce oxygen as do higher plants. Other bacteria are **chemosynthetic**, meaning that they derive energy from chemical bonds rather than from sunlight. **Ocean hot springs** or **geothermal vents**, at depths of 2½ kilometers below the ocean's surface, teem with life of various sorts, including species that are new to science. Heat from the vents with temperatures of 175° to 350°C heats the surrounding water, providing an ideal environment for the rapid growth of bacteria that use hydrogen sulfide to provide energy for food. These chemosynthetic bacteria are the base of the food chain in this system. Although it may seem that this is an exception to the rule that the sun's energy is necessary to drive all ecosystems, it is not, because the hydrogen sulfide is produced as a waste product of bacteria that metabolize organic matter that was formed through photosynthesis and eventually drifted to the ocean bottom. In any case, bacteria are crucial to the existence of these ocean hot springs ecosystems.

As Nitrogen Fixers

As we noted earlier, bacteria play a significant role in enriching the soil with nitrogen compounds. The best-studied case of this phenomenon is the legume–*Rhizobium* symbiosis. As you recall, legumes are members of the pea family that have nodules on their roots which house nitrogen-fixing bacteria. Since plants cannot use atmospheric nitrogen, nitrogen-fixing bacteria in the soil are essential in order to convert nitrogen from the air into nitrogen compounds that can be used by plants (Figure 12.3). Several types of bacteria fix nitrogen only when associated with higher plants, and a number of nonleguminous plants have root or leaf nodules containing nitrogen-fixing bacteria. But in the legume–*Rhizobium* relationship, the plant and the bacterium can fix nitrogen only when grown together. As the legume begins to grow in the soil, it secretes chemical substances conducive to *Rhizobium* growth, attracting bacteria from the soil, which then rapidly colonize the plant. The rhizobia infect the plant roots and produce nodules that are colonized by millions of bacteria. Growth of the bacterial colonies is stimulated by hormones produced by the plant's roots. The plant produces carbohydrates, which provide the bacteria with carbon and energy, and the bacteria fix nitrogen, which is used by the plant. The plant benefits by increased growth, which is especially marked in nitrogen-poor soils. Because of their higher nitrogen content, le-

Figure 12.3 *Rhizobium leguminosarum* in Nodules of the Garden Pea, *Pisum satiuum*. Normally rod-shaped, they form clusters of **bacteroids** in the nodules. Magnified 12,500×.

gumes are higher in protein than nonlegumes. Soybeans are well known for their high protein content.

Agricultural land is usually fertilized with nitrogen and/or phosphorus-containing chemicals. Legumes are sometimes grown between other crops or are grown alone and plowed under to increase the soil's nitrogen content. Twenty-five to 50 kilograms of nitrogen may be added to a hectare of soil by legumes. This is equivalent to adding 250 to 500 kilograms of commercial fertilizer!

Nitrogen-fixing is not restricted to bacteria or to associations with higher plants. Some blue-green algae and some bacteria capable of photosynthesizing can fix nitrogen in the absence of other plants. This is mainly true for aquatic ecosystems.

Other Bacterial Functions and Associations

Many types of bacteria (as well as other microscopic organisms) are associated with the roots of plants. The metabolic activity of the bacteria causes changes in the environment around the roots, altering the pH, temperature, oxygen concentration, and moisture. Roots excrete many organic substances that are used by bacteria and then converted into chemicals that other organisms can use.

Humans carry ten times as many bacterial cells in their intestinal tract as they have cells in their body. Most of these bacteria are found in the mouth and in the large intestine. Perhaps 30 million bacteria are concentrated in a small amount of saliva. Different types of bacteria colonize different portions of the digestive system and perform different digestive functions. They are so abundant that 20%–50% of the solid matter of feces is bacteria. All these bacteria are essential to digestion in humans and many other animals. Without them, we would become undernourished and die. People receiving antibiotics to combat an infection by pathogenic bacteria typically have intestinal upset because their normal bacterial population is considerably decreased and food cannot be efficiently digested.

Bacteria are especially important to mammalian herbivores, which depend on a diet of plants full of hard-to-digest cellulose. Cows, goats, deer, sheep, antelope, giraffe, and others have a special part of the stomach called the **rumen** in which bacterial digestion takes place. In mammals that have no rumen, bacterial digestion is not as efficient; rabbits eat their feces and thus pass food through their digestive system twice.

Some bacteria are luminescent, producing light in special organs of deep-sea fish. The fish are somehow able to regulate the amount of light these bacteria emit, suggesting that the "lights" are used by the fish to find their way in the dark waters, attract prey, or communicate with each other.

Human Uses of Bacteria

Bacteria are used in the production of foods, chemicals, and clothing; the decomposition of waste; and most recently as producers of medicinal drugs.

Most sewage is composed of organic wastes. Secondary sewage treatment, used by 75% of U.S. communities, consists essentially of bacterial degradation of the sewage. The material is decomposed, leaving a layer of dead bacteria and fine-particled debris called **sludge**. This sludge is dried and disposed of in landfills or used as compost or fertilizer in agricultural lands. Sometimes, under **anaerobic** (lacking oxygen) decomposition, methane gas is produced, which can be used to produce heat or as a source of fuel for machines.

The decomposition of carbohydrates by bacteria is used to produce food such as sauerkraut, pickles, yogurt, cheese, vinegar, and soy sauce. Alcohol for drinking or for fuel is produced by bacterial fermentation of organic waste such as corn husks and rice stalks. Some bacteria help to strengthen fabrics by combining chemical molecules. Bacteria in the ocean that assist in degrading oil spills are being tested for their ability to produce food from petroleum products.

In recent years scientists have been able to change the genetic content of microorganisms to enable them to produce useful chemicals. This "genetic engineering" has produced bacteria that efficiently manufacture growth hormone, insulin, and interferon, a potential cancer-fighting drug. In the past, insulin has had to be carefully extracted from the pancreas of slaughtered cattle, a tedious process. Bacterially produced insulin, certified by the U.S. Food and Drug Administration in 1982, is the first bacterially produced drug to be approved for human use. Purer drugs will be produced less expensively through genetic engineering, and bacteria may become the most important producers of medicine in the future. At present, they produce 20% of prescription drugs. Bacteria are obviously important to natural ecosystems and are rapidly gaining in importance in human systems.

THE BLUE-GREEN ALGAE

The blue-green algae are unicellular or filamentous organisms closely related to bacteria. In fact, microbiologists are tending more frequently to consider blue-green algae as a type of bacteria (Figure 12.4). These algae are common in fresh water, marine intertidal zones, in the soil, on tree bark, on wet rocks and ledges, and even in hot springs at temperatures up to 74°C (bacteria may survive at

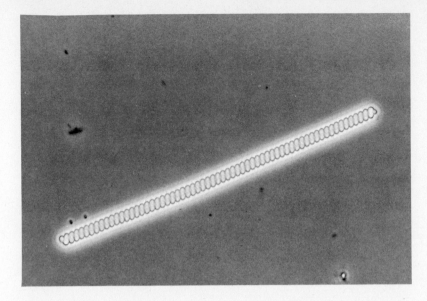

Figure 12.4 Blue-green Algae. The blue-green algae are the most primitive of all algae and are closely related to bacteria. They are found in a wide variety of habitats, both terrestrial and aquatic. Some species are dried and used for food. *Spirulina,* seen here, is being touted—with considerable commercial exaggeration—as a high protein wonder food.

temperatures up to 88°C). Some blue-green algae grow symbiotically with other organisms such as lichens, ferns, fungi, some flowering plants, sea anemones, and a variety of other animals. One species even grows in the hairs of the tree sloth, helping to camouflage this slow-moving mammal from its predators.

The blue-green algae, which contain the same type of chlorophyll as higher plants, are capable of photosynthesis, producing oxygen in the process. Having evolved perhaps 3 billion years ago, these algae were undoubtedly significant in enriching the earth's atmosphere with oxygen, making it suitable for oxygen-consuming forms of life. Some blue-green algae are able to fix atmospheric nitrogen and are common in the nitrogen-poor soils of some tropical areas. They grow well in rice paddies, for example, thus reducing the need for nitrogen fertilizer. They are often the first colonizers after a volcanic eruption. Spirit Lake, which was heated and polluted by the effects of the Mt. St. Helen's eruption in 1980, was devoid of any life until recolonized by blue-green algae and bacteria.

Algal "blooms" in freshwater lakes and ponds with high nutrient contents are due primarily to blue-green algae. In the spring, as the water warms and becomes conducive to growth, the algae photosynthesize, utilize nutrients, and reproduce rapidly. This results in a thick mat of green "scum" on the water, often thick enough to prevent sunlight from reaching the lower depths; when it is this thick, a die-off of other aquatic plants results. As the algae die and sink, processes of decomposition deplete the oxygen supply, killing fish and other animals. Over a period of years the lake will fill in with the organic matter produced. So-called biodegradable detergents are high in phosphates, and although they do not produce as many suds as conventional detergents, they add nutrients to the water and thus contribute to algal blooms. Most sewage is treated in such a way as to break down organic matter, but the treated water coming from the sewage plant is high in nitrogen, thus feeding blue-green algae in whatever body of water the treated effluent goes. Only 5% of all sewage treatment facilities are capable of removing nitrogen compounds. In situations where human communities derive their water from supplies containing blue-green algae, these tiny organisms clog filtration systems.

The scum of green slime you may find floating on a bird bath, growing on the side of your flower pots, or making your favorite swimming hole look less than inviting was very important to the evolution of higher forms of life and still plays a significant role in both natural and human ecosystems.

THE PROCHLOROPHYTA

The prochlorophyta are very similar to the blue-green algae except that the prochlorophytes have a photosynthetic pigment system closer to that of the eukaryotic green algae. In some ways they are intermediates between the blue-green algae and higher plants. They were discovered only in 1976, growing in association with a marine invertebrate (tunicates), and very little is known about them. As is true for many organisms that are not well known, they have no common name.

SUMMARY

The kingdom Monera is composed of the prokaryotic organisms. The three major groups are the bacteria, blue-green algae, and the prochlorophytes. This group contains producers, consumers, decomposers, parasites, mutualistic symbionts, and organisms of commercial importance.

STUDY QUESTIONS

1.   Define prokaryote, eukaryote, pathogen, chemosynthetic, sludge, anaerobic.
2.   Compare prokaryotes to eukaryotes.

3. Why do you think prokaryotes have been so successful, having evolved hundreds of millions of years before the eukaryotes, and still surviving essentially unchanged?

4. In what ways do bacteria interact with humans?

5. Explain the process of bacterial decomposition. What environmental factors affect the speed of decomposition?

6. Discuss the role and importance of nitrogen-fixing bacteria. Give specific examples.

7. What are the differences between bacteria and blue- green algae? Do you think there is reason to consider blue-green algae as simply a type of bacteria?

8. Give some examples of symbiosis among the prokaryotes.

9. Explain how bacteria are used to treat sewage wastes.

10. Antibiotics are used as medicines to kill disease-causing bacteria. What side effects can their bacteria-killing ability cause?

Chapter 13

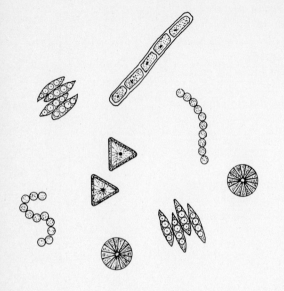

Kingdom Protista

The kingdom Protista consists of unicellular and simple colonial eukaryotic organisms. The membership of the Protista is not generally agreed upon, as almost all authors include a different set of organisms. One of the reasons for the existence of this kingdom is to have a place for organisms that are clearly neither plant nor animal, such as the slime molds. For present purposes we will consider the kingdom Protista as being composed of the following:

Division Chrysophyta: yellow-green and golden-brown algae
diatoms
Division Pyrrophyta: dinoflagellates
Phylum Protozoa: protozoa
Division Myxomyceta: slime molds

This is a very conservative and simplified classification scheme; other authors may include 15 or more categories here.

Major categories in the old system of two kingdoms, Plants and Animals, were termed Divisions and Phyla, respectively. Even though these groups have been rearranged into different kingdoms, the older terms are still used.

THE CHRYSOPHYTA

Yellow-green and Golden-brown Algae

The photosynthetic yellow-green and golden-brown algae are so named and are classified in a different group from other algae because of their characteristic pigmentation. They are mainly inhabitants of fresh water, but some are found in terrestrial and marine habitats. They grow primarily on the muddy shores of streams, lakes, and ponds; on the mud flats of estuaries; and on the bottom mud of tidal flats. They may also grow on damp soil in agricultural fields, on rocks, or in the form of mats floating on ponds where filamentous forms provide habitats for many small organisms. Although there are significant morphological differences between the two groups, they are ecologically very similar (Figure 13.1).

Diatoms

Diatoms are encased in rigid silicon shells that look like fragile sculptures of glass. They are found in both fresh and salt water, growing on the bottom mud, or encrusted on plants, pier pilings, mollusc shells, and even whales. They may become so dense as to cover everything in a tidal zone with a brownish growth. And they are so diverse that hundreds of species may be found in one sample of mud (Figure 13.2).

Figure 13.1 Yellow-green/Golden-brown Algae.

Exercise 13.1. Identifying Phytoplankton.
Virtually all organisms can be identified through the use of a dichotomous (branching in two parts) key. Plankton, which contains many kinds of small organisms, especially algae, can be collected from fresh or salt water with a fine-mesh plankton net. Your instructor can guide you through the following simple key to some phytoplankton.

Green, yellow, or brown pigment present:
1. Bluish-green pigment scattered through cells—blue-green algae

2. Pigment yellow-green or brownish:
a. Cells crystallike, composed of two halves—diatoms

b. Cells not crystallike, not composed of two halves—golden-brown algae

3. Pigment green, brown, or red:
a. Transverse groove present, two flagella present—dinoflagellates

b. No pigmentation present—zooplankton

Figure 13.2 Diatoms.
Diatoms secrete a two-part capsule of silicon around their bodies. If all the earth's diatoms could be weighed, their combined weight would exceed that of the biomass of any other group of organisms, even though diatoms are only fractions of millimeters in length.

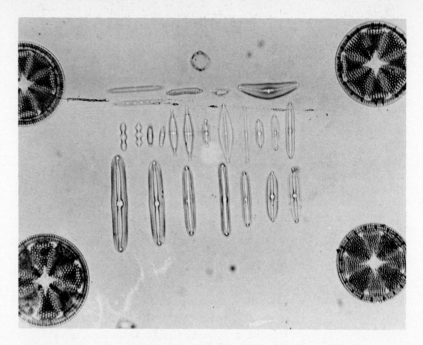

When diatoms die, their shells sink to the bottom and may accumulate in large concentrations. This aggregation of shells is commercially "mined" as diatomaceous earth and used as a mild abrasive in shoe polish, toothpaste, furniture polish, and window cleaning preparations; as insulation; and in filters and paint remover. They may also have played an important role in the formation of oil deposits below the ocean floor.

All algae are important in their role as producers of oxygen and organic matter, but the extremely high abundance of diatoms (perhaps a million per liter of seawater) makes them a significant part of the food chain.

THE DINOFLAGELLATES

Dinoflagellates are small, usually one-celled organisms. Most are photosynthetic but some species are heterotrophic, feeding on small pieces of organic matter. A few species are parasites on marine organisms. They are abundant in both fresh and salt water and are nearly as important as diatoms in the role of producers (Figure 13.3).

Dinoflagellates are perhaps best known as the cause of "red tides," which kill enormous numbers of fish. The numbers of dinoflagellates are so great during these peaks of abundance that the

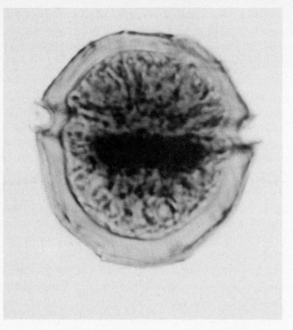

Figure 13.3 Dinoflagellate. Some dinoflagellates are enclosed in a silicon or calcuim shell and others are enclosed in a cellulose "envelope." They move about with flagella that extend out grooves in the shell.

ocean appears red during the day and phosphorescent at night. These organisms contain a toxin that is poisonous to fish and that accumulates in shellfish, killing them or making them inedible. Red tides are most common in the Gulf of Mexico, but they also occur on the California coast and elsewhere.

THE PROTOZOA

There are many thousands of species of protozoa, usually single-celled, although many forms are colonial. They are usually small but some can be seen by the naked eye. They are found virtually everywhere and fill a wide spectrum of ecological niches. Some are producers, some consumers, and some decomposers. They serve as predators, parasites, and mutualistic symbionts. They are found in fresh and salt water, in the soil, and in and on other organisms. Protozoa are probably the best known of the Protista, as well as the most diverse. A few drops of stagnant pond water will contain a wide variety of protozoa. They are also very abundant; a kilogram of soil may contain 1 to 10 million individuals.

Protozoa are well-adapted creatures and much more complex than they seem. They can react to light, temperature, and chemical stimuli. Some reproduce sexually, some asexually, and many are capable of both. Under unfavorable conditions of drought, extremes of temperatures, and food shortages, they can become dormant cysts. Their classification and ecological roles are largely based on their method of locomotion: flagella, cilia, amoeboid movement, or lack of locomotory organelles.

The flagellates are protozoa that possess one or more long, whiplike structures that push or pull them through the water. *Euglena* and *Volvox* (Figures 13.4a and b) are examples. The ciliates possess many short fibers called cilia, which work like many tiny oars to move the organism or produce a current to draw food particles into the protozoan's mouth. *Stentor* uses its cilia for feeding (Figure 13.4c). Members of the amoeba group move by flowing via extensions of their bodies' cytoplasm. These extensions, called **pseudopodia**, or false feet, also engulf prey. Some amoeba are encased in shells, the pseudopodia extending through holes for feeding and movement. The group that has no specialized organs for movement is the entirely parasitic sporozoa. Members of this group move by being transmitted from one host to another, traveling through the circulatory system of the host and absorbing nutrients through their cell membranes.

As Producers

Some protozoa contain **chloroplasts**, green-colored bodies containing chlorophyll, which enable the organisms to photosynthesize. *Euglena* is a photosynthetic protozoan that also feeds on organic matter; this latter ability enables it to live without light. It is also equipped with a flagellum for movement, and was once claimed by both botanists and zoologists because it contains features of both plants and animals. Most *Euglena* live in fresh water, but some live in the soil or even in the intestines of animals. As primary producers, *Euglena* and other photosynthetic protozoa are of minor ecological importance because they are not very abundant.

As Consumers

Most protozoa are heterotrophs, feeding on particulate organic matter, on smaller organisms, or as internal parasites of larger organisms. Typical foods include bacteria, algae, and smaller protozoa.

Protozoa are also responsible for many diseases. Sleeping sickness is caused by *Trypanosoma,* malaria is caused by various species of *Plasmodium,* and several intestinal disorders are due to some species of *Entamoeba, Giardia,* and *Trichomonas.* Like some bacteria, some protozoa (*Entamoeba coli,* for example) are normal and harmless inhabitants of the human digestive tract and aid in digestion. Since single-celled organisms evolved much earlier than the forms they parasitize, they must have been free-living at first, and

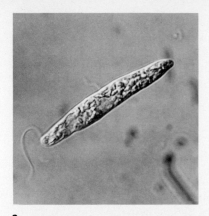

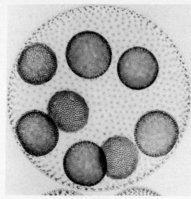

a   b

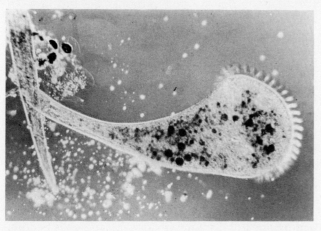

c

Figure 13.4 Protozoa. (a) A one-celled, flagellated photosynthetic protozoan, *Euglena*. (b) A multicellular, ciliated, photosynthetic protozoan, *Volvox*. (c) *Stentor*.

only later evolved into a parasitic niche. Think of a harmless form of protozoa, living in pond water and being constantly ingested by larger animals drinking the water. If some individuals survive in the animals' guts, they may ultimately adapt by evolution to live there as symbionts. Many species of protozoan parasites are very specific, inhabiting only particular organ systems of certain animals. Their role as disease-causing organisms is not insignificant and may play a role in the population control of the animals they infest.

As consumers, protozoa form a major part of some food chains. We should make a distinction here between saprozoic and heterotrophic means of nutrition. **Saprozoic** consumers are animallike heterotrophs that ingest particles of organic matter and other nutrients from the abiotic environment. The term heterotrophic usually implies that the organism is obtaining its food directly from another organism. Many protozoa are capable of saprozoic nutrition and thus help to recycle matter back into the food web.

Some species of protozoa have formed mutualistic relationships. Some contain green algae, which produce carbohydrates for the protozoan and receive shelter in return. Protozoa living in the gut of termites convert wood cellulose, which the termites cannot digest, into sugar for these insects and benefit in turn by having a protected home.

THE SLIME MOLDS

The slime molds have been a mystery to many biologists because of their combination of animal- and plantlike characteristics. Thus, they are prime candidates for inclusion in the kingdom Protista (Figure 13.5). Slime molds are heterotrophic and function primarily as decomposers of dead organic matter in moist environments. Their body is essentially a slimy-appearing mass of protoplasm that streams over the substrate at a rate the naked eye can detect (up to 1 millimeter/second). As they stream, they can ingest bacteria and protozoa as well as nonliving nutrients.

Slime molds, like bacteria, are found virtually everywhere. The spores that land in a suitable habitat begin to form the mass of protoplasm known as a plasmodium. Plasmodia of slime molds are particularly abundant on damp forest floors, especially on the bark of dead trees. Some are quite colorful—yellow, orange, or blue—but many are white or black. The spores become amoebalike structures that move like protozoan amoebas, wiping up bacteria from the surface of rocks, twigs, or soil. As the bacteria become scarce and the environment begins to be unsuitably dry, the amoebas become attracted to each other, fuse, and form a plasmodium that forms spores to begin the life cycle when the environment becomes favorable again.

Although some slime molds cause plant diseases (such as cabbage clubroot) and feed on bacteria, their major role is as nutrient recyclers. Compared to the other groups of decomposers, however, slime molds are much more interesting than they are ecologically important.

SUMMARY

The kingdom Protista consists of unicellular and simple colonial eukaryotic organisms. Major groups of Protista are: yellow-green and golden-brown algae, diatoms, dinoflagellates, protozoa, and slime molds. The algae and diatoms are photosynthetic, the protozoa may be producers, consumers, decomposers, or pathogens, and the slime molds are decomposers and predators.

a

Figure 13.5 Slime Mold.
Slime molds have some
characteristics of animals,
such as active movement and
feeding by engulfing food
particles, and some
characteristics of fungi, such
as spores. (a) Spore-
producing bodies
(sporangia) of *Lepidoderma
tigrinum* magnified 1000✕.
(b) Slime mold mass.

b

STUDY QUESTIONS

1. Define pseudopodia, plasmodium, saprozoic, chloroplast.
2. List the distinguishing characteristics of the yellow-green and golden-brown algae, the diatoms, and dinoflagellates.
3. List the characteristics of the protozoa. With what characteristics is the protozoan group subdivided?
4. Give examples of the various ecological roles that protozoa fill.

5.   List the characteristics of the slime molds. Could they perhaps be considered a type of animal?

6.   *Euglena* is photosynthetic but also feeds on organic matter. Being both an autotrophic and heterotrophic organism would seem to be advantageous. If it is, why aren't more organisms filling both roles?

7.   In what ways are protozoa more specialized than bacteria? In what ways are bacteria more specialized than protozoa?

8.   Describe the life cycle of a slime mold.

9.   Why is the parasitic life style so common among protozoa?

10.   What is the "red tide"?

Chapter 14

Kingdom Fungi

Because of their immobility and plantlike form, the fungi were classified as plants for many years. But because their bodies are made of chitin rather than cellulose (chitin is a combination of protein and sugar that makes up the exoskeleton of insects as well as structures in other animals), because their cell walls are absent or incomplete, and because they are not photosynthetic, fungi are fundamentally different from the green plants and are now considered as a separate kingdom. All are heterotrophic and most are saprophytic, but a large number are parasitic, a few are predaceous, and some are economically important. Most require a moist environment and are particularly abundant in humid habitats. Some forms are economically important. Fungi include not only mushrooms, but mildew, mold, rust, smut, blight, yeast, and together with algae they form lichens (Figure 14.1)

The saprophytic fungi release digestive enzymes onto the organic matter on which they are growing and absorb the nutrients as they are released. Most fungi produce threadlike structures called **hyphae** that grow rapidly to produce a fuzzy-appearing structure called a **mycelium**. As the hyphae grow and absorb the nutrients, the fungus grows so rapidly that its growth may sometimes be observed by our eyes. Fungi are perhaps equally as important as bacteria in recycling the nutrients contained in dead organic matter.

Like bacteria, fungi are ubiquitous, and as soon as organic matter dies it is infected with fungal spores, or the fungi already on it

Figure 14.1 Shelf Fungus. This fungus is named after its shelflike arrangement. The "shelves" are woody and flexible when moist, and stiff when dry. It is most often found on dead wood, from which it absorbs nutrients. The concentric lines each represent a year's growth.

begin to grow. Fungi reproduce asexually by spores (but also sexually by other means), producing enormous numbers of spores—in the millions or billions. The spores are small and light, are often carried long distances by wind or water, and can be found virtually anywhere.

MAJOR GROUPS OF FUNGI

The fungi can be grouped into four major categories:

Class Phycomycetes: bread molds
Class Ascomycetes: sac fungi
Class Basidiomycetes: club fungi
Class Deuteromycetes: imperfect fungi

All of these are called "true fungi" to distinguish them from the slime molds. The classes of true fungi are divided according to their method of reproduction.

Phycomycetes

The phycomycetes (or zygomycetes), the algal fungi, reproduce sexually or asexually. Spores are produced at the ends of filaments called hyphae and are dispersed by the wind. If the spores land in a warm, moist environment, they germinate and produce hyphae. *Rhi-*

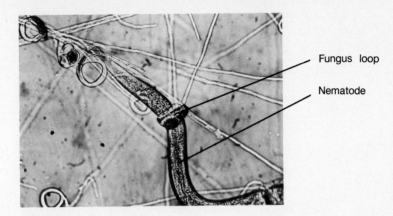

Figure 14.2 **Phycomycetes**. Photomicrograph of nematode trapped by fungus (×150). See text for explanation.

zopus nigricans, the black bread mold, is the best known member of this group, and grows on cheese, fruits, vegetables, and a variety of other foods besides bread. The mycelium grows rapidly through the food, absorbing nutrients in the process. Keeping foods in the refrigerator reduces the chances that spores will land on the food and inhibits the growth of spores that do. Many foods, especially processed breads, are treated with a fungicide to inhibit the growth of mold.

One species of phycomycetes feeds on soil nematodes (round-worms) that are parasitic on the roots of plants. This mold produces loops on its hyphae, and when a nematode accidentally wriggles through the loop, the cells of the loop expand and trap the worm. Small fungal fibers then invade the worm and digest it. Interestingly, this fungus will not produce the loops if no nematodes are present in the soil (Figure 14.2).

Ascomycetes

The ascomycetes are called sac fungi because their spores are produced in long sacs. This group includes the yeasts, molds, mildews, truffles, ergot, and cup fungi (Figure 14.3). *Aspergillus* is a common ascomycete that causes rotting of various fruits, tobacco, and even clothing under humid conditions. *Penicillium* fungi are also very common, being found on fruits, paper products, clothes, and other organic matter. This fungus is also the original source of the antibiotic penicillin. *Ergot* is a fungus that attacks rye grain, producing toxic chemicals in the process. Humans eating infected rye contract a condition called **ergotism**, which is characterized by convulsions and hallucinations and may be fatal. Centuries ago demented women who were accused of being witches may in fact have been victims of ergot fungus.

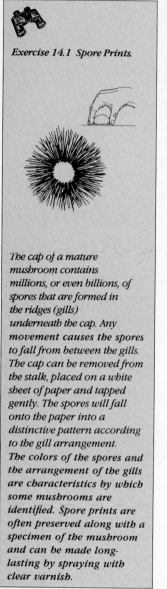

Exercise 14.1 Spore Prints.

The cap of a mature mushroom contains millions, or even billions, of spores that are formed in the ridges (gills) underneath the cap. Any movement causes the spores to fall from between the gills. The cap can be removed from the stalk, placed on a white sheet of paper and tapped gently. The spores will fall onto the paper into a distinctive pattern according to the gill arrangement. The colors of the spores and the arrangement of the gills are characteristics by which some mushrooms are identified. Spore prints are often preserved along with a specimen of the mushroom and can be made long-lasting by spraying with clear varnish.

Figure 14.3 An Ascomycete: Peach Leaf Curl Fungus. This fungus infection causes the leaves of some fruit trees to curl up, form knobby protrusions, and ultimately fall off. This reduces the growth and fruiting potential of the tree and will eventually kill it. It can be controlled by several sprayings of a copper-containing fungicide applied before the leaves appear.

Basidiomycetes

The basidiomycetes, or club fungi, produce spores in club-shaped cells (Exercise 14.1). This group includes the smuts, rusts, mushrooms, puffballs, bracket fungi, and jelly fungi. The smuts and rust cause diseases of agricultural crops, while the others are harmless but essential decomposers. Some mushrooms and all puffballs are edible (Figures 14.4a and b).

Deuteromycetes

The fungi imperfecti are so called because they apparently have no form of sexual reproduction as do the other groups. Some are soil organisms, and others are parasitic, causing a number of diseases such as ringworm and dermatitis in humans. Some scientists include *Aspergillus* and *Penicillium* in this group (Figure 14.5).

a   b

Figure 14.4 Basidiomycetes. (a) This shaggy mane mushroom is common along roadsides, in meadows, open forest, and grassy lawns. Although it is an edible mushroom, it may cause severe illness if eaten with alcohol. (b) Upon maturity, the outer skin of a puffball ruptures and releases millions of spores. Drops of rain, wind, or the touch of an animal create ''puffs'' of spores. This puffball was tapped by the author's finger shortly before the photo was taken. One species of puffball may grow to a weight of 10 kilograms and produce several trillion spores. All puffballs are edible if picked while young and white inside. They are prepared by steaming before frying in butter.

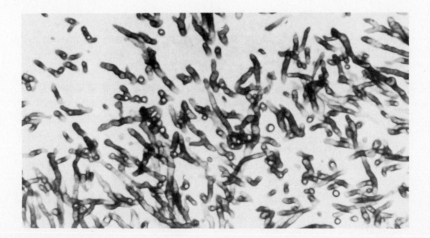

Figure 14.5 Fungi Imperfecti. Imperfect fungi reproduce by means of asexual spores. One of these fungi causes the common foot ailment known as athlete's foot, which causes itching of the skin and discolored and malformed toenails.

AS PARASITES AND PATHOGENS

Fungi are the cause of many plant and several animal diseases. Their spores may be distributed by the wind or carried by animals. They may parasitize their hosts either internally or externally. Occasionally, fungal parasites have had significant ecological or economic effects. The potato famine in Ireland in the 1840s was caused by fungal blight that destroyed the potato crop, for example. Chestnut blight and Dutch elm disease have had devastating effects on trees in the eastern United States, and there is no effective treatment for either. Corn smut appears as dark blisters on the ears, stems, and leaves of corn, and infects other grains. Some smuts can be prevented with mercury compound coatings on the seeds. Rust, so-called because of its appearance, may affect pine trees, apples, and grain. Wheat stem rust has been partially overcome by developing resistant wheat varieties, but new varieties of wheat rust are evolving that can infect the resistant wheat.

Athlete's foot, jock itch, ringworm, and a variety of other minor skin irritations are due to fungal infections. Most are minor and are easily cured. In rare cases, however, when a person's immune system has been weakened or destroyed by disease or heavy chemotherapy, normally harmless fungal spores can multiply rapidly and cause rapid debilitation or death. Bread mold, corn smut, and wood rot fungi have all been implicated in these unusual, and untreatable, infections. Generally, the fungi that decay our food can simply be scraped or cut off with no harm. There are a few species of fungi that are normally pathogenic to humans, however; some cause respiratory infections, and one type of yeast infects the nervous system and causes meningitis.

AS MUTUALISTIC SYMBIONTS

Fungi live in association with a number of insects, in arrangements that are either very loose or highly organized. In one case a flylike midge lays her eggs in the stem of a plant, causing the plant to form galls. These galls contain fungal spores, probably deposited along with the eggs. The developing insect larvae do not feed on the fungi, but on the plant tissue, which is broken down by the fungus. Each midge species is associated with only one fungus species. One fungus forms a thick mat that is inhabited by scale insects. The scale insects receive protection in the mass of fungal filaments, and the fungal filaments attach to some of the insects and suck nutrients from them. Some termites and ants "cultivate" fungi by bringing spores into their nest and lining the fungus garden with saliva and dirt. Unlike termites that have protozoa in their gut, these termite species have none, and so rely on the fungi to break down the cellulose in the wood they eat. The fungus itself is also eaten. The fungus-growing ants collect plant and animal debris, cut it into small pieces, and add them to the fungus garden. When the queen ant flies elsewhere to establish a new nest, she takes with her a pellet of fungus with which she begins a new garden. Fungus-gardening insects, which also include flies and beetles, secrete or excrete substances that prevent the spores of nonsymbiotic fungi from growing in their gardens.

Some fungi live in close association with the roots of higher plants. The fungi apparently increase the plants' uptake of both water and nutrients. An attempt is being made to mass produce these fungi in commercial laboratories. After culturing, they will be injected into the soil around tree roots. It is expected that the survival of pine and ornamental tree seedlings can be increased by 35% with this method.

HUMANS AND FUNGI

Fungi are eaten and used to prepare drugs, foods, and chemicals. Roquefort and Camembert cheeses are produced by bacteria, but the flavor of many other cheeses is due to species of *Penicillium* fungi. Species of this fungus are also used to produce the antibiotic penicillin, the hallocinogenic drug LSD, industrial alcohols, and food dyes. Yeasts are used in brewing, wine production, and baking. (Dough rises because the yeasts produce carbon dioxide as they respire.)

Of about 5000 species of mushrooms in the United States, only about 100 are known to be poisonous. A number are nonedible because of their taste or partially poisonous effects on some people. A few, such as some *Amanita* species, are extremely toxic, causing death in 24 to 48 hours (Figure 14.6). In spite of many old wives' tales, there is no sure way to tell poisonous from nonpoisonous mushrooms. (There is no botanical difference between "toadstools" and mushrooms.) Every mushroom fancier has a particular method to tell the edible from the poisonous, but the only sure way to avoid eating the wrong mushroom is to be skilled enough to identify each species. There's something to be said about the phrase, "There are old mushroom hunters and bold mushroom hunters, but there are no old, bold mushroom hunters."

Most mushrooms are not particularly nutritious, but they are comparable to most garden vegetables and tend to be high in vitamins B and C, copper, and iron, and low in calories. By weight,

Figure 14.6 The Amanita Mushrooms. Shown here is *Amanita muscaria,* or the fly amanita. This particular mushroom is seldom fatal but can produce hallucinations if ingested. In fact, some natives of northern Russia eat this mushroom for its hallucinogenic properties. Its close relatives, such as *A. phalloides* and *A. verni,* are extremely toxic, however. Less than a teaspoonful of these mushrooms causes paralysis of the heart, liver, and kidneys, resulting in death within two or three days in over half the people so poisoned. Many amanitas are aptly called "death angel".

however, 100 times more protein is yielded by a hectare of mushrooms than a hectare of cattle. Unfortunately, recent evidence indicates that mushrooms tend to accumulate environmental pollutants, so we may be making even the edible varieties nonedible.

One of the most sought after, hard to find, and thus expensive edible fungus is the truffle. Truffles grow slightly below the ground in association with oak tree roots in the forests of southern France. They are hunted by the French with pigs trained to detect the truffle's odor. Truffles contain large quantities of a chemical that is also found in boar's saliva, and this serves to elicit a sexual response in the sow. This same substance is secreted by the armpit sweat glands of men and in the urine of women. The ability of pigs to locate truffles and the human predilection for them may thus have a sexual basis.

Fungi are deserving of more respect than the average person bestows upon them. Rot, mold, and mildew seem like unpleasantries, but they are simply part of the fungal world that plays a crucial part in decomposing, digesting, and recycling matter.

Figure 14.7 Foliose Lichen. This lichen, growing on a dead branch of an oak tree, exhibits the foliose growth form. The threadlike parts of the lichen (*hyphae*) attach it to the branch and produce the acids that decompose the branch. A lichen can reproduce by fungal spores that later combine with algal spores, by parcels containing both types of spores, or by fragmentation—a piece of a lichen can be broken off and land in another suitable environment.

LICHENS

Lichens are difficult, if not impossible, to classify because they appear to be one kind of organism, yet they are actually a composite of two—an alga and a fungus. They are a very diverse group, widely distributed, have many distinctive characteristics, but are not very well known. They are found on rocks, tree bark, hanging from tree branches, and on hard soil. They are abundant in the arctic tundra, some species being called "reindeer moss."

Lichens may be found in a leafy form (**foliose**), a crustlike form (**crustose**), or hanging or erect forms (**fruiticose**) (Figure 14.7). These morphological forms are not indicative of their taxonomic relationships, and closely related species may differ morphological-ly. The body of the lichen, called a **thallus**, is the fungus. The alga grows in a layer on the fungus. Typically, one species of alga is associated with a particular fungus, indicating that there has been coevolution between the organisms. Most lichens are gray or brown when dry, but turn a greenish hue when the algae are wet. Many lichens contain colored pigments—green and yellow are common, and red, blue, and violet colors are less common.

In nature and in the lab, both the fungal and algal species can exist separately, but in nature the fungus becomes "lichenized" when it is colonized by spores of algae. The algae photosynthesize and some of the food produced is used by the fungus. The fungus provides a moist surface on which the alga can grow.

Lichens were found to be something other than individual plants only about 100 years ago, and about 60 years ago scientists were speculating that lichens may represent a parasitic relationship, with the fungus feeding on the alga. Experiments have shown that some fungi will infiltrate and kill alga of a species not normally associated with that fungus. It appears that the usual algal symbiont has evolved some resistance to the fungus with which it associates, in a sort of controlled parasitism.

The lichen grows at a very slow rate, increasing in diameter by perhaps only 2 millimeters per year. Lichens are very resistant to extremes in temperatures and desiccation. Experiments have shown that lichens may be completely dried out for over a year and begin to metabolize again when moistened. In their environments they may be exposed to temperatures well below freezing, and they have been experimentally subjected to temperatures of 100°C without permanent damage.

Given their hardiness, it is no surprise that lichens are often the first organisms to colonize barren rocky areas and are most noticeable in the high latitudes of the Arctic. The only plantlike organisms that grow in the Antarctic interior are lichens. Lichens have the ability to penetrate bare rock with hyphae and break down the rock by producing various weak acids (Figure 14.8). Decomposition of the rock provides nutrients for the lichen, allows the lichen to gain a better foothold, and also begins the soil-making process. Many pioneer lichen species are capable of fixing nitrogen because the thallus houses nitrogen-fixing algae. Lichens, being such slow growers, are easily outcompeted by mosses and higher plants and ultimately

Figure 14.8 Crustose Lichen. This photo shows a flat crustose lichen growing on a granite rock—certainly a very harsh habitat.

covered. There are a number of species that are epiphytes on trees. These lichens are not pioneer plants and their habitat is not nearly as severe as that of their colonizing relatives.

Lichens harbor a number of creatures, among them protozoa, roundworms, insects, and snails, all of which use the lichen for food or shelter. Ants, snails, and butterflies are known to aid in dispersal of the spores of the fungus and alga.

Some insects, frogs, and salamanders are colored so as to look like a piece of bark-covered lichen. One species of lichen grows on the shells of tortoises that live on the Galápagos island of Santa Cruz. Dozens of bird species use lichens in building their nests, often providing an effective camouflage. The caribou of North America and the reindeer of Europe (which are the same species) rely heavily on lichens as a winter food course, scraping away snow with their hooves to expose the lichens. Deer, mice, and marmots have also been known to eat lichens. Some species of bats roost in hanging lichens.

Humans and Lichens

Under adverse circumstances, people of the Middle East and the northern latitudes have eaten lichens for sustenance. Lichens are considered a delicacy in Japan. Lichens have been used by many cultures worldwide to produce dyes for clothing or cosmetics. Litmus paper, used to test the acidity of solutions, is made from litmus extracted from lichens. For hundreds of years, lichens were the base for some medicines, their effectiveness substantiated recently by the discovery of an antibiotic substance they contain.

In addition, lichens have the ability to accumulate radioactive materials and metals put into the environment by industry; thus, they are reliable indicators of air pollution and are used to monitor pollution levels in some industrial areas. In addition to measuring air quality deterioration, lichen recolonization of an area denuded by air pollution is being used as an indicator of air quality improvement.

SUMMARY

Fungi are nonphotosynthetic, nongreen plants that are characteristic of moist habitats. The major groups of fungi are the bread molds, cup fungi, sac fungi, and the imperfect fungi. All are saprophytic or parasitic and some are important as food, the cause of diseases, or commercial products. Lichens are an organism formed by the close association of an alga and a fungus.

STUDY QUESTIONS

1.   Define mycelium, ergotism, hypha, foliose, thallus.
2.   Describe the chemical and physical means by which fungi operate to decompose organic matter.
3.   List the characteristics of the four major groups of fungi.
4.   Describe some of the diseases caused by fungi.
5.   Describe some of the symbiotic relationships in which fungi are found.
6.   What are some of the human uses of fungi?
7.   What are the characteristics of lichens?
8.   What evidence is there to indicate that the alga and fungus forming a lichen have a mutualistic relationship? What evidence indicates that this is a parasitic relationship?
9.   Describe the three growth forms of lichens. How is each adaptive?
10.   What are some of the human uses of lichens?

Chapter 15

Kingdom Plantae

The organisms in this group are considered true plants by most biologists, but again there is not total agreement on the inclusion of all the groups. Sometimes all the algae, fungi, and the slime molds are included, sometimes not. But whatever the classification, this kingdom includes an enormous variety of organisms that are found in virtually every habitat (Figure 15.1).

Often, we refer to the "lower" plants and the "higher" plants. The lower plants include the algae, mosses, liverworts, and their relatives; and the higher forms, the coniferous trees and the flowering plants. Intermediate groups such as ferns, horsetails, and club mosses are usually, but not invariably, considered lower plants. The terms "lower" and "higher" give a misleading impression, as they imply that lower organisms are somehow less specialized or less well adapted to their environment than higher ones. Actually, they are only lower in the sense that they evolved earlier and are lower down on the "family tree" (Figure 15.2). Algae and bacteria may be simpler organisms than orchids and trees, but they are nonetheless successful. We might even consider lower plants greater successes than higher ones because many lower plants existed for hundreds of millions of years before the higher plants came on the scene.

There have been some major steps in the evolution of plants from lower to higher forms, primarily adaptations to terrestrial environments. For example, spores, which need to land in optimal habitats and conditions to germinate, were replaced by seeds, which

Figure 15.1 Plants and Habitats. Plants are fundamental components of the environment. They provide food and shelter for animals, modify and add to the soil, prevent erosion, and buffer the effects of wind, precipitation, and temperature. Habitats are often defined by the type and structure of plants which in turn determine the kinds of animals that will be found there. This photo shows smoke trees in a desert wash in southern California.

contain a food supply and a protective seed coat to help them become established under less than optimal conditions. Also, spores often require water for fertilization or dispersal; seeds do not need water for fertilization and have a variety of dispersal mechanisms. Higher plants have a true **vascular** (circulatory) system to absorb and distribute water and nutrients throughout the body of the plant; most lower plants depend on diffusion through their external surfaces. The vascular system, along with "woody" stems, has allowed plants to reach enormous heights and girths; giant sequoia trees, for example, are the largest living organisms (Figure 15.3). The most recently evolved plants are the flowering plants. Flowers serve to attract birds, insects, and other animals that pollinate plants, and thus pollination is typically not dependent on wind and water as it is in the lower plants.

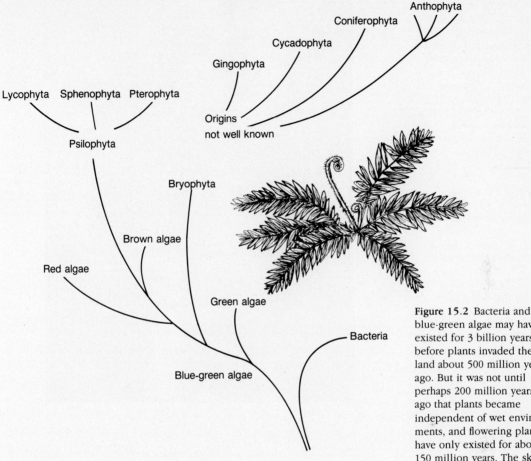

Figure 15.2 Bacteria and blue-green algae may have existed for 3 billion years before plants invaded the land about 500 million years ago. But it was not until perhaps 200 million years ago that plants became independent of wet environments, and flowering plants have only existed for about 150 million years. The sketch indicates supposed relationships of major plant groups that exist today.

We will consider kingdom Plantae to be organized as follows:

Division Chlorophyta: green algae
Division Phaeophyta: brown algae
Division Rhodophyta: red algae
Division Bryophyta: mosses, liverworts, and hornworts
Division Psilophyta: whisk ferns
Division Lycophyta: club mosses
Division Sphenophyta: horsetails
Division Pterophyta: ferns
Division Cycadophyta: cycads
Division Ginkgophyta: ginkgoes
Division Coniferophyta: conifers
Division Anthophyta: flowering plants

Figure 15.3 Giant Sequoia. The giant sequoia is the largest living thing in the world (by volume) but not the tallest. The tallest giant sequoia, the General Sherman tree of Sequoia National Park, California, stands 84 meters tall but is exceeded by a coastal redwood 30 meters taller. The General Sherman is about 2500 to 3000 years old; most of recorded history has occurred while this tree was growing. Of the numerous tiny (1/100 gram) seeds produced, only one in a million germinates; perhaps one in a billion takes root. Since each sequoia produces about 60 million seeds in its lifetime, that means the population of sequoias is declining, and probably has been for perhaps thousands of years.

Note that the division names end in **-phyta**, meaning plant. Some classification schemes consider all the groups from Psilophyta to Anthophyta to be included as subgroups within a single division, the Tracheophyta, those plants with vascular systems. Let's briefly examine each of these groups (Exercise 15.1).

ALGAE

The Green Algae

Algae are the most widespread of the photosynthetic plants. They are found in a wide range of colors although they all depend upon chlorophyll for their primary photosynthetic activity. The green algae contain a diverse array of organisms. They may live as individuals, colonies, filaments, mats, or large sheets of "seaweed" (Figure 15.4a).

Chlamydomonas is a widely distributed green alga, found in fresh and salt water, on tree bark, in the soil, and even in acid-polluted waters. One species, with a red pigment, grows on snow and ice, giving a patch of snow a pinkish hue. (If this snow is eaten, *Chlamydomonas* can cause severe intestinal disorders.) Other green algae are found attached to animals, on frogs' eggs, as pond scum, and growing on the hulls of boats. A few species have no chlorophyll and live as internal parasites in other plants. These parasitic species infect magnolia, citrus, and avocado trees, tea plants and a large variety of other tropical and subtropical plants. They are red in color and cause the stems to crack, which leads to a condition known as "red rust."

The Brown Algae

Almost all of the brown algae are saltwater plants and are most abundant in the tidal zones of colder oceans, where they are exposed to the surging surf and drying wind. Adaptations to this severe environment include a strong, rootlike **holdfast** to attach to rocks, and tough, leathery, flexible bodies covered with slime to retard drying. This slime absorbs and retains water, protects the cells from freezing, and helps to buffer the physical effects of the surf (Figure 15.4b).

None of the brown algae are one-celled, almost all being large plants. The brown algae are, in fact, the largest of all algae. Giant seaweeds (kelps) grow in deep waters with their upper portions floating on the surface 60 meters above.

The Red Algae

More than 90% of the red algae live in marine habitats and are most common in warmer waters. They may be found at low tide level; some species are found at greater depths than any other type of algae. A few species are terrestrial, forming red films on the soil. Most are colored pink to deep red, but some are brown or greenish-gray (Figure 15.4c).

These algae contain green chlorophyll, but many species appear red because of a high concentration of red pigment (**phycobilins**). These red pigments absorb blue light best, so those algae that live at great depths in the ocean where blue light penetrates best can efficiently photosynthesize. (Red light, although high in energy, is rapidly filtered out by seawater and thus provides little or no energy to deep marine plants.)

Figure 15.4 Algae. (a) Green algae. (b) Red algae. (c) Brown algae. The brown algae are primarily inhabitants of shallow and cold marine waters. They provide food and shelter for smaller animals, although the giant kelp (up to 50 meters long) compose virtual underwater forests. This is *Fucus,* a common brown alga of tide pools.

a

b

c

The Ecology of Algae

Algae are subjected to a number of changing environmental conditions, which determine the type of algae that can survive at a particular location. On rocky coasts we sometimes see zonation of algae, a reflection of the different environmental conditions from shore to open water. Waves and tides buffet the algae about and only species that have strong holdfasts survive. The degree of exposure to drying conditions and warm air temperatures at low tide also determines the presence or absence of certain species. The substrate, such as rocks, boulders, pebbles, mud, or sand, also determines kinds of algae that can anchor. Water temperature also influences the existence of certain species; some are restricted to temperate or arctic waters and others to tropical waters. Some can tolerate temperatures of up to 40°C. Even more important is the fluctuation in water temperature. In shallow pools and salt marshes the temperature may change considerably in a day's time. But the algae growing in these pools are tolerant of changes ranging up to 12°C.

Light intensity also has an effect on algae. It can vary considerably according to seasons, water depth, and water clarity. The clearer the water, the deeper the depth at which algae can grow. Light is reduced considerably with depth, however, and only 25% of it penetrates below 2 meters. The clearest water will allow algae to grow at a maximum depth of up to 200 meters.

Algae also exist in vast numbers as **phytoplankton** (free-floating plants). Phytoplankton is found in oceans, lakes, ponds, streams, and rivers, and is especially abundant in nutrient-rich waters such as marshes and sewage disposal ponds. The plants are not attached to any substrate and float at various depths, buoyed up by pockets of gas and oil droplets. Plankton blooms are sometimes caused by the upwelling of nutrients from the lower depths, and since phytoplankton is the major producer in most aquatic food chains, upwellings can cause a body of water to become quite productive. Upwellings along the Peruvian coast produce one of the greatest fisheries areas in the world.

Human Uses of Algae

Various types of algae are used commercially as thickening agents for paint, starch, medicines, soups, sauces, and as stabilizers for many food products such as ice cream and milk (for example, to prevent the chocolate from settling out of chocolate milk). Brown algae are a source of iodine and have been incorporated into foods as supplements. They are used in the manufacture of rubber products, plaster, waxes, polishes, cleaners, insecticides, and are used in medium on which to grow bacteria. Dense kelp beds off the Pacific coast shelter and provide food for many species of commerically caught fishes. Algae are also eaten, especially in Japan, and may become an important food in the future throughout the world.

THE MOSSES, LIVERWORTS, AND HORNWORTS

These plants, collectively called bryophytes, are typically small, low-growing plants that grow on the soil, in rocks, or as epiphytes on tree trunks. They are terrestrial and found from polar to tropical regions; they are most abundant in moist habitats and rare or absent in dry habitats. They often form mats of dense vegetation. To those unfamiliar with them, these plants look like moss, the name bryophyte meaning "moss-plant."

For about 3 billion years, life was concentrated in the oceans, but about 500 million years ago the terrestrial environment was invaded by plants. This invasion required the plants to make some drastic adaptations. A water environment provides a medium full of nutrients and supports plant bodies. In this environment there was no need for a circulatory system within the plant, or for structural support. Being surrounded by water, there was also no danger of drying out. In addition, sexual reproduction in algae involves a swimming sperm, and thus water is necessary for it to swim in to reach the ovum. Terrestrial environments, on the other hand, pose problems for plants of nutrition, support, desiccation, and reproduction. But as plants invaded the land, adaptations that overcame these obstacles evolved.

The bryophytes are intermediate between the truly aquatic and truly terrestrial plants. They are terrestrial but are restricted to wet environments. They absorb moisture through structures similar to roots and leaves, but are not "true" roots and leaves because they do

Exercise 15.1 Collection and Preservation of Plants. *Plants can be preserved for many years of study, if the preservation is done carefully. Smaller plants can be preserved whole, while only selected parts of larger plants, such as twigs, flowers, and fruit, should be preserved. Small plants should be collected with their roots, if possible. They should be laid on several sheets of newspaper, and the plant arranged as desired. Most petals and leaves should be facing you, but some should show their opposite sides. Lay several newspapers on top and press gently to flatten plant.*

Sandwich the newspaper and plant between two sheets of blotter paper and then between two sheets of corrugated cardboard. The entire "sandwich" should then be put in a plant press bought from a commercial supplier, or into one made from 1"-×-¼" hardwood slats that have been made into a latticework. Putting a weight on top or lashing the bundle tightly with straps will also press the plants. The press should be put in a warm, dry environment and the newspaper changed every few days if the plant specimen was moist. After a week or so the dried plants can be mounted on herbarium sheets or other heavy paper with straps of white glue laid over the

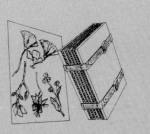

plant's stem. The sheets should be labeled and kept in insect-proof cabinets with mothballs. A simpler method is to mount the dried plants on 5-×-8 cards and cover with clear contact paper. Dried plants are essential, as any moisture under the contact paper will allow rapid decomposition.

not have vascular, or conductive, tissues. A swimming sperm is present and the plants tend to be short and broad, avoiding the need for strong supporting tissues. The bryophytes produce tiny spores that are dispersed by wind and water.

Mosses

Mosses tend to grow in mats rather than as individual plants. They are often found on stream banks, rocks, tree barks, and other areas with a plentiful water supply (Figure 15.5). Their leaflike structures are short and closely packed, helping to hold moisture, as well as being only one cell thick so that moisture can be readily absorbed from the environment. They often follow lichens in the process of succession, helping to form soil on top of rock. Successive layers of moss on rocks produce a mat of soil in which grasses and other flowering plants can grow.

Sphagnum moss is common in bog areas and will eventually cover the bog. In northern latitudes, particularly in Scotland and Ireland, the organic matter formed by peat mosses is cut into bricklike chunks and used for fuel. Sphagnum moss is also used to package live plants for shipping, as a mulch in gardens, and as a medicinal herb in the Orient.

Figure 15.5 Moss. Mosses are common in wet areas, more likely growing on rocks or tree bark rather than on the soil, and often grow alongside lichens and horsetails.

Figure 15.6 Liverworts and Hornworts. The cuplike structures on the leafy body of the liverwort *Marchantia* contain asexual structures called gemmae. When raindrops fall into the cups, the gemmae are splashed out and away from the parent. Each gemma may form a new plant.

Liverworts and Hornworts

Liverworts also grow in mats, either growing flat on the ground or extending a few centimeters upward. Liverworts have wide, leaf-like structures, unlike the small, needlelike structures of the mosses (Figure 15.6). As in other bryophytes, liverworts are dependent on water for dispersal of their spores. Some liverworts have cuplike structures on their leaves that contain spores that are splashed out by raindrops and dispersed. Liverworts get their name from "wort," an archaic word meaning plant, and the fact that someone once thought the plant was shaped like an animal's liver.

Hornworts superficially resemble liverworts, but under close examination their external structure is quite different, and they often contain nitrogen-fixing blue-green algae. Hornworts grow in moist areas on the ground, on trees, or on rocks. Their reproductive structures resemble horns—thus the name.

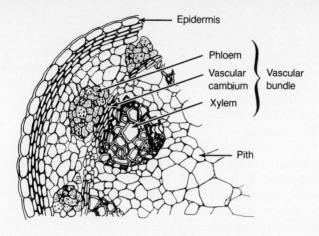

Figure 15.7 **Cross section of a Vascular Plant.** The vascular system is composed of xylem tissue that transports water and ions, and phloem tissue that transports carbohydrates throughout the plant. The **vascular cambium** is tissue that produces xylem and phloem. Xylem, phloem, and the vascular cambium together are termed a **vascular bundle**.

VASCULAR PLANTS

The vascular plants, or **tracheophytes**, are so named because they possess a vascular (circulatory) system to transport water and minerals from the roots (**xylem**) and photosynthetic products from the leaves to other parts of the plant (**phloem**). These tissues make it unnecessary for the plants to absorb nutrients across their body surfaces; materials are absorbed by roots or rootlike rhizomes and transported to other parts of the plant's body. Thus, vascular plants can be thicker, taller, and live in drier habitats than the bryophytes (Figure 15.7).

Roots or rhizomes anchor the plant as well as absorb nutrients. Besides a circulatory system and anchorage, many vascular plants have a supporting system of fiber cells to keep up the plants' bulk. (These fiber cells are used for paper products, cloth, and rope.) Wood is a mixture of supporting and vascular cells.

The ferns and their lower tracheophyte relatives reproduce via spores. The higher tracheophytes have seeds, a considerable advancement that will be discussed later in this chapter.

Lower Tracheophytes

Club Mosses, Horsetails, and Allies

Club mosses, horsetails, whisk ferns, spike mosses, scouring rushes, quillworts and other such plants are vascular plants that, like the ferns, lack seeds. Once dominant plants in the Devonian period, they are now of minor ecological importance. For the most part, they too are inhabitants of wet or moist areas.

Club mosses, ground pines, spike mosses, and the like resemble mosses or small pine trees that creep over the ground. The ancient club mosses grew to 30 meters. Modern species, only a few centimeters long, grow well in sterile soils that allow little competition from other plants. The spike moss spreads out when it has sufficient water, but curls up into a ball when dry; it is sometimes used as a moisture indicator in a terrarium. Ground pines are used in Christmas wreaths.

Horsetails are plants with jointed stems found in poor but moist soil or sand. The branching underground parts help to stabilize the soil. The stems are rough due to a high concentration of silicon; for some obscure reason they also accumulate gold from the environment in their tissues. Horsetails, or "scouring" rushes, were once used by American pioneers to clean pots and pans (Figure 15.8).

Quillworts look much like tufts of grass with swollen bases. They are found in pond, river, and stream habitats, where they may be eaten by muskrats or grazing cattle. Ancestral quillworts of 130 million years ago are essentially identical to those alive today.

Figure 15.8 Horsetail. Most horsetail species are short, but some species exceed 4 meters in height. The name comes from the taillike appearance of the branched species. Some species are unbranched, but in all species most of the photosynthesis occurs in the stems and not the branches.

Ferns

Ferns are one of the most well known of the lower tracheo-phytes, some species commonly being used as house plants. Frequently considered tropical plants, as the majority indeed are, ferns are also found in aquatic, desert, deciduous, and coniferous forest environments. Generally, they are restricted to areas where they receive sufficient moisture for reproduction. Their spores are borne on **sporangia** on the underside of the leaf, or **frond**, and are dispersed by wind or water; the spores must land in a damp environment to germinate. There are about 9000–10,000 species of ferns, which vary in size from tiny aquatic species to the 20-meter tall tree fern found in tropical Pacific areas.

Ferns and their closely related ancestors were very common plants in Devonian times (about 400 million years ago), and many fossils from that period contain ferns. Forests were then composed of ferns that would tower over our present-day houses. Much of our fossil fuel supply was formed by ancient ferns and plants during that time and the following 200 million years of the Carboniferous period. The present-day uses of ferns by humans are mainly ornamental, but ferns have been used for food, medicinal purposes, and even fertilizers (Figure 15.9).

Figure 15.9 Bird's-Foot Fern. The bird's-foot fern is so-called because the fronds are often divided into parts at the tip, resembling a bird's track. This fern commonly grows in dense clumps in rock crevices and is able to withstand the hot summer sun. Indians once used the long rhizomes for basket making.

Plant Reproduction

Figure 15.10 Generalized and Simplified Life Cycles. (a) The Liverwort, a Bryophyte. In the bryophytes the gametophyte generation is the largest generation. See text for explanation. (b) The Horsetail, a Tracheophyte. In the tracheophytes, the sporophyte generation is the largest generation. Figures 15.7 and 15.8 show only the sporophyte generation. See text for explanation.

Alternation of Generations

The lower plants have no seeds and are restricted to wet environments because their sexual cycles involve swimming sperm that need water. The male and female may be far apart, but it is not always necessary for the sperm to swim the distance between them; splashes caused by raindrops can carry the sperm to the egg. These plants typically grow close to the ground to facilitate sperm dispersal. When the sperm reaches the egg, fertilization occurs and the embryo begins to grow.

The fertilized embryo grows to form a stalk that produces hap-

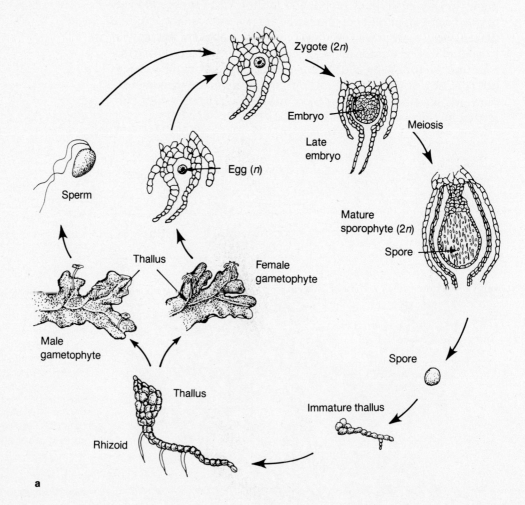

a

loid spores that, in turn, produce the males and females. So there is a diploid form (the **sporophyte**) that produces haploid spores by asexual means, and a haploid form (the **gametophyte**) that repro- duces sexually by gametes. This **alternation of generations** is charac- teristic of plant reproductive cycles. In the bryophytes (mosses and allies), the gametophyte is the largest of the two generations. In the tracheophytes (ferns and allies), the sporophyte generation is the largest, most obvious, part. See Figures 15.10a and b for cycles of bryophytes and tracheophytes.

The evolutionary trend in plants has been a reduction in the size of the gametophyte generation. In the seed plants, the gameto- phytes have been reduced to small structures called the **pollen** and **ovule**.

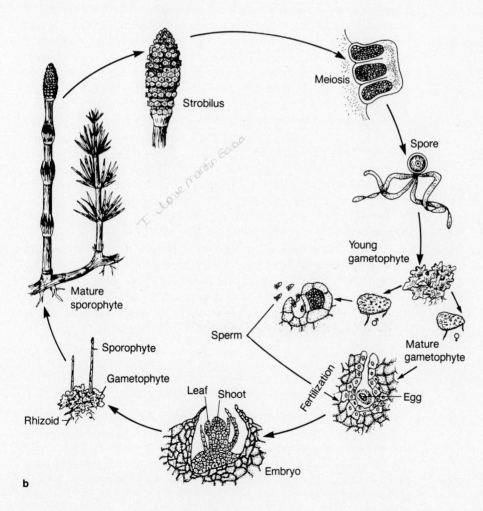

b

Exercise 15.2 Seed Germination.

Seeds of different plant species require different sets of environmental conditions to germinate. Some require periods of darkness and/or cold, which they would be exposed to under winter conditions. Seeds such as those from tomatoes or radishes can be used to compare the effects of various environmental conditions on seeds. Simply placing a few seeds on a piece of wet blotter paper in a dish and covering it with glass will induce germination. Compare germination times and the rate of growth of dishes placed in the dark, in the light, under cold conditions, in warm temperatures, and so forth. Be sure to test only one variable at a time, however. If you place one in the dark and one in the light, for example, be sure they are at the same temperature.

Seeds

The pollen from the male plant, or part of the plant, enters the female ovule to produce an embryo that grows to become a seed. The **seed** actually consists of (1) a matured ovule, (2) **endosperm**, a tissue that surrounds the ovule and provides food for it, and (3) the seed coat, which serves to protect the seed.

A seed is a marvelous adaptation that allowed the rapid proliferation and present abundance of seed plants over the nonseed plants. The seed is surrounded by a protective coat and supplied with nutrients so that it can survive long periods before germination begins; germination is relatively independent of the need to obtain energy or nutrients from the environment (Exercise 15.2).

Seeds may live for many years before germinating. Seeds of pioneer plants are particularly hardy and may remain viable over 100 years. A few species, taken from archeological digs of known age, have germinated after 1700 years. How does a seed "know" when to germinate? The burying of a seed by wind, animals, or other such activity makes some seeds light-sensitive. When these are exposed to light again they will germinate. Thus, when trees fall or are taken down, the seeds in the soil below them receive more light than they did when shaded by the trees, and they begin to germinate. When a farmer plows his corn or wheat stubble into the soil, he brings weed seeds to the surface and exposes them to the sun, encouraging their germination. Those seeds that are not light-sensitive germinate more randomly and may sprout deep beneath the soil and never become mature. It is only because enormous numbers of seeds are produced that these species are successful; some soil holds 100,000 dormant seeds per square meter (Figure 15.11).

In addition to having a protective coating and built-in food provisions, pollen is transported by wind or animals, both more effective and less restrictive than the wet conditions required by the swimming sperm. And the long dormancy of seeds allows them to be transported great distances by wind, water, or animals. For these reasons seed plants are the dominant terrestrial plants.

The Seed Plants

The seed-producing plants include the most successful and dominant plants as well as two uncommon groups. These are the cycads, ginkgoes, conifers, and the flowering plants. All but the flowering plants (**angiosperms**) are called **gymnosperms** (naked seed) because their seeds are not encased in a fruit (Figure 15.12).

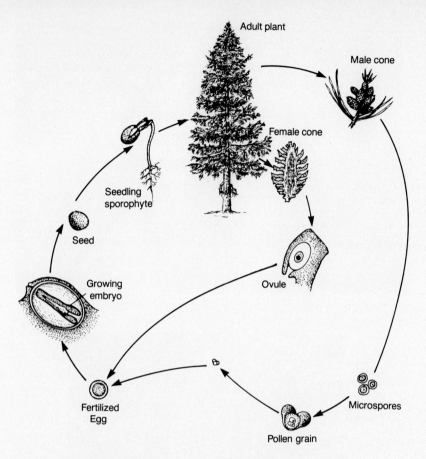

Figure 15.11 **Generalized Life Cycle of a Seed Plant.** See text for explanation.

Cycads

The cycads were abundant and dominant about 200 million years ago during the Mesozoic era. The resemblance of cycads to ferns caused that time period to be called the "age of ferns." Only about 100 species of these palmlike/fernlike plants still exist, all in tropical or subtropical areas. Some species are hosts to nitrogen-fixing bacteria. The sexes are separate (Figure 15.13).

Ginkgoes

A number of ginkgo species existed 200 million years ago, but only the maidenhair tree survives today. It has survived practically unchanged for perhaps 350 million years and is sometimes referred

Figure 15.12 Red Mangrove Seed. In a few species, seeds have no dormant period at all. In the red mangrove, the seed begins to germinate while it is still on the tree. When the root is 20 to 30 centimeters long, it falls off the tree and plants itself in the muddy water of coastal estuaries in which the trees grow. The red mangrove is an angiosperm as the seed is covered by a fruit.

Figure 15.13 A Cycad: Mexican Horncone. Native to tropical forests of Mexico, this palmlike cycad gets its name from the two horns at the ends of the cone scales. This is the cone of a male plant. The horncone is sometimes grown in greenhouses or outdoor gardens in mild climates.

to as a "living fossil"; it may be the oldest of all living seed plants. Once distributed throughout the northern hemisphere, it is now native only to China although it has been naturalized as an ornamental in many locations. The sexes are separate and the fruits borne on the female tree have an unpleasant odor, causing male trees to be planted preferentially. It also seems to be resistant to insects and air pollution, making it a preferred choice for an urban street tree.

Conifers

Conifers contain the most well known of the gymnosperms: junipers, pine, cypress, hemlock, spruce, fir, redwood, larch, yew, and others. The terms conifer and evergreen are often used interchangeably or in conjunction, but by no means are all evergreens conifers, and some conifers are deciduous (for example, dawn redwood, bald cypress, larch). Most, however, are evergreen, meaning that they do not drop all of their leaves at once. These shed and replace their leaves continuously, any one leaf remaining on the tree for 2 to several years.

There are about 500 species of conifers, many of which comprise large expanses of boreal forest as well as smaller stands in other areas of the world. All are trees or shrubs, and most grow in the typical shape of an inverted cone with one main trunk. They range in size from short shrubby junipers to the redwoods that reach over 100 meters in height and the giant sequoias that measure over 30 meters in circumference (Figure 15.14).

Coniferous trees possess a variety of leaves: scalelike, linear, flat, needlelike, and so on. Most are adapted to dry environments, part of the adaptation being waxy leaves. Waxy leaves take 3 to 5 years to decompose, leaving coniferous forest floors with a thick layer of leaf litter but a thin humus layer. The resin in the leaves produces a weak organic acid as it decomposes, slowing down the activity of decomposers and making the soil acid.

The thin soil is easily leached and is thus poor in minerals, but due to a mutualistic association with fungus, many if not most conifers are able to survive in poor soil. Hyphae of fungi cover the roots of the conifer, absorbing some nutrients from the roots. The fungi, in turn, pass nutrients from the soil into the conifer roots. Both organisms benefit. All the dominant trees in boreal forests are involved in this mutualistic association with fungi.

Conifers are perhaps the most commercially important of all trees. Often called softwoods, many are easily cut and worked, are straight-grained, and take paint, varnish, and glue easily. They are used for construction, furniture, and firewood. Distillation of their resin produces turpentine and rosin. Juniper berries provide the flavoring for gin. And conifers provide us with Christmas trees.

Figure 15.14 A Conifer: Brazilian Pine. Native to southern Brazil and adjacent Argentina, this is an important timber tree in that region. The young leaves are larger than the mature ones. Separate male and female cones are found on the same tree.

Ecologically, conifers provide habitats for innumerable other organisms (see Chapter 8). Pine nuts, juniper berries, fir cones, and even the short needles of some conifers are eaten by grouse, jays, squirrels, wood rats, and a variety of other birds and mammals.

The most common and familiar of the conifers are the pines. With their conical shape, needlelike leaves in groups of two to five, and their woody cones, they are easily identified. They are one of the major sources of commercial products such as lumber and paper, and are the dominant trees in many coniferous forests. The seeds of many pines are eaten by animals and once served as a major food for some American Indian tribes. Pines are resistant to desiccation, and many species are found in very dry areas. The ponderosa pine, distributed throughout western North America from British Columbia to Mexico, is extremely drought- and fire-resistant. It inhabits high elevations where precipitation is in the form of snow, and lower elevations where summer heat may be intense. The loblolly pine, however, is characteristic of wet bottom lands of the

Figure 15.15 Tree Rings and Dendrochronology. **Dendrochronology** is the science of dating by the use of tree rings. For example, portions of trees found in excavated archeological sites can be matched with old living trees from the same area. If any part of the two tree ring paterns overlap, the age of the archeological site may be estimated. With enough specimens, a series of overlapping tree ring patterns can be constructed going back thousands of years, even if any one tree is not more than several hundred years old. Radioactive carbon dating gives the approximate ages of organic matter, but the system can be calibrated against a series of tree ring patterns (or one from an old bristlecone pine) to give more exact ages. Many archeological sites have been shown by dendrochronology to be hundreds of years older than determined by radiocarbon dating.

southeastern United States and is not resistant to fire. The lodgepole pine of the western United States and Canada is not only resistant to fire, but its cones typically do not open and release their seeds unless burned; dense stands of lodgepole pine follow a few years after a severe burn. One of the most common Christmas trees in the United States is the Scotch pine; although orginally native to Europe and Siberia, it is now distributed more widely than any other pine.

The bristlecone pine deserves some special mention because it lives longer than any other organism; the oldest bristlecone pine tree in the White Mountains of southeastern California is more than 4600 years old. These trees grow in the subalpine zones (2500–3500 meters high) of six western states. They are almost totally resistant to insect, fungal, and microbial infestations because of their dense, resinous wood. Lack of soil moisture, cold temperatures, and drying winds allow these pines to grow only very slowly—they may add only 2 centimeters to their diameter every hundred years or so.

An examination of the growth rings of long-lived trees gives us an indication of the climate over the past hundreds or thousands of years. Tree rings can be counted on a tree cross section. The rapid spring growth produces a light-colored ring, and the slow winter growth produces a dark ring. Each pair of rings, then, indicate a year's growth. Wet, mild years produce considerable growth while dry, cold years produce little or none (Figure 15.15).

Welwitschia

Welwitschia and about 30 other unfamiliar or uncommon plant species make up the **Genophyta**, whose members are found only in the xeric regions of the world. This group contains herbs, shrubs, small trees, and vines. For several reasons (stem structure, seed covering, and so on) they seem to be somewhat intermediate between the gymnosperms and angiosperms (Figure 15.16).

Angiosperms

The flowering plants, or angiosperms, are very much a part of both natural ecosystems and human culture. Of approximately 300,000 species of plants, 92% are flowering plants. They are the dominant plants in terrestrial ecosystems and constitute virtually all of our agricultural crops. The angiosperms arose in the Cretaceous period about 150 million years ago and have radiated into a large variety of species in almost all habitats and niches. They grow in the form of trees, shrubs, herbs, vines, floating aquatics, and emergent plants. Most are photosynthetic, but a few are saprophytic, some are parasitic, and a few are carnivorous. The characteristic that is common to all of them, is, of course, flowers. There are microscopic flowers on grasses and aquatic duckweed and large showy flowers on dogwood trees and orchids (Figure 15.17 and Exercise 15.3).

Figure 15.16 Welwitschia. A very unusual conifer, belonging to a small group known as the Genophyta, this tree grows only in the Namib desert of southwestern Africa, where rainfall averages 2.5 centimeters per year, with some years having no rain at all. The plant obtains water from dew, which it absorbs through its leaves. The trunk of this tree never grows taller than several centimeters, although it may become a meter in width. Only two leaves are produced, which may grow to a length of 3 meters. *Welwitschia* grows very slowly and may live to be 1000 years old.

Figure 15.17 Milkweed. The milkweed is a pioneer plant characteristic of disturbed areas along roadsides and farmed fields. The root system may be 5 meters in length and serves to anchor soil. Seeds formed in pods have feathery tufts attached for wind dispersal; the tufts were once used for life-preserver fillers. The plant is fed on by monarch butterfly larvae, which pick up chemical poisons (cardiac glycosides). The butterflies are protected by this chemical, which makes them extremely distasteful to predators and may cause the predators to vomit. This particular species is called the "horsetail milkweed."

Exercise 15.3 Phenology of Flowering.
Phenology is the study of periodic biological phenomena such as bird migration, leaf fall, and fruit production. One easy observational exercise is to determine the dates on which various species of plants flower. Once or twice a week, from early spring to mid-summer (or during other appropriate periods), note the species of plants in bloom. Also, gather information such as amount of rainfall, temperature, and macro-and microhabitats of the plants. Although some speculations can be made after perhaps a month, several months of data are preferable. Can you relate time of flowering to individual or sets of environmental factors? What clues the plant to bloom? Do different habitats show different phenologies? Does the same species of plant bloom at different times in different habitats? After doing this study, determine how you would change it to obtain better and more detailed data. What other information would you need to make valid

conclusions as to the causes of flowering? What hypothesis would you make and how would you test it?

Angiosperm literally means "vessel-seed," referring to the seeds being enclosed in a fruit, another characteristic of this group of plants. Flowers did not evolve to grace our windowsills, but to attract animals to serve as dispersers of pollen and seeds.

Flowering plants may have separate sexes, there may be separate male and female flowers on the same plant, or flowers with both male and female parts may be present. Figure 15.18 is a diagram of a typical flower. The typical life cycle of a flowering plant is shown in Figure 15.19.

Pollen and Pollinators

For seed plants to be successful at reproduction, pollen from the anther must reach a pistil. In plants capable of self-pollinating, this is not such a problem since the pollen needs to travel only a short distance. But for an exchange of genes to occur, cross-pollination between plants is necessary. In many plants, self-pollination is in fact prevented. In the apple tree, for example, pollen from the same tree or variety is prevented from growing through the pistil by a chemical inhibitor, but the inhibitor is neutralized by chemicals produced by pollen from a different variety of tree. In other plants, mechanisms prevent pollen from a flower from landing on the pistil of that flower (Figure 15.20).

Movement of pollen by wind is common in gymnosperms but considerably less common in the angiosperms. Wind-dispersed pollen may be carried long distances—even mid-Atlantic Ocean air has pollen—but its chances of landing on a pistil and producing a seed are one in a million. Thus, a great deal of pollen must be produced. The Austrian pine produces 1½ million pollen grains per male cone and the common juniper 400,000, for example. In general, wind-

Figure 15.18 The Flower. A generalized flower with both male (**stamen**) and female (**pistil**) parts is shown here. Flowers differ in the presence or absence of various parts, the arrangement and number of the parts, and the arrangement of flowers as single entities or in clusters. In the ovary is an ovule that, after fertilization, becomes a seed. The ovary becomes the fruit. After fertilization, the petals and reproductive parts degenerate and the fruit and seeds grow.

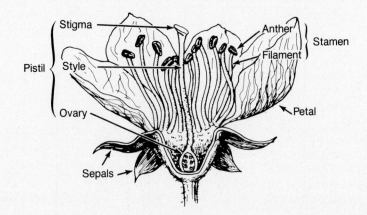

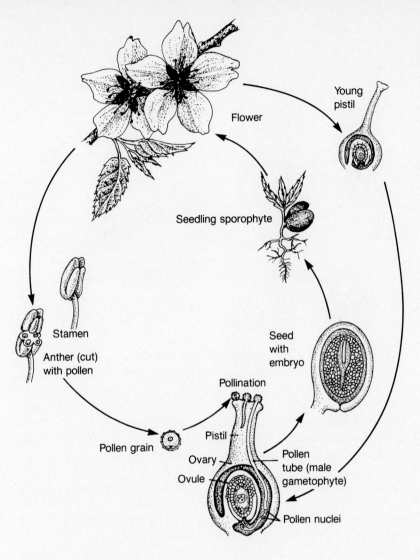

Figure 15.19 Life Cycle of a Flowering Plant. Pollen from the anther is deposited on the pistil. The pollen grain produces a pollen tube that grows down through the pistil to the ovule. The haploid pollen nucleus fuses with the haploid ovule (egg) nucleus to form a diploid zygote. The zygote develops into an embryo (seed), which germinates to become a plant.

pollinated plants grow in open areas (grasses) or in the upper portions of forests (conifers), where they are exposed to the breeze.

Pollination by animals is presumably more efficient than by wind since animals, such as insects, fly from flower to flower (Figure 15.21). One out of 6000 pollen grains will reach a pistil and fertilize an ovule—even if 90% of the pollen is eaten by the insect! Thus, flowers and nectar evolved to attract pollinators such as insects, birds, and bats. Many plants and their pollinators have coevolved close relationships. In areas of the world where insect pollination is

Figure 15.20 Pollination of an Orchid. Pollination of the orchid, *Stanhopea grandiflora,* by the bee *Eulaema meriana*. The bee enters from the side and brushes at the base of the orchid lip (A). If it slips, the bee may fall against the pollen sac, which is placed on the end of a column (B), and the pollen sac becomes stuck to the hind end of the thorax (C). If a bee with an attached pollen sac falls out of a flower, the sac may catch in the stigma, or pollen receptor organ (D), which is so placed on the column that the flower cannot be self-fertilized.

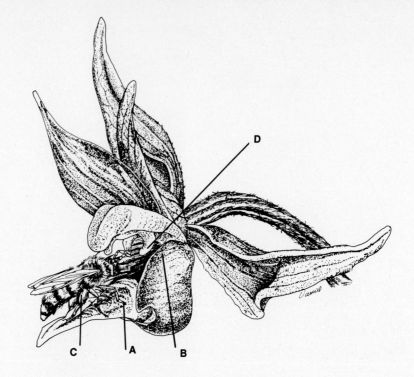

most common, blue-colored flowers, attractive to insects, are dominant. In tropical areas where birds are more important pollinators, red flowers are more common.

Many plants are adapted to accept only a narrow variety of pollinators. Hummingbirds pollinate "hummingbird flowers," which are tubular-shaped to allow the entrance of hummingbird bills but not insects. Bat-pollinated plants open at night, are light in color, and have a distinctive odor. Nectar-feeding bats have larger eyes than other bats, and a gap in their front teeth allows the protrusion of a long tongue into the flower. Some orchids have petals that form to resemble a female wasp; a male wasp, in attempting to copulate with the mimic female, picks up pollen, which is then transmitted to another orchid (Exercise 15.4).

Most flowers have both male and female parts, as does the jewelweed. But when the jewelweed's flowers first open, only the male parts are exposed. A visiting hummingbird or insect can pick up pollen, but the flower cannot fertilize itself because the female parts are still concealed. A day or so later the male parts drop off and the female parts are exposed so the flower can be pollinated by pollen from another flower; cross-fertilization is ensured.

It is advantageous for plants to have their pollen dispersed by

Figure 15.21 Bee on Rose Flower. Since insects detect ultraviolet light better than visible light, they are attracted to blue and yellow flowers, while red flowers appear black to them. This bee is gathering nectar from a yellow rose. Many flowers are adapted to attract only a narrow range of animals for pollination; for example, flowers with thin petals formed into a long tube allow only some birds, long-tongued bees, and long-tongued butterflies and moths to gain access to their nectar. This system helps to ensure that pollen finds its way to another flower of the same species. A number of flowers, however, accept almost any pollinator as does the rose shown here. Not as efficient as the system just described, it is still considerably more efficient than the wind. Bees will visit 250 to 1500 flowers per trip, or about 500 flowers per 25 minutes. One kilogram of white clover honey represents visits to 19 million flowers and 7221 hours of bee labor!

animals, but what benefit do the animals receive? Nectar, being primarily a solution of sugar water in a 15% to 50% concentration, is an excellent source of energy. It also contains amino acids, proteins, fats, and minerals in small amounts. Pollen does not seem to serve as a source of energy, but does provide protein for bees, at least.

Fruits

Another major evolutionary advancement was the development of fruits. With flowers attracting animals, the ovules were vulnerable to predation. A covering, the fruit, evolved to enclose the ovule and protect it. Fruit (and seed) production is stimulated by hormones from the fertilized ovule. The fruit develops from the ovary of the flower's pistil. Fruits are of two main types: fleshy and dry. Fleshy fruits are those with soft tissue surrounding the seeds, such as apples, raspberries, peaches, and tomatoes (Figure 15.22). Dry fruits are those such as peas, beans, peanuts, and acorns.

Fruits not only protect the seed but often attract animals that eat the fruits and disperse the seeds. The fruits of mistletoe, a parasitic plant that roots in trees and taps their nutrients, are eaten by birds. The seeds, after passing through the digestive tract, have a sticky

One of the recent areas of interest in ecology is pollination ecology. The goals of this science include determining what pollinators visit which species of plants, what nutrients the pollinators receive from the plant, how far the pollinators travel from the plant, and other such factors. Observe a flowering plant, or population of plants, and try to determine which pollinators are most important to that plant. How much time does each individual and species of pollinator spend at each plant? What characteristics of the plant make it more attractive to some kinds of pollinators and not others?

coating and are deposited on another tree, where they germinate. Often, seeds that pass through a bird's gut germinate better than those that do not. The animals, of course, derive food energy from the fruit or enclosed seeds. Seeds of some plants are eaten by animals, and although some are chewed up or digested in the process, other seeds survive and are dispersed. Squirrels may find only one-third of all the acorns they bury, ensuring that some will become oak trees. Earlier in the text we discussed other mechanisms of seed dispersal such as winged fruits, tufted seeds, and fruits that attach themselves to animals. As with flowers, fruit color is important; red fruits, for example, are common because of their attraction to birds, which are excellent dispersal agents.

Flowers and fruits have evolved to exploit the dispersal abilities of animals. Animals have evolved to take advantage of the food sources angiosperms were providing. It is then no surprise to see that when the angiosperms underwent a burst of evolution about 130 million years ago, birds and mammals did the same.

Seeds and fruits are eaten by many animals—a price flowering plants have to pay for dispersal. As long as the advantages of dispersal outweigh the disadvantages of predation, the system will continue. However, where the potential exists that too many seeds will be eaten, plants have evolved mechanisms to reduce predation: hard shells, reduced palatability, difficult access, and so on. Bamboo plants have evolved an interesting defense. A grass native to Asia, bamboo is one of the fastest growing of all plants (over a meter in 24 hours). Bamboo flowers once in its life and then dies, but some species do not flower for as long as 100 years. Others live 10, 20, or 60 years before they flower and die. Any animal that tried to depend on the seeds of bamboo as a significant source of food would rapidly go extinct. When the bamboo does flower and go to seed, so many seeds are put into the environment that animals cannot eat all of them so many are left to begin the bamboo cycle again. As a result, vast forests of bamboo may disappear in one season and begin again the next year. Many other plants seem to use this same strategy of saturating the environment with too many seeds to be eaten.

Plants Provide Habitat and Food

The major terrestrial ecosystems of the world are dominated—indeed created—by flowering plants. Besides seeds and fruits, animals eat the buds, flowers, stems, leaves, twigs, and bark of angiosperms. Beetles burrow through the trunks of trees, and aphids suck the sugary sap from leaves. Raccoons, nuthatches, mosquitoes, and others use the crevices or holes formed by plants to probe for food, escape the weather, hide from predators, or raise their young. Plants provide food and shelter in some form for all animals.

Figure 15.22 Bananas. Bananas are considered to be a type of berry because they develop from a compound ovary and contain more than one seed. However, bananas grown for the marketplace develop from unfertilized ovules and thus have no seeds. A native of West Africa, bananas are now grown in tropical areas throughout the world.

As immobile organisms with a large store of energy and nutritive material, plants have evolved defenses against being eaten. Some plants are thorny, distasteful, fibrous, resinous, toxic, and/or noxious smelling. Some trees produce a hormone that arrests development in larval insects that feed on them. Leaves and stems of the jack-in-the-pulpit flower contain calcium oxalate, a mechanical irritant. The passion-flower vine has a rather unique defense. Butterflies of the genus *Heliconius* lay their eggs on the leaves of passion-flower vines. The newly hatched caterpillars are voracious eaters and can easily devour the plant. The passion-flower leaves, however, produce tiny white bumps on their surface. A female butterfly, looking for an appropriate place to deposit its eggs, rejects the seemingly occupied leaves.

Colors of Angiosperms

Flowers and the rest of the plant are the color they are due to one or more chemical pigments, of which chlorophyll is the most common. Carotene is also abundant, causing an orange color (as in carrots); xanthophyll is yellow, anthocyanin red and purple, and so on.

Colors evolved for a purpose; the green of chlorophyll in leaves and stems is obviously the most important since it is essential to photosynthesis. The color of flowers and fruits serves to attract animal pollinators and dispersers. The brilliant colors of fall forests are the results of trees' preparation for the winter. As the daylength shortens, nutrients in the leaf are absorbed by the tree and stored for the next season. Photosynthesis diminishes and chlorophyll is no longer produced. As this green pigment disappears, the pigments that were previously masked appear, turning the leaves to brilliant reds, oranges, and yellows. In the United States the autumn colors begin in the far north and at high elevations and move southward at the rate of about 800 kilometers per week.

Brilliant fall colors are ultimately replaced by shades of brown as the leaves wither and fall from the trees. The brown color is caused by **tannin**, a chemical common to many plants; it is found in the bark, leaves, or fruits such as acorns. Tannin inhibits attack by insects, the probable reason for its presence. American Indians put acorns in running water to leach out the bitter tannins before eating the acorns; tannins have been used for many years to tan leather.

The Rhythm of Plants and Blue Pigment

Flowers are the first real signs of spring to most of us, and the fall of leaves, shedding of acorns, the appearances of lilies at Easter and poinsettias around Christmas indicate other seasons. Since temperature is associated with seasons, it was once assumed that temperature initiated blooming. But when plants are put in greenhouses and the temperature kept constant, flowering occurs at the same time as those plants grown in the natural environment.

Further experimentation showed that flowering was brought on by a particular photoperiod. Some plants bloomed with 10 hours of light per day but others needed 15. Those plants that bloomed early in the spring, before the daylength reached 12 hours, were termed **short-day plants** (examples are asters, soybeans). Those that required more than 12 hours (such as hollyhocks and clover) were called **long-day plants**. Continued experimentation revealed that the blue pigment **phytochrome** controls the time of flowering. Phytochrome exists in one chemical state in the day and another at night. And it is the relative proportions of these two phytochrome states that determines the time when the plant will flower. (Some, such as tomatoes, react to other environmental factors and may flower under differing light conditions.)

The Meat Eaters

There are several species of plants that are carnivorous. Pitcher plants are vase-shaped, with slippery walls down which unwary insects slide to be dissolved by the plants' digestive juices. The Venus flytrap closes shut around an insect to trap it (Figure 15.23). The sundew and butterworts secrete sticky droplets that adhere to any insect that touches them. The bladderwort, a tiny aquatic plant, has pods with a trapdoor opening through which insects are sucked. These plants became meat eaters because they live in bogs, marshes, depressions in granite rocks, pools of water formed by bromeliad leaves, and in other habitats that are low in nitrogen, phosphorus, and other minerals. The insects they eat provide these nutrients and a source of protein.

Some insects live in a symbiotic relationship with the pitcher plant. The pitcher plant mosquito can lay its eggs, and the larvae develop, in the digestive juices of the plant. The mosquito derives protection from the pitcher plant and benefits the plant by attracting predatory insects into the pitcher plant's innards. A species of flesh-eating fly also lays its eggs in the pitcher plant, and its larvae (maggots) feed on the dead insects caught by the plant. Both the mosquito and the fly apparently secrete enzymes that neutralize the digestive enzymes of the pitcher plant.

Figure 15.23 The Venus Flytrap. Native to the swamps and marshy areas of North and South Carolina, the Venus flytrap is known worldwide for its insect-trapping ability. Its two hinged sides of a modified leaf close, trapping an insect within. To avoid being closed by a raindrop or falling leaf, there are three trigger hairs on each side; two trigger hairs must be tripped in succession before the closure occurs.

Flowering Plants Without Chlorophyll

The vast majority of plants are photosynthetic, but a few kinds of flowering plants make their living as parasites or saprophytes. Dodder grows as a vine on many types of plants. It germinates in the soil but loses its contact with the ground at maturity, tapping its host (alfalfa, clover, onion, and so on) for water and nutrients. Indian pipe takes its food from the roots of living plants. It may also receive nutrients from decaying plants by means of a mass of fungal hyphae wrapped around the Indian pipe's roots. The fungus decomposes dead matter, and the Indian pipe takes nutrients from the fungus. Pine drops and snow plants are also saprophytes (Figure 15.24).

Economic Uses of Flowering Plants

Flowering plants have been put to an enormous number of uses by humans. Plants supply us with food (rice, beans, wheat, corn, barley, oranges, apples, bananas, and so on), paper products, lumber, fiber (cotton, hemp), spices (cinnamon, cardamon, pepper, and so on), and the list goes on. Perhaps 50% of all medicines were first discovered in some form in a plant. Fuels such as wood and

Figure 15.24 Snow Plant. The snow plant is found in the late spring, poking its stem and flowers through the decaying litter of the coniferous forest floor. Its bright red color stands out strikingly against the usually present blanket of white snow. It absorbs nutrients from decaying organic matter, but there is some evidence that it is also parasitic, tapping into the roots of pine trees.

charcoal have been used for centuries but research is presently being conducted for ways to convert plants into "fossil" fuels in the laboratory. Genetic engineering and selective breeding techniques are creating faster growing and higher yielding food crops, tomatoes that grow in seawater, rice with higher protein content, and a plant that produces both tomatoes (above ground) and potatoes (below ground) (Figure 15.25).

Fruits such as apples, peaches, cherries, oranges, plums, and bananas, which are borne on trees, are of economic importance and provide us with essential nutrients as well as tasty treats. Artificial selection of the most favored varieties have given us seedless grapes, freestone peaches, bananas with no seeds, and disease-resistant varieties of many fruits. Only the crab apple is native to the United States, but over 7000 varieties of apples are now grown in this country.

Viruses, fungi, bacteria, and insects destroy or damage many fruit crops. Some of these pests cause only cosmetic damage, but since consumers prefer attractive fruit, the value of a blemished fruit is decreased at the supermarket. In spite of the known or suspected detrimental effects of pesticides, some are necessary to produce a higher yielding crop that will sell.

Figure 15.25 Triticale. Triticale is a hybrid of wheat and rye that combines the wheat's high-yield characteristics with the rye's resistance to disease and adverse climate conditions.

In recent years scientists have discovered many plants that contain chemicals that are distasteful or toxic to insects. In some cases, these chemicals, which can be extracted or synthesized, can be used as pesticides specific for the pest, and thus they are not likely to harm other organisms. Some strains of agricultural crops are being hybridized with wild plants that have resistance to pests, a trait that was apparently lost or diminished as crops were selected for their fast-growing capacity or high yield.

SUMMARY

The kingdom Plantae includes the "lower" plants such as algae, mosses, and liverworts and the "higher" plants such as the conifers and flowering plants. Major steps in the evolution of plants included many adaptations to the terrestrial environment, the development of seeds, and the origin of flowers. The major groups of plants are the green algae, brown algae, red algae, liverworts and allies, whisk ferns, club mosses, horsetails, ferns, cycads, ginkgoes, conifers, and flowering plants. Most are photosynthetic, but some are saprophytic and some carnivorous.

Seeds, flowers, and fruits allowed the higher plants to become quite successful and dominate terrestrial environments, while the lower plants continue to dominate aquatic habitats. Many plants have coevolved adaptations with animals.

STUDY QUESTIONS

1.   Define vascular, holdfast, plankton, frond, gametophyte, sporophyte, pollination.

2.   Discuss the adaptiveness of sexual reproduction over asexual reproduction in plants.

3.   Differentiate between pollination and fertilization.

4.   Discuss the adaptations that plants evolved as they moved from aquatic to terrestrial habitats.

5.   What are the general characteristics that distinguish "higher" plants from "lower" plants?

6.   In what ways are seeds evolutionary advancements over spores?

7.   In what ways are flowers adaptive? In other words, why have flowering plants become so successful?

8.   How do dispersal mechanisms differ between flowering and nonflowering plants?

9.   Why have some plants evolved the habit of eating insects?

10.   Explain and discuss the concept of alternation of generations.

Chapter 16

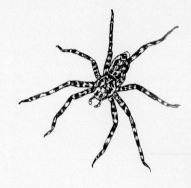

Kingdom Animalia—
The Invertebrates

The organisms in this group are the animals, although some seem much more animallike than others. Sponges and a few other groups superficially look so much like plants that they were classified as such for many years. Today we see such statements in the popular press as "birds and animals," a reflection of the common perception that "mammals" and "animals" are synonymous, while worms, spiders, and squids seem to remain unclassified. But mammals are only a small fraction of the 1 million plus animal species. The vast majority of animals appear to be a vague mystery to most people although many are ecologically very important.

Animals, with some exceptions, are typically more motile than plants; they crawl, swim, burrow, hop, run, and fly. They are heterotrophic, most ingesting their food through some sort of mouth rather than absorbing materials from the environment through their body surface. There are differences at the cellular level also, the most obvious being the lack of a cell wall, the presence of which is characteristic of the other kingdoms.

Having evolved in the varied habitats created by plants, animals radiated into a wide variety of niches: herbivores, carnivores, omnivores, scavengers, and parasites. They live in, on, and above the ground and from the tops of mountains to the deepest parts of the ocean. They may produce a few young in their lifetime or millions of offspring in a year. The animal kingdom is not only as old as the other kingdoms, it is at least as diverse and certainly as interesting (Figure 16.1).

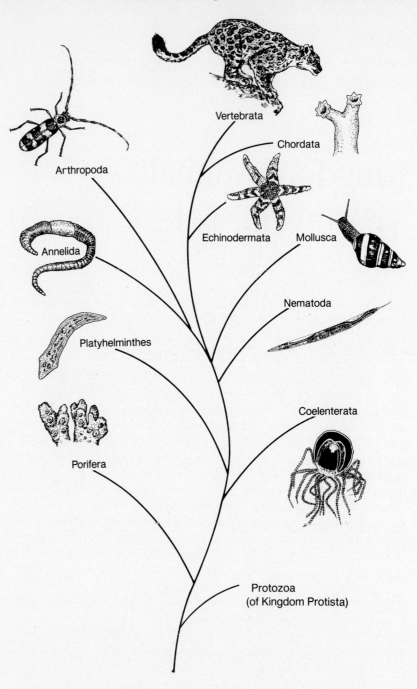

Figure 16.1 Family Tree of Kingdom Animalia. The major groups (phyla) of animals showing their approximate evolutionary relationships.

EVOLUTIONARY TRENDS

Like the plants, the animals have demonstrated evolutionary trends as they adapted to changing environmental conditions and diverged into various forms. Aquatic forms were abundant and dominant for hundreds of millions of years before animals invaded land. Being surrounded by water, gas exchange, nutrient absorption, and the elimination of body wastes were easily accomplished by simple diffusion across the body surface. As long as the animal's body is small and thin, this system will work. But as animals increase in size, their volume increases faster than their surface area, making diffusion into and out of cells deep in the body impossible. Specialized organs that perform particular functions solved these problems. Lungs and gills provide for gas exchange, a digestive system processes food, kidneys eliminate waste, and the circulatory system transports materials throughout the body. A larger size also requires more support, a problem solved by a skeletal system.

Although extant animals range in size from the microscopic single-celled animals to the enormous blue whale, most animals seem to be fairly small; perhaps some sizes are more adaptive than others. For example, there are more species of beetles than there are species of any other animal group; the number of fly species comes in second. Together, flies and beetles have more species than the remainder of the animal kingdom. This may say something about the adaptiveness of size.

Again, like plants, animals evolved adaptations that allowed them to invade and occupy the land. The relative lack of moisture reduced the ability to respire, absorb nutrients, and eliminate wastes; and water no longer surrounded and buoyed up animals' bodies. Adaptations for support and for avoiding desiccation were therefore necessary. As with problems of increasing size, internal organ systems solved many of the problems posed by terrestrial habitats, and external coverings such as plates, scales, or a thick skin reduced the hazard of drying out as well as prevented abrasion as the animal moved along the ground.

Homeothermy evolved in the birds and mammals. Poikilothermic animals were restricted to either warmer environments or to only seasonal activity in colder habitats; they were excluded entirely from the coldest habitats. Homeothermic animals could thus expand into unfilled niches. The ability to maintain a constant body temperature seems to have had its beginning in the dinosaurs, perhaps 200 million years ago. To reduce the energy required to maintain body temperature at a high level, insulation evolved in the form of fat, fur, and feathers. Mammals and birds are the most recently evolved animals (they are about 150 million years old). Success in the biological world, however, is not equated with newness but with perpetuation. Some animal groups have existed essentially unchanged since their evolution; an example of this is the jellyfish, fossils of which have been found to be 600 million years old. That is certainly a indication of success.

KINDS OF ANIMALS

There are several dozen phyla of animals, some of which contain commonly recognized forms such as fish, reptiles, birds, mammals, octopuses, earthworms, butterflies, snails, and others. There are also many rather obscure groups that contain animals that few people are familiar with. One rarely reads of tube or arrow worms, tusk shells, tunicates, sea cucumbers, or flukes. In general, the animals that are the most familiar are those that seem the most abundant, important, or largest in the ecosystem, or those that have some economic value. Endangered species are well known because of the attention they receive and the problems they present. Thus, if you ask someone to name five animals that live in the ocean ecosystem, the response will most likely include whales, seals, porpoises, tuna, and salmon rather than sponges, corals, palolo worms, viperfish, and chimaeras. With that in mind, and recalling that our purpose is not a comprehensive survey of all the forms of life, we will examine some of the familiar, important, or unique animals.

Animal Phyla

Phylum Porifera: sponges
Coelenterata (or Cnidaria): jellyfish, corals, and relatives
Platyhelminthes: flatworms, flukes, and tapeworms
Nematoda: roundworms
Mollusca: molluscs
Annelida: segmented worms
Arthropoda: arthropods
Echinodermata: spiny-skinned animals
Chordata: chordates
   Subphylum Vertebrata: backboned animals
      Class Pisces: fishes
      Class Amphibia: amphibians
      Class Reptilia: reptiles
      Class Aves: birds
      Class Mammalia: mammals

In this chapter we will consider only the invertebrate animals, those without backbones.

THE SPONGES

Mature forms of sponges are sessile, spending their entire lives attached to a rock or other object. This immobility and their plant-like shape prompted Aristotle and other early natural historians to consider them as plants. Careful observation in the mid-eighteenth century established their true animal nature (Figure 16.2).

Almost all sponges are marine; they are common in all oceans, where they attach to rocks, shells, coral, wood pilings, or the sandy bottom. They range from thumb to barrel size and come in a variety of shapes and colors. The three major kinds of sponges are distinguished by the composition of their skeleton, which may be made of calcium, silicon, or a protein called spongin. The natural sponges we use to wash with have skeletons of spongin (Exercise 16.1).

The skeletons of sponges are formed by cell secretions. These secretions are formed into crystallike shapes called **spicules** and may be composed of silicon, calcium, or a network of protein fibers. Sponges are the most primitive of the multicellular animals and possess no true organs. Different types of cells are specialized to perform different functions although a special type of cell, the amoebocyte, has the ability to transform into any other type of cell. The skeletons of live sponges have been crushed experimentally and the cells separated from the skeletal material. Ultimately, the cells clump together and form a new sponge. If cells from two species of sponges are intermixed, the cells first separate into two

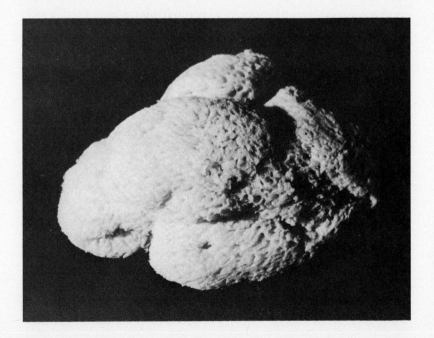

Figure 16.2 Sponge. The bath sponge is commercially harvested and is used for bathing, washing cars, and other such uses. They are most abundant in shallow tropical ocean waters.

Exercise 16.1 Supporting Structures of Sponges.

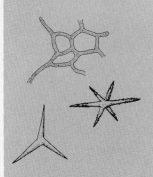

With simple preparation, the supporting structures of sponges can easily be seen. Examined under a microscope, the ordinary bath sponge can be seen to be a fibrous network. The calcium and silicon sponges have rod or crystallike spicules embedded in their gelatinous body for support. Boil a fragment of these sponges in household bleach to dissolve the tissue. Rinse with water and examine the two- or three-pointed calcareous spicules or the six-pointed siliceous spicules. Some species of bath sponges also contain spicules of calcium.

species and then congregate by species, indicating species recognition at the cellular level.

Sponge Ecology

Sponges feed by filter feeding, a common mode of nutrition in the animal world. Flagellated cells draw water into the sponge along with algae, bacteria, and small particles of organic matter, which are then trapped by special cells. Waste matter is discarded also via water currents through the "mouth" of the sponge.

Many plants and animals live in symbiotic relationships with sponges. Crabs, algae, and other small plants and animals may live in the crevices of the sponge skeleton. Sponges are often found growing on corals, oysters, clams, and other animals. One of the silicon "glass" sponges harbors a pair of a species of shrimp. The shrimp enter the fine meshwork of the sponge while they are still young, but become trapped as they grow larger. The Japanese once used these sponges with the imprisoned shrimp in wedding ceremonies to symbolize a long-lasting bond. Sponges do not seem to be a major source of food for other animals but they are eaten by starfish and sea slugs, and are grazed on by some fish.

Human Uses

Before the introduction of manufactured cellulose sponges, bath sponges were used for bathing and household cleaning. Live sponges are collected by hooks dragged from boats or by divers in shallow tropical seas such as in the Mediterranean and off Florida and the West Indies. The live sponges are laid in the sun to dry and to allow the cellular material to decompose, leaving only the skeletons. Then they are beaten, washed, bleached, and dried. Sponges may be cultured by being cut into small pieces, attached to rocks, and dropped back into the ocean to grow.

THE COELENTERATES

The coelenterates (or cnidarians) are grouped into three categories: (1) the *Hydra,* Portuguese man-of-war and allies; (2) the jellyfish and relatives, and (3) the corals and sea anemones. Their diversity in size, shape, and color is enormous, but they all share one characteristic: **nematocysts.** Nematocysts are hairlike structures that are ejected from **cnidocytes** (stinging cells) in response to mechanical or chemical stimulation and are used for defense, prey capture, or holding onto a substrate. Different types of nematocysts penetrate with spear points, adhere with sticky tips, or wrap around an object in whiplike fashion.

Coelenterates are mostly marine, but there are a few freshwater species such as *Hydra* and some jellyfish. They occur as either sessile or slow-moving cylindrical "polyp" forms or as motile umbrella-shaped forms called **medusae**. Many have a life cycle with both forms present. The majority are found in the coastal regions of oceans, primarily in warmer waters (Figure 16.3).

All coelenterates are predators on other animals and feed either with nematocysts, which sting and paralyze large prey or entangle smaller prey, or by sweeping small food particles into their mouth with tracts of hairlike cilia. They eat a variety of foods from protozoa to large fish.

Symbiosis is common among coelenterates. Algae inhabit some forms of hydra and coral, but whether this is commensalism or mutualism is not clear. Some small species of damselfish inhabit the tentacles of some sea anemones without causing the anemone's stinging cells to discharge. The fish thus receive protection from predators, which would be stung if they approached, and the fish presumably act as lures to entice larger fish to enter the tentacles of the anemone. Some hermit crabs pick up and attach small anemones to their backs or claws as protection against octopuses. The anemones perhaps benefit by the crab's motility and/or protection. The relationship is fairly specific, as only certain anemone species will attach to certain species of crabs.

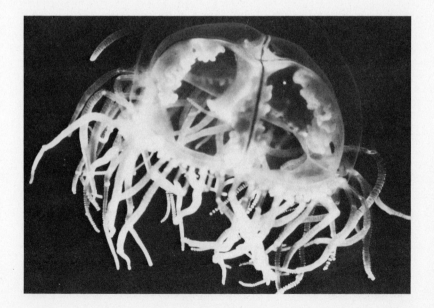

Figure 16.3 Coelenterate. All coelenterates are carnivorous and feed primarily on small crustaceans. Their nematocysts paralyze and entangle the prey and pull it towards the coelenterates' mouth. After digestion, undigestible materials are forced out of the mouth.

Corals

Corals probably exert the greatest ecological influence of all the coelenterates. The living bodies of plantlike corals poke out of their hard skeletons to feed. It is the skeletons made by the corals' secretion of calcium carbonate that form islands, reefs, and ridges that provide habitats for a vast array of other organisms. The skeletons of dead corals provide a place for other corals to grow, and off the coast of Australia, for instance, they form the Great Barrier Reef. This reef is nearly 2000 kilometers long and 100 kilometers wide, and provides habitats for an enormous variety of other animals (Figure 16.4). An individual coral animal is small, but these animals live in colonies, where their skeletons fuse to form coral formations. These self-forming corals are called "stony" corals, to distinguish them from "soft" or "gorgonian" corals that are fan- or tree-shaped and rather flexible.

Some corals contain symbiotic algae that photosynthesize and provide all the coral's food. The algae use the carbon dioxide produced by the coral, provide oxygen for the coral, and increase the coral's ability to produce calcium carbonate. The corals not only provide a habitat for other organisms, but grow rapidly in the warm waters and serve as food for other animals. In addition, the mucus they produce to protect their soft bodies sloughs off and is apparently a source of nutrition to many plankton species.

Corals are eaten by various kinds of animals such as parrotfish, whose strong jaws and razor-edged lips allow them to nip off pieces of the coral skeleton in order to reach the soft-bodied animal inside. A population explosion of the crown-of-thorns starfish some years ago on the Great Barrier Reef caused some concern because of the starfish's taste for coral. Besides these starfish and the wave action of the ocean that constantly wears away and breaks up coral reefs, the greatest threat to this habitat is its alteration and pollution by human activities.

Human Uses

Some jellyfish and sea anemones have been used as a minor source of food in some areas of the world, but human usage of coelenterates is, for the most part, restricted to decorative purposes. Coral is made into necklaces, earrings, lamp bases, and various knick-knacks. In tropical areas it may even be incorporated into road beds and air strips. But perhaps the greatest ecological as well as esthetic contribution of coelenterates is their bright colors and varied shapes, which provide character and structure to shallow tropical waters, much as trees provide those attributes to forests.

Figure 16.4 Coral Reef. An island of the Great Barrier Reef off the northeastern coast of Australia. Formed entirely of corals, the reef islands provide habitats for all sorts of marine animals. The white ring around the island indicates a zone of shallow water.

THE FLATWORMS

The flatworms are subdivided into three groups, each of which has a distinctive morphology: the free-living flatworms, the parasitic flukes, and the parasitic tapeworms. None of the groups commonly comes to mind as fulfilling important heterotrophic roles because of their inconspicuous lifestyles, but the flukes and tapeworms are common parasites of higher animals and affect the populations of these animals, posing severe health problems to both domestic animals and humans in many parts of the world (Figure 16.5).

The Free-living Flatworms

The free-living flatworms, or turbellarians, are mostly marine, but a few species are found in fresh water or in moist areas on land (Exercise 16.2). Some swim, but most dwell on the ocean bottom where they crawl over the mud or sand. Most are shorter than a centimeter but a few tropical terrestrial species are over 60 centimeters in length!

Figure 16.5 Tapeworm. All tapeworms live in the gut of their host, almost all of which are vertebrates. The tapeworm body is protected from being digested by means of a thick covering. This thick covering, however, allows the absorption of nutrients from the host, a necessary process as tapeworms have no digestive system; most of their innards are filled with reproductive organs. The head of the tapeworm is armed with hooks that allow the animal to hold onto the gut lining of the host.

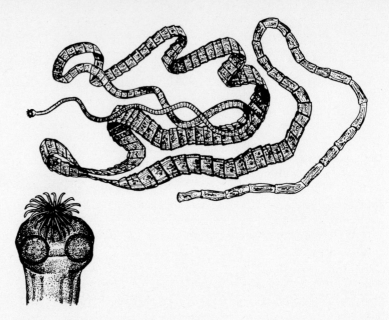

The turbellarians are carnivores and eat smaller worms, arthropods, and other small invertebrates. Some turbellarians are green, due to symbiotic algae, but whether or not the flatworm benefits from the algal photosynthesis is not known. Presumably the green color provides camouflage.

The Flukes

Adult flukes are all parasites of vertebrate animals, but the intermediate stages almost always develop in intermediate mollusc hosts, most commonly snails. (See Chapter 10 for a diagram of a fluke life cycle.) Some immature forms of flukes are capable of swimming, but the adults are essentially immobile and attach themselves to the digestive tract, eyes, lungs, liver, or other organs of the host by means of strong suckers. Blood flukes live in the bloodstream of vertebrate hosts, including humans.

In the United States and other industrial countries with high standards of sanitation and good waste disposal systems, the infection of humans and domestic animals by flukes is fairly rare. But because the eggs of flukes are passed with waste products, flukes pose health problems in some areas of the world with poor sanitation systems. In some parts of Asia, for example, workers in rice fields defecate and urinate in the water, releasing fluke eggs. The eggs hatch in the intestines of snails and eventually transform into free-swimming larvae that burrow into the legs of another worker to start the cycle again. Schistosomiasis, one of the major diseases of the world, is caused by a blood fluke transmitted in this manner. Other fluke infections can be had by eating raw fish or vegetation such as wild watercress, which may harbor encysted forms of flukes.

The Tapeworms

Like the flukes, the tapeworms are totally parasitic; the intermediate host is an invertebrate and the host for the adult is the digestive tract of a vertebrate. Tapeworms often attach themselves by means of hooks on a rounded head and may grow to enormous lengths—perhaps to 20 meters in the gut of a cow. Like the flukes, infection by tapeworms is most common in areas where proper sanitation is not practiced.

Humans may be infected by pork, beef, or fish tapeworms if the flesh of those animals is not adequately cooked before being eaten. The adult tapeworm is covered by a cuticle resistant to digestive juices. Occasionally, tapeworm eggs are ingested and the larval forms encyst in the muscle or even the brain, sometimes causing death.

An old saying referred to someone who ate large helpings at the dinner table as having a tapeworm. However, although low-level infections of tapeworms produce few or no symptoms, larger infections actually cause a loss of appetite. In fact, early in this century, female models who wished to stay or become slim ate the heads of tapeworms to induce a loss of appetite; needless to say, this is not a recommended procedure.

Life As a Parasite

Although most groups of invertebrate animals contain some parasitic species, the flukes, tapeworms, protozoa, roundworms, and arthropods are probably the most well known and important of the groups. Parasites are smaller than their hosts and depend on one or more hosts to varying degrees. Ticks may remain on a coyote's back for days and finally drop off when full of blood, only to find another host later. Adult tapeworms, however, must spend their entire life inside a host, their reproductive organs comprising most of their body, to the exclusion of a digestive and circulatory system. Parasites depend on their hosts not only for food but for dispersal of eggs, larvae, and adults. Parasitism indeed requires a close relationship between organisms.

Such a dependency on one or more hosts may seem to imply a simple, even primitive existence. Just the contrary. Parasites are usually considered the most advanced of the animals in their phylum. Many specializations such as hooks, suckers, the ability to burrow through skin, and so on have evolved. The blood flukes even

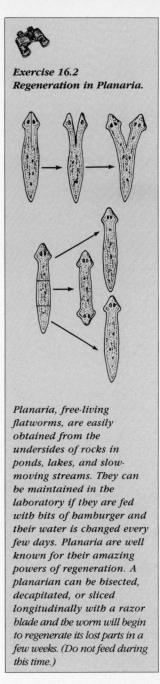

Exercise 16.2
Regeneration in Planaria.

Planaria, free-living flatworms, are easily obtained from the undersides of rocks in ponds, lakes, and slow-moving streams. They can be maintained in the laboratory if they are fed with bits of hamburger and their water is changed every few days. Planaria are well known for their amazing powers of regeneration. A planarian can be bisected, decapitated, or sliced longitudinally with a razor blade and the worm will begin to regenerate its lost parts in a few weeks. (Do not feed during this time.)

have the ability to coat themselves with blood proteins so the host's body does not attack them.

Parasites are highly specialized predators, often restricted to one host species (or set of species). But parasites have different effects on their hosts than do typical predators. Parasites may occasionally kill an individual host, but since killing the host kills the parasites or at least interrupts their life cycle, natural selection has provided for prudent parasites, which feed on but do not normally destroy their hosts.

Parasites do, however, lower the overall health of their hosts, making them more susceptible to diseases and harsh weather, and probably reduce their life spans and reproductive rates. In dense populations of hosts, parasites reproduce and disperse more readily and thus more severe infections occur. As the host populations decline, the parasite "load" does also. Parasites, like predators, are thus density-dependent factors.

THE ROUNDWORMS

The roundworms, more commonly called nematodes, are one of the most abundant of all multicellular organisms. New species are being discovered continually, and estimates of the total number of nematode species approach one-half million, although only about 50,000 are known. They are so widespread and numerous that it has been said if the earth, except for nematodes, was dissolved, nematodes would form an outline of the world and its inhabitants. An exaggeration, but it does convey an impressive message: Nematodes are everywhere in great numbers. They inhabit the soil, fresh water, salt water, and are found from the poles to the equator. Many are free-living, but many also are parasitic on plants and animals (Figure 16.6).

Roundworms are successful partly because of their ability to withstand severe environmental extremes. "Vinegar eels" are sometimes found in vinegar, which has an acid pH of 1.5. They are capable of withstanding long periods of desiccation and extremes of temperatures. Experimentally, nematodes have even been frozen in liquid nitrogen and have survived.

Ecological Roles

Free-living nematodes are decomposers and predators, feeding on bacteria, algae, fungi, various small invertebrates, yeast, waste matter, plant juices, and even other nematodes. Nematodes themselves are prey for various insects, mites, and spiders. One species of fungus traps nematodes by forming a lasso-like loop in which the nema-

Figure 16.6 Roundworm.
Nematodes are extremely
abundant. One square meter
of ocean bottom muck may
yield 5,000,000 worms. A
decomposing apple may
have 100,000. Undoubtedly
they are one of the most
successful animal groups.

todes are trapped as they attempt to wiggle through; the fungus then digests the nematodes (see Figure 14.2).

Nematodes parasitize plants by burrowing into their roots. The plants react by forming "root knots" that are harmful to and may eventually kill the plant. Sometimes marigold flowers are planted to rid the soil of nematodes, as marigolds apparently secrete a chemical that is toxic to the worms. Most soil nematodes are harmless to plants, however, and some even benefit plants by parasitizing plant pests such as caterpillars and beetle grubs.

Parasitic infections of animals, especially humans, by nematodes are much more severe. Hookworm and pinworm are two of the most common parasitic infections in the United States. Before proper sanitary procedures were initiated, trichinosis, a sometimes fatal disease caused by eating insufficiently cooked pork infected with trichinella worms, was not uncommon; it is still found to some degree in about 15% of the U.S. population. Large nematodes of the genus *Ascaris* grow to 18 centimeters and often infect horses, cattle, pigs, and humans. Unlike the tapeworm and flukes, many parasitic roundworms require only one host. The eggs are passed out in the feces and ingestion of the eggs by another host begins the cycle again. Thus, improper sanitation is related to the incidence of nematode infections.

In North Africa, portions of the Middle East, and parts of Asia, nematode and other parasitic infections are common. **Elephantiasis**, the enlargement of appendages due to the accumulation of fluid, is caused by nematodes blocking the lymph ducts. "Devil" or "guinea" worms live just under the skin and cause ulcerated sores. The African eye worm lives in the cornea of the eye and may cause blindness. In North America, although pinworms and hookworm are not uncommon in people, much of our concern is centered around worming our pet puppies, virtually all of which are infected with nematodes.

THE MOLLUSCS

The molluscs are known as the shelled animals although not all are shelled. Most are marine, but quite a few are freshwater species and a number are terrestrial. There are five major groups of molluscs: (1) the snails, slugs, limpets, and allies—the largest group,

Figure 16.7 Molluscs. This group is one of the new invertebrate classifications that have received much attention by amateur biologists. Shell collecting has been a popular hobby for many years and, as a result, molluscs are known almost as well as insects, birds, and mammals. Pictured are empty shells of the most well-known group of molluscs, the gastropods.

numbering about 35,000 and inhabiting freshwater, marine, and terrestrial habitats; (2) the chitons, inhabitants of tidal areas; (3) the clams, oysters, mussels, and allies, of fresh and salt water; (4) the octopuses, squids, and relatives, of marine habitats; and (5) the tusk shells, ocean floor dwellers. All of these groups existed in the Cambrian period over 500 million years ago. Extant species, numbering about 80,000, closely resemble their ancient ancestors. They have a worldwide distribution and fill a variety of niches (Figure 16.7).

Molluscs are carnivores, herbivores, and filter feeders that feed on both live and dead organic matter. Octopuses and squids catch and eat fish and other large animals. Giant squids have even been known to attack whales, their large size and undulating movement perhaps accounting for some "sea serpent" sightings. Small squids are themselves important prey, being food to whales, sharks, and some fish; and many molluscs are eaten by humans.

Snails, slugs, chitons, and their relatives possess a tonguelike organ that resembles a file or rasp. With this **radula**, a structure unique to molluscs, they scrape algae from rocks, eat leaves, consume dead and decaying animals, or even attack live animals. Small snails feed on clams by inserting their radula into the clam and tearing the flesh. Seashells of the genus *Conus* are carnivorous, and their radula contains a toxin that paralyzes their mollusc prey; the toxin may be fatal to humans. Slugs, which resemble snails without shells, are frequently cannibalistic during times of the year when food and shelter are scarce—a form of territoriality.

The radula and teeth of some molluscs are so powerful that they can bore into rocks to form their own shelters. The teredo clam (sometimes called "shipworm") is very destructive to boats and wharves, as it burrows into wood and digests the cellulose. A few species are parasitic. One snail species lives inside of a sea cucumber and feeds on its internal organs.

Clams, mussels, oysters, scallops, and their allies feed primarily by filter feeding. Water is drawn in and excreted through openings where their two shells meet. The water is passed over the gills and organic particles filtered out. Some clams accumulate particular chemicals, and an analysis of their body tissues may indicate water contamination when the pollutants are too diluted to detect in the open water. Other species are sensitive to water pollution and their disappearance from a stream may indicate contamination.

We find some symbiotic relationships among molluscs and other creatures. The shells of molluscs are colonized by algae, barnacles, and sea anemones. The empty shells of dead molluscs are used as homes by hermit crabs (a posthumous symbiosis, perhaps). And, as mentioned earlier, snails serve as hosts for most parasitic flukes. But perhaps one of the most spectacular forms of symbiosis (if indeed it can be called that) occurs with some large marine, unshelled molluscs called sea slugs. Large (up to 30 centimeters) and very colorful, some sea slugs are covered with tentaclelike projections on their back that contain stinging nematocysts. However, the sea slugs do not produce their own nematocysts; they obtain them from coelenterates. They eat the coelenterates, and the nematocysts of the latter are then passed to the tentacles of the slugs (a process that may take as little as 20 minutes). For some reason the sea slugs do not cause the coelenterates to discharge their nematocysts during the process.

Molluscs reproduce sexually, the eggs and sperm usually meetings outside the body in open water. In some filter feeders, the sperm is drawn into the body to fertilize the eggs, which then develop in the body; this happens with clams. Since external fertilization is not effective on land, terrestrial species such as snails and slugs have internal fertilization.

Humans and Molluscs

The mollusc group has undoubtedly been exploited to a greater extent than any other group of invertebrates except perhaps the arthropods. We eat squid, octopus, clams, mussels, snails, oysters, scallops and abalone. Each year over 500,000 tons of squid are caught in waters off the coast of Japan. The Japanese even raise squid in "squid ranches." We culture oysters for pearls and make buttons and decorative inlays from mollusc shells. Mollusc shells

Figure 16.8 An Edible Snail.
This snail is an edible pest, but most Americans would be loathe to eat it. Snails can be trapped in shallow trays of stale beer, but many people prefer to poison them. Recently an entrepreneur in California has been offering 1¢ per snail; he keeps the snails on snail farms and then ships them to France.

are high in calcium and are used to make calcium food supplements for humans and pets. Mussels and other shells attach themselves to rocks, ships, wharf pilings, fishing floats or other structures with a gluelike secretion; this glue is so waterproof and strong that a synthetic version of it is being considered as a dental adhesive to glue in fillings, and so on. Seashell collecting is a popular hobby, and some seashell varieties are worth thousands of dollars. Unfortunately, like most hobbies involving wild organisms, it may be having some effect on their populations.

Molluscs can cause a considerable amount of damage. They burrow into the hulls of wooden ships or into the supporting pilings of piers and eat vegetable crops. In California, a species of edible snail that was imported in the 1800s from France to provide escargot escaped and became a garden pest (Figure 16.8).

THE SEGMENTED WORMS

The familiar earthworms and leeches belong to this group, as do the less well known marine worms. These three groups collectively inhabit a wide variety of environments and niches. They are terrestrial or aquatic, both marine and freshwater. They are predators, omnivores, detritivores, and decomposers (Figure 16.9).

Earthworms

There are lots of animals called worms, so termed because of their long cylindrical shape and flexible body. Earthworms are *the*

Figure 16.9 A Segmented Worm. The earthworms, marine worms, and leeches have very different habits and habitats but they are similar in many other ways. The major structural similarity is their segmentation—their body is divided into similar parts from front to back, their internal organs being divided as well. Pictured here is a leech, the most advanced and specialized of the segmented worms. Thought of as bloodsuckers, some species are predators and feed on other animals, even other leeches.

worms to fishermen, gardeners, and most everyone else. Earthworms burrow through the soil by gripping the walls of the burrow with tiny bristles that occur on their sides and slide through easily with the help of a mucus-covered body. They feed on or near the surface of the soil on decaying organic matter, helping to break down large pieces into smaller ones and forming humus. Burrows help to aerate the soil and allow for the percolation and drainage of water. One hectare of soil may contain 200,000 worms, which help produce a centimeter of new soil every two years.

Earthworms range in size from nearly microscopic forms to the nearly 4-meter long terrestrial worms of Australia. They are susceptible to heat and desiccation and will burrow deep into the soil during dry and hot seasons. During the rainy periods they stay near or on the surface—hence the "night crawlers" that are found in lawns or on the sidewalks after a heavy rain. They are sensitive to vibrations and will come to the surface when a power mower is run over a lawn or the top soil is subjected to vibration in some way. This may be an adaptation for sensing rainfall before rainwater begins to fill their burrows and threatens the worms with drowning. In the southern United States, fishing worms are collected commercially by rubbing a car spring against a piece of dense ironwood tree set on the ground. The vibrations cause the worms to come to the surface, where they are picked up.

Worms, of course, serve as prey for larger animals such as robins, shrews, jays, moles, snakes, and lizards. Earthworms are very tolerant of a variety of pesticides. Pesticides used to control weeds, insects, or even the earthworms (which make mounds of soil on lawns and golf greens) are ingested by the worms and concentrated in their bodies to six or more times the soil concentration. Animals eating the worms thus get a potent dose of pesticides. One study showed a close relationship between the use of pesticides to control weeds and the decline of robin populations that ate contaminated worms.

Leeches

Leeches are flattened worms with a sucker at each end. Most are found in fresh water, but some are marine and a few are terrestrial. A common belief is that leeches are inhabitants only of swampy tropical areas. In reality, they range from deserts to mountaintops; there are more leeches in Antarctic waters than in the tropics! They are carnivorous and feed on the blood of vertebrates by attaching to and slitting the skin or inserting a tubelike tongue. An anticoagulant in their saliva prevents the blood from clotting. Leeches have also evolved an anesthetic chemical in their saliva so that they may feed on their host without attracting attention. When the leech is full of blood, it drops off and swims away. They are predators but not true parasites since most depend on another animal only when feeding and are rarely species-specific. A few do, however, have a specific host; one species feeds on young penguins in the Antarctic and another feeds only on bats in New Guinea caves. Terrestrial species eat earthworms, slugs, snails, insect larvae, and even other leeches.

For centuries, bloodletting or bleeding was a treatment used by many medical practitioners. A vein was sliced open and some blood, presumed to hold the cause of illness, was allowed to flow out. In the 1500s leeches were used to draw the blood out and their posteriors were sometimes cut off to encourage more flow. Charles II of England in the 1680s and Adolph Hitler in the 1930s were treated with leeches in this manner. Pharmacies once used to stock "medicinal" leeches, one species being appropriately named *Hirudo medicinalis*. This therapy presumably did more harm than good and fell to disfavor. However, it has recently been discovered that leeches can be applied after microsurgery on fingers, toes, ears, and so on to prevent blood clotting (due to the anticoagulant in their saliva) and draw out excess blood around the site of surgery. They may also be good for removing concentrations of blood around black eyes. Leeches may never enjoy the popularity in medicine they once had, but a drug store in Chicago still imports leeches from "leech farms" in Russia and Hungary and does a pretty good business at $10 a leech.

Marine Worms

The marine worms are odd-shaped worms; many have paddle-like extensions and long bristles on their sides and bottom. Some are free-swimming and predaceous, feeding on fish and invertebrates captured with pincerlike jaws. Other species build burrows in the sand by cementing sand particles together, and spend their lives flapping their paddles to produce a current through the tube. A sac of mucus traps organic particles and, when full, is moved to the mouth and swallowed. Some species move around carrying their tubes of sand particles with them.

Marine worms, or polychaetes, are extremely abundant and many thousands may inhabit a square meter of mud bottom. They play an important role in the recycling of nutrients because they ingest so much organic matter and then often become prey to fish, crabs, coelenterates, and other animals. Humans occasionally use them for fish bait. One species of polychaete lives in the shell of the keyhole limpet, a mollusc. If the limpet is attacked by a starfish, a common predator, the worm sticks its head out from under the limpet shell and bites the fleshy underparts of the starfish, causing it to retreat.

The palolo worm swarms to the surface in enormous numbers on only two nights in October and two nights in November each year. These reproductive swarms allow the usually sedentary animals to mate and lay eggs. The South Seas natives consider the worms a delicacy and eat them raw right from the sea, like a handful of wriggling spaghetti.

THE ARTHROPODS

The arthropods include a myriad of familiar groups, including crabs, centipedes, thrips, spiders, ticks, lice, barnacles, lobsters, and insects (Exercise 16.3). Arthropods are more numerous in numbers and species than any other group of animals, and they occupy the widest variety of habitats and niches of any animal group. Most invertebrate groups are largely aquatic, but many arthropods are terrestrial because they have succeeded to a greater extent in adapting to the land environment. Their jointed legs allow them considerable mobility, and their exoskeleton of chiton prevents abrasion and desiccation (Figure 16.10). There are approximately 1 million species of arthropods. The largest and most well known groups are the insects (butterflies, beetles, moths, bugs, termites, flies, and relatives), the arachnids (spiders, scorpions, mites, and ticks), and the crustaceans (crabs, lobsters, barnacles, shrimp, and allies).

Exercise 16.3 Collecting Terrestrial Invertebrates.

soil
mesh

alcohol

Larger invertebrates, primarily arthropods, can be easily collected from terrestrial habitats with sweep nets or in soil samples. Sample trees, shrubs, herbs, and grass for invertebrates by briskly sweeping durable sweep nets of canvas back and forth against them. Then anesthetize the sample with ethyl acetate or ether while it is still in the bag and identify the animals in the field, or preserve the sample in 70% ethanol for later study. For soil and litter animals, the Berlese funnel or similar apparatus is used. Place soil above the funnel spout and illuminate it. The heat and light drive the invertebrates downward through the soil to eventually fall into a container of alcohol.

Figure 16.10 An Arthropod. Spiders have the exoskeleton and jointed appendages as do all arthropods, but spiders are a unique group because they have eight legs, are capable of producing silk, and have excellent vision.

The Insects

The most well known of all the arthropods are the six-legged insects. With the greatest number of species, insects fill the roles of decomposers, omnivores, herbivores, carnivores, predators, prey, parasites, pollinators, and disease carriers. Insects comprise the only group of organisms besides bats and birds that is capable of flight; as such, the insects disperse easily and some are migratory. There are a number of insect groups, but nearly 70% belong to five major groups: the beetles; the butterflies and moths; the flies; the bees, wasps, and ants; and the true bugs.

Some insects such as the true bugs suck plant juices. Termites eat dead wood with the help of symbiotic protozoa. Beetles eat the live wood of trees, decaying vegetation, and a variety of other organic matter. Many bees eat nectar and pollen, grasshoppers eat leaves, and female mosquitoes feed on blood.

The sensory environment of the land is more variable than that of water, so insects have evolved many refined sensory structures. Most insects have two kinds of eyes: a simple eye for detecting light and a compound eye for seeing images. Some insects can see ultraviolet light or even carbon dioxide, both of which are invisible to the human eye. Insects have a pair of antennae that can smell or detect vibrations. Others hear via eardrums or sensory hairs on the sides of their body. A number of insects can apparently detect polarized light. It has been demonstrated that bees orient to polarized light, and many migratory insects may use polarized light as a navigational clue in the same way we use a compass.

Insects, along with humans and other primates, are one of the few animal groups that have evolved complex social systems. Termites, bees, and wasps are social insects, but ants have evolved the greatest degree of sociality (Figure 16.11). Army ants, for instance, have colonies that consist of a queen and various kinds of workers, which feed the young and protect the nest. Large columns of worker ants stream out of an underground nest each morning to capture prey, usually other arthropods. (Contrary to popular belief, army ants do not devour every living thing in their path, though they will occasionally eat a lizard or snake during their raids.) One group of small workers has the job of feeding the larval ants. The intermediate-sized workers do most of the prey capturing, and the largest individuals have enormous heads and jaws and protect the colony against intruders. All their activities are conducted in well-coordinated groups.

A number of animals undergo **metamorphosis**—a change in body form during their life cycle; the changes from caterpillar to butterfly, maggot to house fly, and grub to beetle are familar to us.

Figure 16.11 Termite Nest. Termites are a very social group of insects and are one of the few groups that makes itself obviously a part of the landscape. Many species live in decaying wood on or above the soil, but some tropical species build mud nests that may reach 10 meters in height. Here the author's father examines a termite nest in northern Botswana.

Not all insects undergo a total metamorphosis (egg–larva–pupa–adult), however, but all go through intermediate growth stages during which they shed their old exoskeleton and grow a new one. For example, young grasshoppers, called nymphs, look like adults but do not have wings. This is **incomplete metamorphosis.** The various stages of complete metamorphosis allow insect forms to overwinter as eggs in a dormant egg case, larvae, or pupae, but they also allow adults and immature forms to avoid competition for the same food. Caterpillars feed on plant parts, for instance, while moths and butterflies eat nectar. And dragonfly larvae are aquatic predators and adults are terrestrial predators.

Insects are prey to many animals (and some plants), the major predators being other insects. As a result, predator-avoidance mechanisms are common among insects. Some treehoppers are shaped like thorns on a rosebush; the inchworm (a moth larva) is colored like a dead twig and, when endangered, positions itself as if it were one. The puss moth caterpillar has bristly antennae and large spots resembling eyes on its head; when threatened, it rears back and presents a posture presumably forboding to enemies. One species of butterfly larva secretes sugar water from its pores. The sugar water attracts ants which crawl all over the caterpillar. This gives the caterpillar protection against predatory wasps and flies that would otherwise lay their eggs in the caterpillar. Ants contain formic acid, distasteful enough to deter some predators. Some birds, however, actually pick up ants and rub them through their feathers, a process called **anting**, apparently to clear their body of lice and mites.

Insects are themselves predators on a variety of other animals. The giant water bug captures fish with its forelegs and injects a paralyzing venom. A wasp will kill katydids and grasshoppers to feed to her young. Water scavenger beetles and diving beetle larvae may feed on tadpoles, and there are a number of insects that feed on the blood of vertebrates.

Insects eat a variety of prey, and some dragonflies, wasps, termites, and others are cannibalistic. One of the more intriguing cases of cannibalism occurs in the praying mantis. As with many insects, copulation takes place and the female ultimately lays eggs. Interestingly, the female praying mantis may eat the male after, or while, copulation takes place. The postcopulation cannibalism seems to occur only when the female is hungry, perhaps due to a shortage of food, and may be adaptive since the female needs the food energy to produce eggs. Even more amazing, however, is that the female may eat the head of the male before copulation occurs. Rather than bringing the intimacies to an end, the removal of his head actually instigates copulatory movements since the nervous system in the male is constructed to do just that. The decapitated male will die after mating, of course, but he has passed on his genes. This odd behavior may have evolved as an adaptation for the females to cause the male to mate when they (the females) are ready to produce eggs and not expend energy in courtship and copulation prematurely.

Major Insect Groups

If you picked any animal species at random, the chances are good that it would be a beetle. Over 300,000 species of beetles are known worldwide. They are found anywhere insects are found and feed on all types of plant and animal matter. Some are carnivores, some herbivores, many are omnivorous, and many are scavengers; a few are parasitic. Many are pests, eating stored foods, especially grain. Others are useful, being predators of crop pests and helping to recycle organic wastes. The incredible variety of beetles is reflected in the common names of their families: wrinkled bark beetles, whirligig beetles, ship-timber beetles, pleasing fungus beetles, carrion beetles, ant-loving beetles, grain beetles, and dung beetles (Figure 16.12).

The butterflies and moths are the most spectacular of the insects. The larvae (caterpillars) have chewing mouthparts for feeding on vegetation, and the adults have a long tubular tongue called a

Figure 16.12 South African Dung Beetles. These beetles are rolling a ball of dung that they have formed from an elephant dropping. The ball of dung seen here is slightly larger than a golf ball but is easily rolled as the beetle pushes it with its rear legs. The beetles lay their eggs in the dung and the developing larvae feed on it. Dung beetles are extremely important in recycling waste material.

proboscis, through which they suck nectar from flowers. A few adults live only long enough to lay eggs and do not feed at all. Some larvae are quite destructive to crops; 90% of the damage to apple trees is caused by the larva of the codling moth. Moths can be distinguished from butterflies by their feathery or hairlike antennae; butterflies have antennae with a knob at the tip.

The bees, wasps, and ants are related, but quite diverse. They feed on a variety of foods, some ants killing their prey with poison fangs. Stingers in the abdomen of female bees and wasps are only for defense. Most have a highly developed social system with work roles well defined within a colony. Sex in most of these insects is determined by whether or not the egg is fertilized. A fertilized egg develops into a female and an unfertilized one becomes a male.

The flies include such animals as mosquitoes, gnats, and midges. Many insects that are called "flies" (sawflies, fireflies, dragonflies, mayflies) actually belong to other groups. The true flies are the only insects with one pair of wings; all the others have two. Flies have sucking mouth parts but some are adapted for biting or piercing. They feed on nectar, sap, blood, other insects, or the flesh of larger animals.

True bugs are distinguished by having sucking mouth parts. June bugs, lightning bugs, and ladybugs are beetles. Most species of bugs are terrestrial but a number are aquatic. Most feed on plant juices, but a number feed on other insects or suck the blood of vertebrates. Many are destructive to crops but many are valuable as predators on destructive insects.

Dragonflies belong to one of the oldest insect groups, having existed for millions of years before the dinosaurs. One primitive dragonfly had a wingspan of nearly a meter, but otherwise they have changed little over the years. Their two pairs of transparent wings, perpendicular to the body at rest (damselfly wings rest parallel to the body), beat independently, allowing great maneuverability. They can fly at 40 kilometers per hour, hover, fly backwards, turn quickly at sharp angles, and stop in an instant. Their large eyes have nearly 30,000 independent sensors and can catch quickly moving prey such as mosquitoes or moths. Male dragonflies establish and defend territories with the help of their bright colors. The dragonfly spends about 75% of its life as an underwater nymph. The nymphs are also predatory, and may eat prey even as large as fish or tadpoles. After several molts, the nymph crawls out of the water on a plant stem and a fully formed adult dragonfly splits the nymphal skin and emerges.

Mimicry, the adaptation of looking like something else, is very common among insects. Some katydids, butterflies, and moths look like leaves, even mimicking the appearance of fungal growth and insect damage (Figure 16.13). A few species of moths have caterpillars with bright colors and large spots that look like eyes, and they assume a threatening posture when a predator approaches. The monarch butterfly has larvae that feed on milkweeds, from which they obtain distasteful chemicals. The viceroy butterfly looks like the black and orange monarch and is avoided by birds that have learned to not eat the unpalatable monarch but have not learned to tell the two similar butterfly species apart. One species of moth looks very much like a bumblebee, and certain flies closely resemble wasps and bees. A predator that has previously been stung is likely to avoid both model and mimic.

Arachnids

This group contains about 60,000 species of spiders, scorpions, ticks, and mites. Most are terrestrial and all have eight legs.

The familiar spiders are closely tied to insects in a predator–prey relationship. All spiders are predatory and equipped with poison fangs. Their prey is injected with venom that digests the doomed insect; the spider then sucks the digested body dry. Their behavior is rare among the animal world since few predators use traps to capture their prey, and their use of silk for webs and traps is unique.

Figure 16.13 A Dead-Leaf Butterfly. This butterfly not only looks like its name implies, but it sits on a twig in an appropriate position and remains motionless.

Spiders have three to seven silk-producing glands at the rear of their abdomen. Each gland produces silk for a slightly different purpose. Some glands produce the silk for the body of the web, some make the adhesive silk that attaches the web to other surfaces, and some produce the silk that wraps the prey. The liquid silk is drawn out into strands by the spider's legs and solidifies. Web-building spiders build a variety of webs in the form of domes, funnels, triangles, and the familiar orb of silk strands radiating out in a flat plane with concentric circles woven among them. Even the most complex web can be built overnight. The design of a web is characteristic of a species and is a behavioral isolating mechanism since males are attracted to a female by the web she builds. Experiments have shown that certain mind-altering drugs (LSD, marijuana, tranquilizers, for instance) fed to spiders will alter the shape of their web, although the basic pattern will remain.

Other methods of prey capture are just as fascinating. Some spiders throw sticky lassos of thread at the prey. Others dig holes, cover them, and hide underneath; when an insect passes over the hole, the spider pulls the hole cover and the insect down. The "trap door" spider builds a hinged lid over a hole, out of which it pops to capture a wandering morsel. Fishing spiders hide in the crevices of trees just above the surface of the water and ambush fish. The spitting spider spits sticky threads at its prey. The tarantula ambushes and chews up its prey. One species of tarantula shares a burrow with a desert toad for mutual protection.

The so-called daddy longlegs are not true spiders but are closely related. Their oval bodies have no waist (as spiders do), their fangs have no poison, and they spin no webs.

Scorpions were perhaps the first arthropods to invade land successfully and they have existed since the Silurian. Scorpions are abundant in tropical and subtropical areas, in both dry and humid environments. They are usually nocturnal, emerging at night from their protective burrows under logs or rocks to prey on insects and spiders. Some sand-dwelling scorpions can detect prey through vibrations in the sand. The stinger on the end of the abdomen contains a toxin that immobilizes prey, but only a few species are dangerous to humans.

Mites and ticks are characteristically parasitic, but not all are parasitic. They burrow into or embed their pincerlike mouthparts in the skin of their host, usually a bird or mammal, and feed on blood. Often they are simply a nuisance, causing itching and discomfort; the itch mite of dogs is an example. They also transmit diseases, however, and in that way can be quite harmful. The cattle tick transmits the protozoan that causes "redwater fever" (so-called because the urine turns red). The spotted fever tick transmits Rocky Mountain spotted fever. Some mites are destructive to greenhouse plants and vegetable crops.

Crustaceans

The crustaceans, like the insects, are ecologically and economically important arthropods. Crustaceans include the crabs, crayfish, barnacles, lobsters, whale lice, shrimp, and pill bugs. But, unlike the insects, the crustaceans are primarily aquatic and most abundant in marine ecosystems. Most are carnivorous but some are filter feeders (for example, barnacles) and some are parasitic, such as the small, flat, crablike creatures that live under the scales of some fish (Figure 16.14).

Species of cleaning shrimp have developed symbiotic relationships with coral, sea anemones, and many marine fish. The shrimp

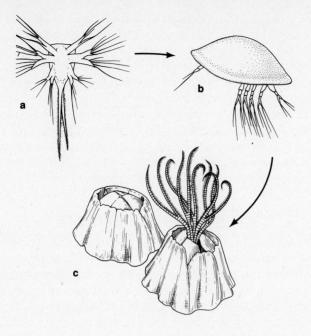

Figure 16.14 Barnacle. The barnacle is one of the more unusual crustaceans. Its featherlike appendages create a current of water that draws food particles into its mouth. (a) and (b) are free-swimming larval stages; (c) is the sessile adult. The barnacle goose of Greenland was once thought to arise from barnacles because of the barnacles' feathery appendages.

live in the crevices of the coral or among the tentacles of the anemones, cleaning these animals and receiving protection from predators in return. (The shrimp are immune to the nematocysts' sting.) The coelenterates are cleaned of debris and parasites. Fish are also cleaned, assuming a particular posture or changing colors to signal the shrimp that cleaning is desired. Thus, the shrimp knows it will get food and protection but not be eaten by the fish. Presumably, any shrimp that tried to clean fish randomly would be soon eaten. The shrimp frantically wave their antennae back and forth to signal the availability of their cleaning services.

Shells are characteristic of the molluscs and a leathery exoskeleton covers the arthropods. But the hermit crab uses the empty shell of a mollusc such as periwinkles, triton clams, and turban snails as a shelter by crawling inside and carrying the shell for protection. Being terrestrial and crawling about on beaches, hermit crabs are exposed to predation by birds and mammals, and a protective shell is necessary for survival. The hermit crab has to find a suitable-sized shell, however, and then exchange that shell for a larger one as it grows. The larger shells are rare, so the hermit crab population competes intraspecifically for large shells, the scarcity of large shells being the major form of hermit crab population control. This leads to aggression between the crabs, and one occasionally pulls another out of its shell in order to occupy it. Sometimes, though, a mutual exchange of ill-fitting shells will occur; a large crab with a small shell will exchange its home with a small crab that has a large shell.

Crustaceans are frequently key elements in ecosystems. Small species such as the common *Daphnia* provide food for many larger organisms in fresh water, and *Calanus* is a major source of energy in the North Atlantic. Crustaceans also play an economically important role in the seafood industry. Crayfish are related to lobsters, but are restricted to fresh water. They are taken from streams or ponds, or "farmed" in the southern United States, and used as food; the boiled tails are very tasty plain or with simple spices.

Economic Impacts of Arthropods

Of all animals besides vertebrates, the arthropods probably have the greatest impact on our lives. They affect our agriculture, health, dwellings, and our general comfort. We spend enormous sums of money and in some cases endanger our environment trying to control insects with pesticides. About 2 billion kilograms of pesticide are applied annually worldwide to control pests—about one-third of that for insect control (the other two-thirds for weed control). Not all "insecticides" are directed towards insects; some are aimed at other arthropods such as mites and spiders (Figure 16.15).

Forests and agricultural fields are sprayed with a variety of pesticides to control corn borers, cotton boll weevils, earwigs, fruit flies, spruce budworms, pine beetles, and so on. Most agricultural insecticides in the United States are used to control the corn borer and cotton boll weevil; most insecticide applications on forests are to control the western budworm and the gypsy moth. Some forms of biological control are also used, such as sterilizing male insects and releasing them to mate with wild females, which then lay infertile eggs. This has been done with mixed success with the screwworm fly and the Mediterranean fruit fly. Some male insects are attracted to traps baited with females of their species, or with sex hormones (**pheromones**) that attract from a distance.

Growth-inhibiting hormones have been used to interrupt the metamorphosis of larval insects. These are species-specific forms of control unlikely to greatly disturb the environment. Other forms of biological control are also species-specific (a certain bacterium attacks only tomato worms) or somewhat broader (the use of ladybird beetles, wasps, or preying mantids). Many other biological methods are being studied as alternatives to pesticides.

Arthropods carry a variety of diseases, including some that are quite widespread and devastating. Fleas are responsible for transmitting the bubonic (black) plague organism, as well as tapeworms

Figure 16.15 Pesticides Pro and Con. The safety of many pesticides and their long-term effects are much argued. Pesticides are, by definition, toxic to living organisms, and although the effects pesticides have vary with the organism, none can be considered entirely safe for humans. Pesticides do have considerable benefits, though, so the dilemma posed to our society is deciding whether the potential dangers of pesticides outweigh their benefits. There is no doubt that millions of lives have been saved by eradicating disease-transmitting insects and increasing the productivity of agricultural lands. But several bird species (brown pelicans, peregrine falcons, bald eagles) have declined in number due to pesticide use, and many fish have pesticide levels so high that, for instance, we are warned not to eat more than a pound of fish per week from Lake Michigan and told to remove the fat from the fish (which stores DDT). It is said 90% of human cancers are environmentally induced, probably some due to pesticides.

Nonspecific pesticides kill insect predators (other insects) and thus open up niches. In Nova Scotia, spider mites were not a problem in orchards until codling moths were controlled by DDT; then the spider mites became pests. Pesticides float through the air, accumulate in the soil, and move through ground and surface water. Persistent ones may last for many years and concentrate in higher trophic levels. In addition to potential health effects, ecosystems are changed and simplified by the reduction or eradication of one or several arthropod species. Unfortunately, there is insufficient evidence to weigh properly the benefits and drawbacks of most pesticides.

and flukes. Mosquitoes carry malaria (caused by a protozoan), yellow and dengue fever (caused by viruses), and filariasis (caused by a nematode worm). Tsetse flies transmit sleeping sickness; deer flies, tularemia; and house flies, typhoid fever, dysentery, and cholera.

Millions of dollars are spent each year to prevent the importation of foreign pests, mostly insects, into the United States, where, in the absence of natural controls, they could destroy crops or spread disease. The Hessian fly helped to destroy the United States wheat crop during World War I, and the Mediterranean fruit fly (on several occasions in Florida, Texas, Hawaii, and most recently, California) has caused millions of dollars of damage to a variety of fruit crops. In the United States, the U.S. Department of Agriculture's Animal and Plant Health Inspection Service employs 700 inspectors at 100 ports of entry. Although they inspect millions of pieces of luggage and the holds of thousands of ships and planes, it is likely that many insects are never detected, and new pests will continue to invade.

On the positive side, arthropods are very useful as food, pollinators, predators of crop pests, and producers of commercial products. Arthropods, especially insects, are important links in the food chain as the major or sole prey of many vertebrates. People eat many kinds of arthropods such as crabs, lobsters, and shrimp, and in many parts of the world locusts, ants, grasshoppers, beetle larvae, and termites are also eaten. Fried caterpillars are a delicacy in Mexico.

Many plants are dependent on insects as pollinators. Beekeepers not only gather honey from their hives but rent them out to orchard owners during flowering time. Apples, pears, plums, lemons, oranges, strawberries, pumpkins, eggplants, peppers, and others depend on insect pollination.

The majority of insect-eating animals are insects, and no doubt do more to control insect pests than pesticides. They may be predators or parasites. Predaceous insects include dragon- and damselflies, tiger beetles, ladybird beetles, some wasps, lacewings, stinkbugs, and robber flies. Parasitic insects attack their host (for example, a caterpillar) only during their larval stage. The eggs are laid in or on the host and when they hatch, they literally eat the caterpillar alive.

Bees produce honey and beeswax, silkworms produce silk, and shellac is produced by the lac insect of the Far East. Dyes have been made from scale insects, and drugs such as cantharidin from the blister beetle, and allantoin (to treat infections) from blowfly larvae. Bee venom has been used to treat arthritis.

The Successful Arthropods

The earliest arthropods evolved over 600 million years ago. Since about 75% of all animals are arthropods, we have to consider them the most successful animals. Perhaps the credit for this success should go to the evolution of a thick exoskeleton, which allows movement but minimizes evaporative water loss and thus allowed invasion of terrestrial environments. Arthropods have a higher proportion of terrestrial species than any other invertebrate group, as they diversified on land in the absence of serious competition from other groups.

THE SPINY-SKINNED ANIMALS

The echinoderms (literally "hedgehog-skinned") are all marine and include the starfish, sand dollars, sea cucumbers, sea lilies, and sea urchins. They have an internal skeleton of bony plates, which project through the skin as spines. Characteristic of and unique to the group is a **water vascular system** of canals throughout the body, which ends in a series of tube feet. Like numerous tiny suction cups, they hold the organisms to the substrate, move the animal, and grip prey. Since the echinoderm skeleton is made of calcium, it preserves readily; hence, echinoderms are one of the few animal groups with an abundant fossil record.

The five major groups of echinoderms are the sea stars, brittle stars, sea cucumbers, sea urchins, and sea lilies. They are found on the floor of marine habitats from shallow tide pools, which are alternately inundated and exposed by the tide, to the ocean bottom 10 kilometers down (Exercise 16.4). Most echinoderms are carnivorous or detrivorous, capturing live animal prey or filter feeding on organic matter. Like many other marine invertebrates, echinoderms produce thousands of eggs, which hatch into a larval form and then metamorphose into an adult. There may be some parental care by the adults—some starfish and sea cucumbers have brood pouches—but very often the eggs are just scattered into the sea.

The colorful starfish have five or more arms and a central mouth. They range in size from a centimeter to a meter, and feed on clams, oysters, and other related molluscs by wrapping their arms around the shells and pulling the clam open, everting their stomach into the clam, digesting the viscera, and retracting their stomach. Some starfish, such as the crown-of-thorns, will also feed on coral. Some starfish filter feed. Starfish can regenerate a lost arm. An arm, if part of the central portion of the starfish remains attached, can regenerate the rest of the body. Sponge fishermen in Florida at one time tried to reduce predation on sponges by capturing starfish and chopping them into pieces; needless to say, they made the problem worse.

The brittle stars have five to dozens of thin arms attached to a circular disc. While the sea stars' arms are rather stiff, the brittle stars have flexible, almost prehensile, arms by which the animals skitter along the ocean bottom. The arms of these starfish break easily, hence their name, and regenerate readily. They feed by raking small animals and bits of debris into their mouth with their arms.

Looking like its vegetable namesake, the sea cucumber is a soft-bodied creature that traps food particles in its tentacles and stuffs them in its mouth. These creatures may slowly crawl along the ocean bottom or remain motionless in a shallow burrow. A species of fish called the pearlfish may spend its entire life in the body cavity of the sea cucumber, coming out only under the cover of darkness to

Exercise 16.4 Collecting Aquatic Invertebrates

Aquatic invertebrates can be collected by hand from tide pools, ponds, and slow-moving streams. but a variety of equipment is required to sample effectively most aquatic environments. Dredges, scoops, plankton nets, sieves, and seines are all used. See Field and Laboratory Methods for General Ecology *by J. E. Brower and Jerrold H. Zar (Dubuque, Iowa: W.C. Brown, 1977), pages 91–97 for detailed information on aquatic sampling equipment and techniques.*

feed, although it may also feed on the cucumber's internal organs. Being eaten from the inside is not as disastrous as it seems, since the cucumber has remarkable powers of regeneration. It can eject its internal organs through its anus or even burst open when disturbed; it can then slowly repair and regenerate its innards. In Asia, sea cucumbers are eaten boiled, roasted, or raw and added to soups.

The sea urchins and related sand dollars are enclosed in a hard shell called a **test**, which has spines that are sometimes very long; the spines of some species contain a toxin. Sea urchins feed on a wide variety of foods, live or dead, and some species are filter feeders (Figure 16.16). They have an elaborate chewing apparatus called **Aristotle's lantern**, with which they can chew seaweed scraped from rocks. Sea urchin eggs are easily fertilized, and for many years they have been used in the laboratory in studies of embryo development. Their gonads are eaten in some parts of the world, and their skeletons can be crushed to produce a dye.

The sea lilies, or crinoids, are the most ancient of the echinoderms, having existed since the Paleozoic era 500 million years ago. Most have a stalk with joined appendages at the end, giving the animal a flowerlike appearance. Some are free-swimming but many are sessile. They feed by filtering organic matter from the sea.

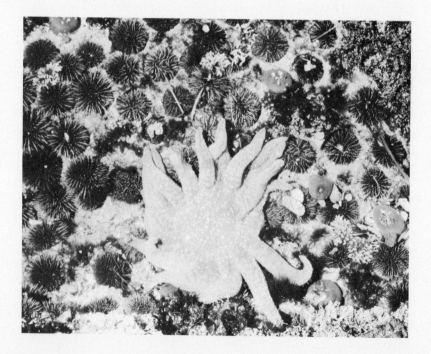

Figure 16.16 Sea urchins. A bed of sea urchins with a sea star moving over them.

THE CHORDATES

The chordate group is ecologically and economically very important because among its members are the vertebrate animals: the fish, amphibians, reptiles, birds, and mammals. Besides vertebrates there are the tunicates, sea squirts, acorn worms, lancelets, peanut worms, and others. These, in an evolutionary sense, are quite important because of their relationship to vertebrates. Ecologically, however, they are relatively small and uncommon, and hold no major role in the marine ecosystems they inhabit.

These invertebrate chordates share with the higher vertebrates three major characteristics: a dorsal hollow nerve cord, a nerve cord, and gill slits. Even humans possess these characteristics early in their development. It is thought that larval forms of tunicates ("tadpole larvae") resemble the earliest fishlike vertebrates, and thus vertebrates may have arisen from the tunicates some 450 million years or so ago.

Although the groups described in this chapter contain many more species of animals in a much greater diversity of form and lifestyles than the vertebrates, the vertebrates have captivated humans to a much greater degree. Thus, Chapter 17 is devoted to vertebrates. Remember, however, that the importance of animal groups in nature is measured not by human interest but by ecological roles. By that measure, invertebrates certainly hold their own.

SUMMARY

Animals are typically more motile than plants and fill more niches. They are heterotrophic herbivores, omnivores, carnivores, and decomposers. They live in, on, or above the ground from the tops of mountains to the deepest parts of the ocean. Like the plants, animals evolved in aquatic habitats and some forms adapted to terrestrial habitats. All plants and most animals are poikilothermic, but the birds and mammals are homeothermic. The major groups of animals are the sponges, jellyfish and relatives, flatworms and relatives, roundworms, molluscs, segmented worms, arthropods, spiny-skinned animals, and chordates.

STUDY QUESTIONS

1. Define spicule, homeothermic, poikilothermic, cnidocyte, metamorphosis, pheromone, Aristotle's lantern.
2. Describe the adaptations that evolved to adapt animals to a terrestrial environment.

3.   The arthropod group seems to be the most successful of all animal groups since it contains the greatest number of species and has representatives in the greatest number of habitats. Why do you think arthropods are so successful?

4.   What invertebrate animal groups do you think have the greatest impact on humans?

5.   Homeothermy seems to be an important adaptation, yet it is restricted to birds and mammals. Why do you think homeothermy did not evolve earlier in the course of animal evolution?

6.   List three ecological characteristics of each invertebrate group.

7.   List as many symbiotic relationships among the invertebrates as you can.

8.   Why are parasitic members of a group considered to be the most advanced members? Which invertebrate groups have parasitic representatives?

9.   Describe the types of metamorphoses that occur in insects. Do any other groups of invertebrates exhibit metamorphosis?

10.   The chordates include the vertebrates. What characteristics do the backboned vertebrates and the invertebrates have in common?

Chapter 17

The Vertebrates

The vertebrates are the backboned animals, which are undoubtedly the most familiar of all animals. If you ask someone to name an animal at random, the chances are very high it will be a vertebrate. Vertebrates hold a special role in the animal kingdom, partly due to their size and visibility; to their evolutionary relationship and similarity to humans; to their close affinity to human society as food, pets, and so on; and to their often obvious role as consumers in many ecosystems.

In number, there are approximately six times as many species of beetles as there are vertebrates. Since vertebrates are often at or near the tip of a food chain, their biomass is frequently much less than that of the invertebrates in the ecosystems, and the energy they contain is but a small fraction of what was trapped by the producers. Thus, vertebrates may be of relatively little importance in some ecosystems; for example, they may often be rare, infrequent, or absent from such habitats as small ponds and streams, and patches of terrestrial habitats. But, among other attributes, their size, color, and the ability of some to make loud sounds, makes them more conspicuous than other animals. Since they are the most well known of all animals, we will discuss them in a little more detail than we did the invertebrate groups.

THE FISHES

There are about 25,000 species of fishes in the world, far more than any other vertebrate group. Some fishes are among the biggest vertebrates; the largest fish, for example, is the whale shark, a plankton feeder that may reach 20 meters in length. The largest freshwater fish is the great paddlefish of China, which may reach 10 meters. The smallest is a goby of the Phillipine Islands whose adult size is less than a centimeter.

The first fish evolved about 500 million years ago in seas that were considerably less salty than now. These early fish, covered with bony armor, had no jaws and fed by filtering organic particles from the water. By 300 million years ago, jawed fishes similar to ones we are familiar with today had evolved. Having existed for such a long time in the watery habitat that covers three-quarters of the earth, it is no wonder that fish have evolved into so many species.

Fish are found in virtually every aquatic environment, although there are no freshwater fish in Greenland or Antarctica. Some fish are adapted to the warm tropical seawaters of equatorial regions, others to stagnant freshwater pools, and others to cold, fast-moving streams. Some live in shallow waters, a few can walk on land (one species can climb trees!), and a variety are found at great depths in the ocean. Most are restricted to either marine or freshwater habitats, but some move between both environments at times during their life cycle. Salmon, for instance, move into fresh water to **spawn** (lay eggs) and die; the young later return to the ocean to grow to maturity. Others live in **brackish** (slightly salty) water all of their lives, killifish and tarpon, for example. Fish are more widely distributed across the earth than other vertebrates, not only across longitudes and latitudes, but vertically as well; they live in mountain streams and lakes and in the deepest depths of the oceans.

What Makes a Fish a Fish?

Fish are poikilothermic animals, with gills for respiration, paired fins for balance and locomotion, air bladders for depth regulation, a lateral line for the detection of vibrations and sound, a flexible internal skeleton of bone, and a covering of scales (Exercise 17.1). There are exceptions to each of these: Lungfish have lungs as well as gills, sharks and their relatives have no air bladder or lateral line, and a number of fish do not have scales. Being poikilothermic, their body temperature is always close to that of the surrounding water; some live in arctic marine waters below 0°C (the salts in the water lower its freezing temperature) and a few species are found in hot springs at 43°C.

Major Groups

There are three major groups of fishes. ("Fishes" is the proper plural when referring to more than one species; for more than one individual of the same species, "fish" is proper). The hagfishes and lampreys are the most primitive living fishes. They have circular, sucking mouths with no jaws, no paired fins, a skeleton of cartilage, and no scales. Hagfishes are marine, scavenging on the ocean bottom. Lampreys are **anadromous**, breeding in fresh water and migrating to the ocean. As adults they attach to large fish with their suckerlike mouth, and suck blood from their host. About a century ago a canal was built to connect Lake Ontario to Lake Erie. Lampreys were part of the natural fauna of Lake Ontario, but when they were allowed into the western Great Lakes, they caused considerable damage to the populations of native fish, especially the trout. Only in the last 20 years, when the lamprey population has been controlled by specific chemicals that kill only lampreys, have the trout populations increased.

The sharks, skates, and rays comprise an ancient group that has changed little in hundreds of millions of years. They have jaws, a cartilaginous skeleton, and tiny pointed scales that resemble teeth and give the skin a sandpaperlike texture. (Shark skin was once used as sandpaper in cabinet making.) Skates and rays are bottom dwellers and sharks may inhabit the ocean floor or open waters. All are marine except for a species of bull shark that lives in Lake Nicaragua.

The bony fishes comprise the largest and most modern of the fishes. They have a bony skeleton, a lateral line organ to sense vibration, a swim (air) bladder to help them adjust their buoyancy, and roundish, overlapping scales (although some, like catfish, are scaleless). In the lungfishes, the air bladder has been modified to serve as a lung, allowing the animals to burrow into the mud to survive the dry season when their habitats dry up.

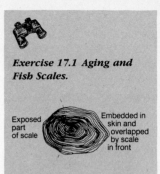

Exercise 17.1 Aging and Fish Scales.

Exposed part of scale / Embedded in skin and overlapped by scale in front

Fish scales show growth rings as do trees. Counting these rings gives an estimate of a fish's age. A set of rings, rather than one ring, represents a year's growth. This set is called an annulus. *Remove scales from a fish and examine under a microscope. Besides the annuli, note the portion of the scale that is embedded in the skin.*

Ecological Roles

Like the insects, fish fill a wide variety of niches, and their feeding habits are quite diverse. There are fish that feed on dead organic matter (carp, catfish, suckers, and some minnows) by sucking up debris from the bottom. Some species filter detritus out of the water by means of fine filaments on the gills: shad, herring, whale and basking sharks, garden eels, and the paddlefish, for example. Herbivorous fish take bites of aquatic plants, algae being a large source of food. Some minnows, carp, guppies, some catfish, and many others feed primarily on plants. Predaceous fish come in a variety of types. Some feed on other fish: sharks, bass, and barracuda swallow their prey whole. Some species of fish feed on larger ones by nipping off their scales or taking chunks of their fins or flesh. The sawfish, a kind of shark with a long snout that bears teeth on its sides (resembling a double-edged saw), violently shakes it head from side to side as it swims through a school of smaller fish. Some of the smaller fish are killed or injured and the sawfish eats them leisurely. The deep sea anglerfish dangles a modified fin spine in front of its mouth as a lure; an unsuspecting fish inspecting the lure gets snapped up (Figure 17.1). The archerfish of some South American rivers spits a stream of water from the surface to an overhanging leaf or branch and knocks its insect prey into the water.

Some fish, particularly sharks, are considered dangerous to humans. Shark attacks do occur, but one's chances of being hit by lightning are much greater. Only 9 of the 250 shark species are considered "man-eaters," and in most cases shark attacks are probably due to the shark mistaking the person for a seal or other usual prey. Recent evidence indicates that sharks can sense weak electrical fields. Virtually every aquatic animal produces a weak electrical field (due to differences in the electrical potential of the skin), and sharks apparently rely on this field much more than on their sense of smell to detect prey. Like the sharks, moray eels have a reputation as ferocious fish. Their snakelike form and large sharp teeth are no doubt the reason for their image, but they are usually quite docile. Some species are even plankton feeders, filtering water for their food. It is also quite likely that the snake eels, up to 3 meters in length, were the actual basis of many sea serpent stories as they undulated across the surface of the ocean.

Figure 17.1 The Anglerfish. The anglerfish is a bony fish and one of a few species that uses a lure to capture its prey. Looking like a rock at the bottom of the sea, it eats unsuspecting fish that are tempted by its lure.

There are a few species of "electric" fishes such as the electric catfish that can produce several hundred volts of electricity to stun their prey. Several species of South American fishes emit continuous weak electric fields for the purpose of echolocation. They live in murky waters and use electric impulses to navigate. Sharks apparently use this "sixth sense" to migrate from one part of the ocean to another.

Fish occupy a number of symbiotic roles. Pearlfish live in the body cavity of sea cucumbers. Some cardinal fishes live in the body cavity of living conchs (tropical molluscs). The damsel and clown fish live among tentacles of sea anemones. The Portuguese man-of-war fish inhabits the tentacles of this large, stinging coelenterate. "Cleaning shrimp" clean the parasites off large fish.

"Senioritas" are small fish that inhabit specific "cleaning stations" in kelp beds; they rid larger fish of external parasites, some gobies, angelfishes, and wrasses may also be cleaners. The parasites they eat are aquatic crustaceans, commonly called "fish lice." The shark suckers or remoras have a suction-type device on their head with which they attach to whales, turtles, and larger fish, especially sharks. They get a free ride and the leftovers from the larger animals' meals.

Fish are parasitized by many organisms and are occasionally parasites themselves. Flukes and tapeworms are common in many fish, and some of these parasites can be passed to humans who eat insufficiently cooked fish flesh. Fungi, bacteria, crustaceans, and leeches also infect fish. The lamprey parasitizes other fish, as does a small South American catfish called the candiru, which sucks the blood from the gills of larger fish.

Reproduction

Fish reproduce in ways as diverse as their feeding habits. They may be **oviparous** (lay eggs), **ovoviviparous** (the eggs develop and hatch in the body and the young are born alive), or **viviparous** (bear live young with only a short egg stage). There may be internal or external fertilization of the eggs, nests may or may not be built, and parental care varies from none to extensive.

Many species lay eggs, usually in great numbers. The eggs are well supplied with yolk to nourish the developing embryos and are encased by a thin shell. (Some skates and rays encase their eggs in a hard egg case called a Mermaid's purse). The shells may be covered with a sticky substance so that the eggs adhere to some substrate and are not tossed about by water motion. Nonsticky eggs are rubbery and resilient during much of their development to resist mechanical damage. Salmon eggs are resilient and become even more elastic when fertilized. At salmon hatcheries, fertilized eggs are separated from nonfertilized eggs by rolling them over a corrugated surface; the fertilized eggs bounce off and the sterile ones roll smoothly downward.

Some fish build nests. Some nests are simply a depression in the sand or mud (bluegill, catfish) or a hollow scooped out from under a rock (darters, some minnows). Others are built from plant materials (stickleback) or bits of vegetation incorporated into a floating nest of saliva bubbles (Siamese fighting fish, gouramis). *Copeina arnoldi,* a fish of the Amazon Basin, lays its eggs on leaves just above the surface of the water and splashes them for several days until they hatch and the **fry** (newly-hatched fish) drop into the water. A number of fishes show no parental care and simply scatter their eggs among rocks or vegetation and abandon them. Some, though, demonstrate extreme dedication to their eggs and young. The female "Egyptian mouthbreeder" incubates the eggs in her mouth, and after the fry are swimming freely, allows them to use her mouth as refuge at any sign of danger. The male seahorse incubates the eggs in an abdominal pouch and ejects the young when hatched. Bass will protect their young against predators for several weeks after hatching.

Importance of Fish to Humans

Fish have been an important part of the human diet since at least prehistoric times. Many cultures, such as the Japanese and some Eskimo tribes, rely on fish as a major or sole source of protein. Fish are a supplementary source of food for nearly every country in the world. They are becoming more popular in the diet-conscious United States because their meat contains less fat than other meats. So-called trash fish (carp, suckers, and so on) have been caught commercially for use in pet food and fertilizers but are now being developed as protein supplements for use in underdeveloped countries.

Most fishes are at or near the top of the trophic level, and our exploitation of them means we are harvesting a low energy source. As a result, many areas of the world are being overfished (indicated by the decreasing average age of fish caught), and competition between fishing fleets of different countries is becoming severe, even leading to hostilities. Pollution by PCBs, pesticides, and mercury has decreased the number or quality of fish in certain areas. Acid rain has killed the fish populations of perhaps 10% of northeastern U.S. lakes.

To avoid the problem of overfishing and/or pollution, fish are increasingly being raised and/or harvested under artificial conditions for commercial or recreational purposes. Since the survival

Figure 17.2 A Fish Farm. Many more fish can be raised in an artificial pond than in the wild, because temperature, food supply, pollution, diseases, and predators can be controlled.

rate of eggs and fry is far higher in hatcheries than under natural conditions, streams and lakes can be restocked rather easily. Fish farms raise fish to sell as food, the fish never having lived in a natural habitat. Although we still rely mainly on wild fish for food, fish farms may be considerably more important in the future (Figure 17.2).

THE AMPHIBIANS

In the middle of the Paleozoic era, about 350 to 400 million years ago, fishes rapidly diversified and the land began to be invaded by the first terrestrial vertebrates, the amphibians. The amphibians, although capable of existence on land, are still tied to water for part of their lives. "Amphibian" means "alternate lives."

In contrast to the fishes, the nearly 3000 known species of amphibians is the lowest number of species of any vertebrate group. Amphibians evolved from a group of lobe-finned fishes that were probably capable of living out of the water for short periods of time. Although the amphibians were a definite advance over these fish, their dependence on water has limited their success. There are three major groups of amphibians: frogs and toads, salamanders, and the lesser known caecilians (Figure 17.3).

Amphibians come in a wide assortment of shapes, sizes, and colors. The smallest are frogs that may reach only 8 millimeters in length as adults; the largest is a 1½-meter long giant salamander

Figure 17.3 Leopard Frog.
The leopard or meadow frog
may be the most widely
distributed amphibian in
North America. It is common
in marshes, streams, lakes,
and grassy meadows from
Canada to Mexico. The
female lays a jellylike mass
of perhaps 5000 eggs, which
hatch in 2 weeks, become
tadpoles in 2 months, and
adult frogs in a year.
Tadpoles eat algae; adults eat
insects, worms, and other
small invertebrates.

from Japan. Many are colored cryptically in shades of brown, green, or gray, but some are very spectacularly colored, especially some of the poisonous tropical frogs that advertise their distastefulness to potential predators.

What Makes an Amphibian an Amphibian?

Amphibians live in aquatic, semiaquatic, or at least moist habitats for part of the year. Most live in water for all or a good portion of their life cycle, but there are amphibians that live in trees and some that live in hot deserts. As a rule, amphibians lay eggs that are fertilized externally and that hatch in winter into gilled larvae that are entirely aquatic. A change in form occurs (metamorphosis), and the gilled larvae become adults with lungs—the tadpole is transformed into a frog, for example. In a few cases, the gilled larvae become gilled adults; these species are aquatic their entire life.

Besides the alternation of life cycles in and out of water, morphological adaptations also limit the success of amphibians. Their thin, scaleless skin must stay moist, and in some cases it is an important respiratory organ; some salamanders, for instance, respire entirely through their skin and the mucous membranes of their throat. Mucous glands help to moisten the skin in the absence of water; that's why amphibians are slimy. Mucous glands may also secrete toxic or distasteful substances to deter predators.

Amphibians are poikilothermic and restricted to warm or temperate environments. They may be active during the entire year in tropical areas but become dormant in cold winters. Generally well adapted for locomotion in water, their legs are weak and clawless and ill-suited for travel across land, although a few frogs can climb trees. The caecilians are limbless and can only burrow or swim.

Major Groups of Amphibians

About 2% of the 2500 amphibian species are caecilians, 7% salamanders and newts, and the remainder frogs and toads.

Frogs and Toads

The frogs and toads are the most abundant and widespread (worldwide) of the amphibians. There are three major groups but the distinction between them is not always clear. Frogs tend to have a smooth skin and spend most of their time in the water or on floating aquatic vegetation. Tree frogs are more terrestrial and spend a good deal of time on land and in terrestrial vegetation, sometimes far from water. Toads have a warty skin and spend considerable amount of time away from bodies of water.

Reproductive patterns vary somewhat, but in general the female lays eggs that are fertilized by the male. Often there is a mass migration and congregation of individuals in the rainy season; continual croaking and chirping function to attract individuals to an area and to identify sexes. Many kinds of calls have been identified, including a "release call" given by a male that is clasped by another male in a mistaken attempt at mating. The eggs are laid in jellylike masses in aquatic vegetation, in nests of foam in the water, frothy nests on land, in water that collects between the leaf stalks of terrestrial plants, in rotting logs, or in underground burrows. The eggs are usually abandoned, but some species will guard their eggs until hatching. The eggs may become free-swimming larvae or develop directly into adults, depending on the species. The eggs of the Surinam toad are deposited by the female on her own back, where they are fertilized by the male. The young develop in the skin of the female's back and ultimately burst out of the skin as small toads. The eggs of the South American marsupial toad are fertilized by the male and deposited on the back of the female as the pair tumble through the water. The eggs develop into tadpoles in the skin of the female's back; she releases them by making an opening in her skin with her hind legs. A few tropical species lay their eggs under logs and carry their larvae to open water when they hatch. The male European midwife toad wraps the eggs in a string around his waist until they hatch. Two species of South American frogs lay their eggs on moist cliffs; the larvae never enter water before becoming adults. Several South African frogs lay their eggs under moist leaves, where the eggs hatch into small versions of the adult, bypassing the larval stage.

In one region of Ecuador, 81 species of frogs exist in a 1 kilometer square area. One reason they are able to coexist is their diversity of breeding patterns; at least ten different modes of reproduction have been identified there.

Salamanders

Salamanders may be totally aquatic, partially aquatic, or terrestrial. A few live underground, many live on the forest floor, and several live right in the canopy of tropical forests (Figure 17.4).

There are about 300 species of salamanders; half of the species are found in the northern hemisphere of Asia, Europe, and North America. The other half live in the tropics of Central and South America. The northern salamanders fit the typical amphibian pattern: Eggs laid in water hatch into gilled larval forms that change into adults. The southern salamanders, in a situation of seasonal rainfall, have become more terrestrial. The eggs are laid in damp places under logs or leaf litter, and the young hatch into fully developed salamanders. Fertilization is internal but contact between the female and male is not necessary. The male deposits a packet of sperm that is picked up by the female.

Some species lay their eggs on land in the autumn; winter rains then wash gilled larvae into a stream. Other species lay their eggs in the water in clumps of moss near ponds, or on land vegetation. Some species are viviparous. The mudpuppy and hellbender are

Figure 17.4 Red-backed Salamander. The red-backed salamander is common to forested areas in the eastern United States, and is found most often under logs, stones, bark, and leaves. It has two distinct color places: the "red-backed" form, with a wide reddish stripe down the back, and the "lead-backed" form, which is all gray above. Other forms of this species, sometimes considered subspecies, have thin or zigzag dorsal stripes. Because of their sluggish habits, salamander populations tend to be somewhat isolated and thus differentiated from one another.

totally aquatic, not only laying their eggs in the water but remaining in the water as gilled adults. They are essentially reproductive individuals in juvenile form.

Caecilians

The caecilians are long and wormlike with no legs and very small eyes. Found in Asia, Africa, and South America, they burrow like earthworms beneath the soil (except for one aquatic form). Their eggs are laid on land but the larvae still have gills.

Ecological Roles

All adult amphibians are carnivorous, preying on small insects, spiders, worms, crayfish, and other aquatic or terrestrial invertebrates. Larger frogs and toads may capture large insects, small fish, rodents, and birds, and are often cannibalistic.

Frogs and salamanders have sticky tongues that can be extended to capture an insect. Some tropical salamanders can extend their tongue a length equivalent to one-third their body length and pull it back with prey in less than 10 milliseconds. Due to their generally small size, paucity of species, and low population numbers, the relative importance of amphibians in most ecosystems is minor. There are some surprising exceptions, however. One study in a New Hampshire forest showed that both the biomass and numbers of salamanders were greater than those of birds and mammals.

During the breeding season, when masses of salamanders or frogs congregate, there may be a temporary impact on their prey, but amphibians are usually not abundant enough to be considered as major predators or in turn to be a significant food item for their predators. Amphibians are eaten by fish, reptiles, birds, and mammals—all the other vertebrate groups. They are occasionally taken by an invertebrate; predaceous water beetle larvae will sometimes eat tadpoles.

As larval forms, amphibians fill the role of herbivores and omnivores. Generally, larval forms such as tadpoles, or "pollywogs," and gilled salamanders feed on algae or other small plants. Not only does metamorphosis involve the development of lungs and the loss of gills (and in frogs and toads the development of limbs), but a change in trophic levels to a higher level consumer. One advantage of metamorphosis from a larva to a very different adult form is that competition between young and adult is reduced because they rely on different food sources. An aquatic tadpole scraping algae off rocks fills a very different niche than the one occupied by the terrestrial insect-eating toad it eventually becomes.

Amphibians and Humans

Amphibians do not seem to play a large economic role in present-day society. Some amphibians are or have been used as food: mudpuppies in the eastern United States, the giant salamander in Japan, and frog's legs in many regions (Exercise 17.2).

Frogs are often used in high school and college biology classes to represent the typical vertebrate, and frog eggs are used in classrooms or scientific research to study embryo development. The unusually large size of the mudpuppy's kidneys make it ideal for studying the vertebrate excretory system. Various newts, aquatic as adults and terrestrial as larvae, are common aquarium pets. Frogs and salamanders are often used by fishermen as bait.

The mucus secreted by the skin of many amphibians is distasteful, irritating, or even poisonous. Handling a toad will not cause warts, but the skin glands of some frogs and toads can cause severe skin irritations. One group of South American frogs produces a very toxic nerve poison; the bright color of these frogs warns potential predators. South American Indians use the frogs' secretions to poison their hunting arrows.

THE REPTILES

Reptiles evolved about 300 million years ago, were the dominant forms of terrestrial animal life for 200 million years, and num-

Exercise 17.2 Captive Amphibians and Reptiles. *The keeping of wild animals as pets is not encouraged, but short-term observation of some species can be interesting and instructive if the animals are maintained properly. Toads and lizards can easily be kept in a glass terrarium with a sandy bottom, some twigs to climb on and bark to crawl under, and a shallow dish of water. Fed a diet of insects or worms and kept in a moderately lighted and heated location, they will do nicely for several months. Frogs and salamanders need such a terrarium also, but the entire habitat should be sprinkled daily to keep it moist. Newts, larval salamanders, and tadpoles need a totally aquatic environment. Turtles are easy to keep if given an environment that resembles their natural habitat: dry, semiaquatic, or aquatic. (Be sure that even aquatic turtles can get out of the water.) If an animal does not adjust well to captivity, it is best to release it.*

Snakes usually do not do well, even under good conditions. See A Field Guide to Reptiles and Amphibians by Roger Conant (Boston: Houghton Mifflin, 1958) for hints on keeping reptiles and amphibians.

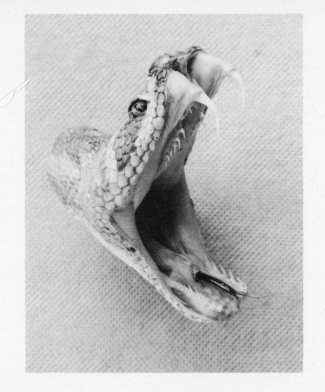

Figure 17.6 Snake's Teeth. The teeth of a snake are arranged in rows in the upper and lower jaws and curve backwards toward the throat. If one is broken or lost, a new one grows in its place. Venomous snakes have grooved or hollow fangs to direct or inject toxin into their prey. This rattlesnake and other "pit vipers" have special sensory pits on the head which detect heat and thus help the snake locate warm-blooded prey. The tongue, flicking in and out, serves as an organ of smell as it takes molecules of air into a pocket in the floor of the mouth where a special olfactory organ lies.

in rattlesnakes. Although not as potent as neurotoxic venom, it is very effectively injected. A rattlesnake's fangs are hollow and sharp, like hypodermic needles. Folded against the roof of the mouth when not being used, they swing into position during the strike. When they pierce a prey, an amount of venom appropriate to the size of the prey is injected. A broken fang is no problem since it will rapidly be replaced by a new one. The most dangerous snakes in the world possess neurotoxins; the boomslang of Africa, the tiger snake of Australia, the krait of India, and others need to inject only 5 milligrams or less of toxin into a human to cause death. Most neurotoxic snakes have a less refined method of delivering their poison. The poison follows a channel down the fangs into the wound made by the teeth. It is not as effective a system as the injection system of hemotoxic rattlesnakes, but considerably less neurotoxin is needed to immobilize a prey.

Statistical evidence of incidents of snakebite is incomplete, but it is estimated that perhaps 1 million people a year are bitten by venomous snakes, 15% dying as a result. Most snakebites occur in the underdeveloped areas of the Far East, Africa, South America, and Australia. In the United States only a few thousand cases are reported each year. Most toxic bites are from rattlesnakes, with perhaps only a dozen resulting in death. Many more Americans die from allergic reactions to bee stings.

Snakes are either oviparous or ovoviviparous. They may guard the young and eggs tenaciously; some snakes, such as the cobra, are particularly aggressive during the breeding season.

Snakes catch their prey with their jaws and either swallow the prey alive, kill it first with poison, or suffocate it by coiling around it. Their eyesight is good, and a special organ at the base of the tongue "tastes" the air as the tongue flicks in and out. In the rattlesnakes and other "pit vipers," a special sensory pit between the eye and nostril is capable of detecting infrared radiation, thus enabling the snakes to find warm-blooded prey in complete darkness. The most well known of the pit vipers are the rattlesnakes. Contrary to an old myth, the age of a rattlesnake cannot be told by the number of rattles. A new rattle is formed whenever the snake sheds its old skin, which may be one to three times a year; rattles also break off. Many snakes that have no rattles and are not poisonous vibrate their tail anyway. In dry leaves, this shaking may create a nose similar to a rattler's buzz. This is mimicry of behavior.

Sea snakes are restricted to the tropical Pacific and are well adapted to a totally aquatic life, with a flattened body, paddlelike tail, and the ability to remain submerged for 20 minutes. Like sea turtles and marine iguanas, sea snakes have special salt glands to rid their bodies of excess salt. Sea snakes feed on fish, which they kill with neurotoxic venom. Experiments have shown that Pacific Ocean fish, normally preyed upon by sea snakes, avoid them as any animal avoids a predator. But Atlantic fish, not having evolved in the same ecosystem as sea snakes, will not try to avoid them and will even try to eat the snakes—with disastrous results for the fish. These experiments were done because of tentative plans to build a sea-level canal that would directly connect the Atlantic and Pacific oceans. The present canal consists of a series of locks with a freshwater lake center, which provides an effective barrier to the dispersal of animals from one ocean to another. A sea-level canal without such barriers, however, would allow sea snakes into the Atlantic with potentially horrible results for Atlantic fish. (The crown-of-thorns starfish, which eats coral, might also be introduced from the Pacific.) Any possible introduction of an exotic species should be studied carefully; there seems to be no doubt that sea snakes do not belong in the Atlantic.

Lizards

Snakes and lizards are distributed worldwide, each with about 3000 species. In many ways, lizards resemble snakes and are indeed closely related. But most lizards have four legs, eyelids, external ear openings, and nonexpandable jaws (Figure 17.7). Their sense of vision is well developed, and their hearing is poor. Lizards live in burrows, swim, climb trees, or live among rock piles. One species with webbed feet can run across the surface of ponds for short distances, and one has weblike extensions of its arms that allow it to jump from trees and glide long distances. Some lizards, such as the anoles of the West Indies and the chameleon of Africa, have skin pigments that allow the animal to change colors to match its background. Lizards are usually either carnivores or herbivores. The horned lizard captures insects on the run, skinks search for invertebrates under rotting logs, chameleons shoot out a long sticky tongue for arthropods, chuckwallas of the U.S. deserts eat creosote bush, and the sea-going Galápagos iguanas dive for algae.

Lizards, like snakes, lay eggs or bear live young. Eggs may be buried in the sand or laid under logs and the newly hatched young left to fend for themselves. The heat produced by decomposition or the warming of the sand by the sun helps to incubate the eggs. A few species of lizards, like several invertebrates, exhibit parthenogenesis—development without fertilization; it seems that only females of these species exist.

Lizards protect themselves from predators—reptiles, birds, and mammals—by cryptic coloration (for example, the chameleon), or speed of movement. The horned lizard (sometimes mistakenly called "toad") can squirt blood from its eyes to repulse predators. The only poisonous lizards are the gila monster and beaded lizard of the southwestern United States and Mexico. They are not particularly dangerous and must hold on to their prey (or enemy) and chew to work the toxin into the wound.

Figure 17.7 This lizard is sunning itself on a log in the early morning cold. Note the claws on its feet and the covering of scales. The external ear opening just behind the angle of the jaws is characteristic of lizards. Note also the more muscular legs and the more vertical orientation of the legs than those of the salamander in Figure 17.4. Lizards are able to move considerably faster.

Turtles

Turtles are freshwater, marine, or terrestrial reptiles encased in bony shells that incorporate their backbone and ribs. They may be carnivorous, herbivorous, omnivorous, or even detritivorous. Instead of teeth, however, their jaws are formed into a beak. The land turtles such as the box turtles, gopher turtles, and tortoises feed mainly on plants and invertebrates. They may not travel beyond a 100-meter radius during a year. Some hibernate in the soil in colder areas and estivate in mud during hot weather. The pond turtles, such as sliders, cooters, terrapins, and painted turtles, spend most of their time in the water, but will sun themselves on logs, lilypads, and the like. They will eat plants or animals in or out of the water, including berries, live or dead fish, worms, and tadpoles.

In many turtle species, the females leave the water at night and dig a pit in the sand or soil. The pit is just under the surface so that the eggs can be warmed by the sun. Perhaps 50 eggs are laid, which hatch in three or four months. During the incubation period predators seek out the nests, and many are destroyed by foxes, skunks, and raccoons. Out of the surviving eggs hatch young turtles, which dig upwards and search for aquatic habitats once on the surface. The mortality of small turtles is high, as they are easy prey for many predators. Those that do reach adulthood may live 25 to 100 or more years.

The marine forms such as the green and loggerhead turtles have paddle-shaped limbs and are totally aquatic, except when the females leave the ocean to lay eggs in beach sand. Some sea turtles, which are the largest of the turtles (up to 400 kilograms), are endangered due to the use of adults as meat, eggs as food, shells as souvenirs, and so on. Some species feed mostly on algae, but others eat animal food such as crabs, jellyfish, and fish. The alligator snapping turtle is more aquatic than most and rarely leaves the water. Normally sluggish, it can attack prey swiftly. It will lie quietly on the bottom of a river with its mouth open, a small red appendage attached to its tongue wriggling like a worm. When a fish comes to investigate this "worm," the snapping turtle rapidly closes its powerful jaws. These turtles grow large and strong enough to kill and eat adult ducks, and large individuals (up to 1 meter in length) have been known to bite through broom handles and snap off fingers. Because the snapping turtle rarely leaves the water, its shell becomes covered with algae, which provides the turtle with camouflage. During winter, snappers hibernate in the mud at the bottom of ponds. Their metabolism is very low then, and they can take in sufficient oxygen across their skin to survive even if the pond is frozen over.

Crocodiles

Crocodiles, alligators, caimans, and gavials are the most closely related living reptilian relatives of the dinosaurs. Once much more abundant, the current 23 species are now restricted to tropical and subtropical areas of the world. All are carnivorous and attack large animals in and out of the water, including humans. They may hibernate or estivate during harsh seasons.

During the mating season, male alligators and crocodiles emit a roar that may be heard for miles in order to attract females and establish a territory in which the nest will be built and guarded. The nest is a mound of decaying vegetation, and the heat of decomposition helps to incubate the eggs. The young hatch and fend for themselves, many being eaten by birds and other animals.

Alligators play a key role in the Florida Everglades by keeping bodies of water open that would otherwise fill in with vegetation. These "alligator holes" provide niches for many kinds of animals.

Humans and Reptiles

Reptiles have long been used for food, clothing, and decorative items. Turtles and their eggs, snakes, lizards, and crocodiles have all been eaten, to the extent that, for instance, sea turtle populations have been drastically reduced. The Galápagos Islands tortoise was almost wiped out by sailors who considered the tortoise an easy source of food. They stacked hundreds of tortoises in the holds of their ships to be used as a fresh source of meat during the voyage. (The Abingdon Island tortoise became extinct in 1962.) The diamondback terrapin turtle was once sold as a delicacy at "$1 per inch" shell diameter. The flesh of boa constrictors, pythons, rattlesnakes, and cobras is eaten in their respective parts of the world, but more as a novelty than as a major food. Shoes, handbags, hats, hatbands, belts, and wall decorations are made from crocodile, lizard, and snake skins. Turtle shells are all too often converted into ash trays or jewelry. For the most part, reptiles do not provide any essential products for humans, although some snake venom is used to produce drugs that may be useful against certain diseases of the nervous or circulatory systems.

Snakes have long been wantonly destroyed by humans. The slithery demeanor, staring, unblinking eyes, and the venomous fangs of some species have caused people to develop an unwarranted fear of snakes (ophidophobia, to use the proper term). But snakes are useful predators of rodents and, in relation to human activities, do much more good than harm.

THE BIRDS

Of all the forms of life, birds are perhaps the most beautiful; certainly they are among the most melodious, most obvious, and most admired animals. Nearly 8700 species of birds are distributed worldwide except in the center of Antarctica. Although there are 27 groupings of birds—and some groups are quite different from others—all birds share many characteristics, and birds are perhaps one of the most homogenous of all groups. As alike as birds are to one another, they are also unique; rarely are birds mistaken for other animals or vice versa (Exercise 17.3).

It has been said that dinosaurs never became extinct—they changed into birds. A bit of an exaggeration, perhaps, but birds did evolve from small dinosaurs and share characteristics with modern reptiles such as egg laying, similar skeletal features, and similar blood proteins. Birds also share some characteristics with mammals (which also evolved from reptiles). Birds and mammals are homeothermic, have a four-chambered heart, and a high, constant metabolic rate. Birds have unique characteristics such as a beak without teeth, the ability to fly, and a feature peculiar to birds: feathers.

Fossils found in 1861 revealed what is considered to be the first bird, a definite transition between birds and reptiles that lived about 150 million years ago. It was lizardlike, with toothed jaws, a slender tail, clawed fingers, and other distinctly reptilian characteristics. Were it not for the unmistakable impressions of feathers, it would have been classified as a reptile. This primitive bird, *Archeopteryx*

Exercise 17.3 *Identifying Birds.*
Of all animals, birds are probably the easiest to identify. Birdwatching is probably the most popular of all nature activities worldwide. All that is needed is a pair of binoculars. A few simple hints will turn the beginning birdwatcher into a skilled amateur ornithologist in a short time. Spot the bird with the naked eye first, then focus the binoculars on it. Look

for as many obvious characteristics as possible: silhouette, flight pattern, wing shape, bill shape and length, plumage pattern and color. Color alone is too often depended upon, but it is not reliable under many conditions in the field. Look for "field marks"—those characteristics that are immediately obvious—a white rump, white outer tail feathers, barred tail, extremely long bill. If the bird sings or calls, that may be a help or even a necessary factor for identification. Often, the

song or call is all the information needed. There are many field guides for birds for every area of the world. You should carry one with you to consult.

(ancient wing), about the size of a crow, probably climbed trees since its anatomy indicates it could not fly.

From this and other early creatures, birds have evolved into a variety of forms. The smallest of all living birds is the Cuban bee hummingbird, which weighs about as much as a penny (2 grams); it drinks nectar, eats insects, and can fly backwards as well as hover. The largest bird is the African ostrich, which weighs 135 kilograms, and eats fruits and small animals. Its lack of ability to fly is compensated by its ability to run at speeds of up to 60 km/hr while taking 3-meter strides. The largest bird that ever lived is the now-extinct elephant bird of Madagascar, which weighed 450 kilograms and laid an 8-liter egg, the equivalent of 183 chicken eggs!

Most birds fly. The ability to fly probably evolved as a means of escape from predators. Most flightless birds either evolved in a location that had no major predators, or were able to escape or avoid predation by running (ostrich), hiding (rail), and/or being nocturnal (kiwi), or by being large and powerful themselves (the ostrich and emu). Besides such well-known flightless birds as the penguins, ostriches, and their relatives, there are flightless cormorants, grebes, ducks, rails, and wrens. Albatrosses, shearwaters, and petrels living on the open sea are the best flyers, spending endless hours just above the ocean. The wandering albatross has the largest wingspan of any living bird—about 4 meters; it is probably the largest flying bird that ever lived. Birds are homeothermic; only mammals share this characteristic. Because they are physiologically adapted to maintain a constant warm body temperature, they are better able to withstand temperature fluctuations and extremes in the environment. Thus, they are active on a year-round basis and can occupy environments that many other animal groups cannot. But being homeothermic requires a high metabolic rate and high-energy food sources. Poikilotherms need less food per gram of body weight, and thus do not need to seek sources of food as frequently as homeotherms. Smaller homeotherms, such as hummingbirds, have a large surface/volume ratio, resulting in a continual high loss of body heat and the necessity to forage almost constantly. But the advantages of homeothermy outweigh its disadvantages: It is one reason why birds and mammals are such successful animals.

Major Groups of Birds

To get an idea of the vast diversity of bird types, let us examine some major groups and their habits.

The herons, bitterns, and egrets are the long-legged wading birds that stalk prey in shallow water. They spear or snap up between their bills frogs, fish, crayfish, and snakes. Often small fish will seek shelter in the shade of one of these birds, only to be eaten by it;

some herons have learned to shade the water with their wings to attract fish or merely to reduce the sun's glare. The related flamingos and spoonbills filter microorganisms from the soft mud by pushing the mud through a sievelike mechanism in their bills. The storks feed on food similar to that of herons, but will also eat young birds and scavenge for dead animals.

The waterfowl are a large and widespread group. These are the ducks, geese, and swans. Some ducks eat fish, but most subsist on aquatic plants and fish; geese and swans graze on terrestrial plants and also take some insects. The Chinese goose is occasionally used to weed out bean fields. Most waterfowl are ground nesters, although some ducks nest in tree cavities. Newly hatched young are flightless, and the young of tree nesters have to jump out of the nest to the ground.

The birds of prey are perhaps the most spectacular and awesome of all birds. Their sharp and powerful talons, keen eyesight, and strong flight, make them effective top carnivores. Most hawks prey on a variety of small mammals, birds, and reptiles, but a few specialize on particular types of prey. The osprey eats only fish, which it catches with its long talons and holds with prickly toes. The Everglades kite of Florida feeds only on one species of snail. Bald eagles will catch fish, but they also eat other animals and even carrion; occasionally they force an osprey to release its catch in mid-air. Some smaller hawks feed primarily on insects. Vultures are related to hawks and eagles, but their weak feet allow them to eat only dead animal matter. They appear to have a sense of smell, which most

Figure 17.8 The common loon's legs are positioned far back on its body, making it an excellent swimmer and diver, but making it very clumsy on land. Found only in the far northern hemisphere, they lay their eggs in floating nests of vegetation in freshwater lakes, although they are common in coastal oceans in the nonbreeding season. They feed primarily on fish and may dive to 50 meters and remain submerged for 15 minutes or more.

birds lack, to detect their food. Their bald head allows them to feed on dead matter without the problem of trying to clean matted head feathers.

The shorebird group includes the gulls, terns, auks, puffins, sandpipers, phalaropes, plovers, and many other types of shore and open water birds of both fresh and salt water. They nest on the ground near shores or on cliffs above the water. Most are fast and agile flyers with long and thin wings. Some of these birds are well known for their long migrations; for example, the arctic tern flies from the Arctic to the Antarctic and back each year—about 40,000 kilometers round trip. The terns, auks, and puffins eat fish, the shorebirds eat various invertebrates, and the gulls are omnivores.

The parrots are found primarily in equatorial areas (Figure 17.9). The only native North American parrot was the Carolina parakeet of the eastern United States; it became extinct in the early 1900s due to habitat destruction. Parrots feed on fruits and seeds and are one of the few birds to use their feet to manipulate food and other objects. Most nest in trees but a few nest on the ground; one rare species even nests in a cave. Many parrot species are endangered due to their desirability as pets.

Figure 17.9 Scarlet Macaw. The scarlet macaw is one of the most familiar of the parrots. Native to South America, it has been threatened by the hunting of adults and nestlings for the pet trade.

Owls were once considered to be related to hawks; however, they are not related except in an ecological sense, as owls are nocturnal birds of prey. Owls have excellent day and night vision, asymmetrical ears that detect the source of sound accurately and frayed flight feathers for silent flight. Their prey is similar to that of hawks: rodents, insects, snakes, lizards, and fish. Indigestible portions of the prey are formed into a ball and regurgitated. Examination of the fur, feathers, and bones in these "owl pellets" allows one to study the feeding habits of owls. Some hawks, kingfishers, and grebes also form pellets.

The hummingbirds are the smallest of all birds, found only in North and South America. Although their main energy source is nectar, they obtain needed protein from insects. They do not actually suck up the nectar, but sop it up with their featherlike tongue. Hummingbird's bodies are so small that they lose heat rapidly and often become torpid at night since they would not survive at their usual high rate of metabolism without feeding. Other nectar-eating birds found in other areas are such species as the honeycreepers of Hawaii and honeyeaters of Africa.

Every forested area in the world, except the island of Madagascar, has one or more species of woodpeckers. Most birds have a toe arrangement of three in front and one in back, but the woodpeckers have two toes forward and two back. This toe arrangement, in conjunction with a stiff tail that is used as a prop, allows woodpeckers to walk up the side of a tree. Their jaws are built to withstand the pounding necessary to excavate a hole for a nest or to reach insect prey. A woodpecker is able to extend its long sticky tongue into cracks and crevices because the tongue is supported by a long muscle-covered bone. Besides building their own home by chiseling a hole in a dead tree, they also communicate by drumming against hollow trees, tin roofs, metal lamp posts, and so on.

The most abundant, most obvious, and most widespread of all birds are the song or perching birds: They comprise over half of all the extant bird species. They have the most complex songs of any bird group, and the males are among the most colorful of all birds. As a group they nest in numerous habitats and may build very simple nests on the ground (for example, some sparrows) or complex woven hanging nests (orioles, bushtits, weaver finches). They feed on a wide variety of food such as nectar (honeycreepers, orioles), seeds (finches, sparrows), insects (warblers, flycatchers), berries (waxwings, robins) or are widely omnivorous (crows, jays, magpies).

Reproduction

All birds lay eggs; all but a few species build a nest; and, with one exception, all incubate their eggs until they hatch and show some degree of parental care. Incubation may last for 60 days or more as in penguins and albatrosses or be as short as 11 days as in some hummingbirds. Most often, it is between 14 and 21 days. Newly hatched birds may be helpless and featherless (**altricial**), or covered with down and ready to follow the parents around within a few hours (**precocial**) such as in the waterfowl. The exception to parental incubation is the Australian brush turkey, which lays its eggs in a pile of decaying vegetation. The heat of decomposition incubates the eggs.

Migration

Their ability to fly has allowed birds to colonize many suitable habitats. The majority of bird species migrate to some extent; some move thousands of kilometers in latitude, others only a few kilometers in altitude. They are able to take advantage of new food sources and nesting sites and avoid the rigors of winter. Few other animals have the migratory habit so well developed.

The ability of birds to navigate on their migratory trips has been the subject of much research. There is still a good deal to be learned, but it seems that birds' ability to navigate and to find their breeding or wintering grounds, although they may have never been there before, is genetic. The mechanisms by which they navigate include orientation to the sun, stars, landmarks, and geomagnetic lines of force. Probably more than one mechanism is used by each migratory species.

Ecological Roles

Birds fit into roles from decomposer to top level carnivores because their habitats and diet are so varied. Their mobility has allowed them to become a part of nearly every habitat in every geographical area on earth. The total number of birds on the earth may be 8 billion, but most species are not very numerous and generally have a minor role in the total ecosystem. Other species, especially sea birds, which congregate during the breeding season in small areas in enormous numbers, or waterfowl or blackbirds, which descend in large concentrations onto wintering grounds during fall migration, may have a tremendous, even overwhelming impact on their habitat.

Birds and Humans

Birds have been used by people for food, clothing, decoration, oil, eggs, and other products. They carry diseases transmittable to people, and may be pests in agricultural fields and airports. Birds provide sport for hunters and are common house pets worldwide.

Perhaps the most abundant bird in the world is the chicken, a domesticated version of the jungle fowl of Asia. Ducks, turkeys, and geese are also commonly eaten. Penguins, ostriches, the extinct dodo of Mauritius, and the extinct passenger pigeon have all been used for food. Songbirds by the trainload were shipped to New York for food in the early 1900s. Songbirds are still captured in great numbers in southern Europe and North Africa for that purpose.

Parts of birds (bills, feet, bones) have been used for decorative purposes, but feather uses are much more common. Feathers to decorate dancers' costumes and ladies' hats once endangered some species of birds. The softer undercoating of feathers (down) is used for insulation in jackets, hats, and comforters. Fortunately, the present demand for feathers is met mainly by domesticated birds rather than wild ones.

In South America, young "oilbirds," which are high in fat content, are killed and melted for cooking oil. Oil from penguins is occasionally used for candles. Besides food, the greatest product provided by birds is guano, which is turned into fertilizer rich in nitrogen and phosphorus.

Many birds have been driven to extinction by the activities of humans (overhunting, introduction of exotic species, and so on), but the loss of habitat is by far the greatest threat to birds and other animals.

THE MAMMALS

Of all the animals that exist in the world today, only 0.5% are mammals—about 5000 species. But compared with most other kinds of animals, mammals are large and some are very numerous. Their size and abundance and their use as a resource by primitive and modern humans for food and clothing has made them very obvious to us. Mammals are a significant part of many ecosystems, but their role as human food, pets, game trophies, means of transportation, and so forth has given them a disproportionate visibility.

Birds are a rather homogenous group. Each bears a distinct resemblance to the others as they share many outward characteristics that quickly label them as birds. Whales, monkeys, pangolins, giraffes, and sea cows are quite a mixed group; they are not of the same mold. The largest mammals are the whales, the blue whale reaching 35 meters in length and 136,000 kilograms in weight. The smallest are the shrews; the pygmy shrew weighs only 2.3 grams—about as much as two raisins. Mammals are diverse in size, shape, habits, and habitats. They live below or on the surface of the ground, in trees, and in fresh and salt water. They burrow, run, crawl, swim, and fly.

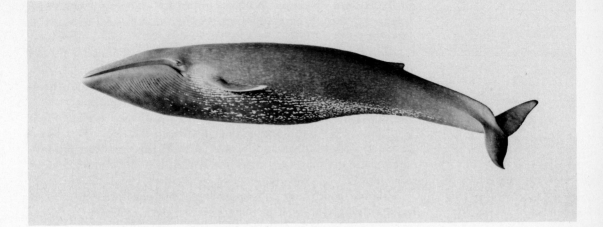

They are carnivores, herbivores, insectivores, fish eaters, and scavengers. Their different anatomical characteristics reflect this wide assortment of lifestyles. (Figure 17.10).

Like birds, mammals evolved from reptilian ancestors about 150 million years ago. As did the feathers of birds, the fur of mammals evolved to serve as insulation. There are a number of major groups of mammals, such as elephants, sea cows, cats and dogs, monkeys and apes, seals, whales, antelopes, kangaroos, and so on. But each of these groups can be put in one of the three major divisions of the mammal class: monotremes, marsupials, and placentals. These divisions are based on reproductive modes and are described in later sections.

Figure 17.10 Blue Whale. The existence of the blue whale, the largest of mammals, is threatened by overhunting. Although all products derived from whales (lubricating oil, perfume bases, dog food) can be easily obtained by other sources, a few countries, especially Japan, continue to exploit this rapidly declining animal.

What Makes a Mammal a Mammal?

Mammals are homeothermic and are covered with fur as insulation to reduce the rate of body heat loss. Some mammals such as musk ox and bear, which inhabit cold environments, are thickly covered. Some, such as deer, giraffe, and lions, which inhabit temperate areas, have less hair. Some, like porpoises, whales, and sea cows are virtually naked, their thick layer of body fat providing the insulation. Fetal whales are covered with hair, however, showing their affinity to more "typical" mammals.

Mammals are also characterized by specialized sets of teeth rooted in their jaws. Incisors are for grasping and cutting, canines for biting and holding, and molars for chewing, grinding, and crushing. Deer use their incisors to browse on bushes and their molars to grind the leaves; they have no canine teeth. The canines and molars of hyenas are large and pointed—both for catching and holding larger prey and for crushing bone. Some teeth may be modified as fangs or tusks.

A wide variety of skin glands has evolved in the mammals, and these are associated with milk production, thermoregulation, and behavior. Mammals nourish their young with milk-producing **mammary** glands; the dependency of the young on the mother for food has necessarily led to a system of parental care. Sweat glands help the body lose excess heat. Oil glands keep the skin moist. Scent glands produce strong and persistent odors that may attract members of the opposite sex, mark territories, identify young, or deter predators.

Mammals are also known for their ability to hibernate or estivate. Although hibernation occurs among birds, it is rare (restricted to some poorwills and swifts) and is more typical of mammals. Dormancy is a way to escape both food scarcity and temperature extremes. Birds' ability to fly to other regions reduces their need to face the rigors of an environment that becomes harsh. Some mammals migrate, such as bats, caribou, zebra, wildebeest, and deer, but a number simply go dormant for various periods of time. True hibernators are dormant for several months, putting on a layer of fat before curling up in an underground den or cave. Their metabolism may drop to perhaps one-fiftieth of normal, and their body temperature falls to just above freezing. Woodchucks, some ground squirrels, some bats, and others exhibit true hibernation. Some bears and squirrels show a dormancy in which their metabolism drops by 50% and their temperature is lowered by 5°–10°C, but they will frequently wake up during the winter. This is not true hibernation.

Most small mammals are nocturnal and most large ones are diurnal. Small mammals reduce detection by predators by being nocturnal. Being homeothermic allows them to be active during the colder nights; many also utilize burrows to modify the effects of the environment. Their sense of smell and their night vision are well developed, as would be expected of nocturnal creatures.

The Three Major Groups of Mammals

There are three large groups of mammals, distinguished by their mode of reproduction. One lays eggs, one bears very undeveloped young, and the third nourishes the young internally until well developed.

Monotremes

The **monotremes** are the most primitive mammals and are found only in Australia, Tasmania, and New Guinea. There are a few spe-

Figure 17.11 Spiny echidna. The spiny echidna or spiny ant-eater has powerful claws for digging up ant nests and for making burrows in which it raises young. The pointed snout lacks teeth and is used to probe for ants. The mammary glands seep milk onto the mother's underside, where it is lapped up by the young.

cies of the spiny echidna and only one species of the duckbilled platypus. Their teeth are small and nonfunctional; a spine that is grooved for the movement of poison is found on the hind legs of males; and, most characteristically, all lay eggs. Egg laying and many skeletal features indicate their relation to their reptile ancestors.

The spiny echidna has a long snout, a long sticky tongue, legs adapted for digging, and a body covered by quills for protection (Figure 17.11). The duckbilled platypus has a leathery ducklike bill, webbed feet, and short hair. The echidna eats insects and the platypus lives in streams, where it eats vegetation and small invertebrates.

Marsupials

The **marsupials**, or pouched mammals, total 230 species and are found in Australia, South America, and one species, the opossum, in North America. The name comes from **marsupium**, the belly pouch in which the young develop. The young are born at a very early stage, and crawl up the mother's belly from the birth canal to the marsupium, in which they attach to a nipple and remain for several months. Birth occurs very soon after conception (eight days in the opossum) and the young are very small (18 newborn opossums can fit in a teaspoon).

Marsupials fill a variety of niches and are represented by such animals as kangaroos, opossums, wombats, and bandicoots. Bandicoots resemble rabbits but eat insects. Phalangers are opossumlike herbivorous tree dwellers. The koala "bear" is arboreal, feeding only on the leaves of the *Eucalyptus* tree. The marsupial cat is very catlike and carnivorous. Kangaroos and wallabies are grassland herbivores (Figure 17.12).

Although Africa, Australia, and South America were once part of one land mass (Pangaea), there are no marsupials in Africa. Biologists have speculated that since mammals evolved after the continents began drifting apart and marsupials evolved in South America, then they must have reached Australia by crossing over the Antarctic continent, which connected Australia to South America. But it was not until 1982, when a fossil marsupial was found in Antarctica, that this theory was confirmed.

Figure 17.12 The Koala. The koala is not a bear but a marsupial. It is a sluggish animal that spends most of its time wandering slowly through Eucalyptus trees in Australia. Its diet consists only of Eucalyptus leaves, and, even more interesting, these leaves contain strong oils that repel most animals.

Placentals

The majority of mammals in the world belong to the **placental** group. The developing embryo is nourished by the mother via a close connection of their circulatory systems: a placenta. Development of a few weeks to several months (the **gestation period**) results in young that are relatively well developed at the time of birth. The marsupials and placentals have evolved in parallel fashions to occupy similar niches, but the placentals are more numerous and widespread. There are 17 major groups of placentals; we will examine a few of them.

The rodents are probably the most successful of all mammals (except perhaps humans). Of the nearly 5000 species of mammals, over half are rodents. Found in nearly every terrestrial habitat, and some aquatic ones, all subsist by eating plants or seeds. Most are fairly small—mouse- and squirrel-sized—the largest being the South American capybara, which is as large as a small pig. Their most characteristic feature is their large, gnawing incisor teeth. The incisors have enamel only on the front, so the rear of the teeth constantly wears away, producing a sharp chisel edge. The teeth grow continually and if they become misaligned or broken, an incisor may grow in a spiral and prevent the mouth from closing, or even penetrate the skull and kill the animal. The rodent group includes mice, squirrels, porcupines, beaver, marmots, rats, voles, gerbils, hamsters, lemmings, chipmunks, and agoutis (Exercise 17.4).

Exercise 17.4 Behavior and Time Budgets of Squirrels. *Most mammals are hard to observe because of their secretive habits. Many are nocturnal and many burrow or hide in vegetation. Tree squirrels, however, are easy to observe since they are diurnal and fairly common in many habitats. Choose an area known to be used by squirrels. Watch the squirrels and try to place each behavior into a particular category, such as jumping, climbing, eating, foraging, and so on. After several hours of observation, calculate the percent of time devoted to each category of behavior. This will be a "time budget." Then try to analyze what aspects of a squirrel's daily activities seem to be most important and why. For example, if 42% of a squirrel's time is devoted to chasing other squirrels, it may be that defense of a territory may be crucial to that squirrel's survival. Compare the time budgets of different squirrels and of the same squirrel at different times of the day. Would you expect there to be differences? Why?*

The carnivores include the cats, dogs, bears, raccoons, weasels, skunks, pandas, badgers, mongooses, hyenas, and others (Figure 17.13). Their canine teeth are well developed for holding, slashing, and tearing. Since they have to catch their food, they need to be fast, agile, and more powerful than their prey, or they have to hunt in packs as wolves and hyenas do. The top consumers in most terrestrial ecosystems are mammals. As is true for all consumers on the top trophic level, they are much rarer than lower level consumers. Thus, mountain lions, tigers, and wolves are uncommon, and human activity has further reduced their numbers.

Bats are the only mammals capable of flight; "flying squirrels" and a few other species can only glide. The bat's wing is a thin membrane of skin that stretches over elongated finger bones and extends rearward to include the legs and tail. It has long been known that bats navigate by sonar, bouncing high-pitched sound waves off objects. They also locate their prey by sonar. As a bat approaches a flying insect, it emits sonar "beeps" at an increasingly faster rate. Some moths can detect bats' sonar and maneuver to avoid being eaten. Many bat species eat insects, but some eat fish, others nectar (pollinating night-blooming plants in the process), and vampire bats lick blood from wounds that they make by slitting the skin with their sharp teeth.

Figure 17.13 Spotted Hyena. The hyena is capable of killing its own prey and does so occasionally, but typically chooses to feed on the leftover kills of other carnivores. The hyena, like most mammals that eat dead flesh, has a good sense of smell. It also has extremely well developed jaw muscles and teeth, adaptations for crushing bone.

Bats, like snakes, have an unearned reputation as dangerous or "evil" animals. Most of the prevailing beliefs about bats are myths, however. Bats are not aggressive, never attack people, and certainly don't land in human hair. Bats are no more prone to rabies than other wild mammals, and only ten people in the United States and Canada have contracted rabies from bats in over 30 years; more people die *annually* from domestic dog attacks.

The ungulates are the hoofed animals such as the camel, peccary, rhinoceros, various antelope species, alpaca, zebra, caribou, and moose. Some of the largest land animals are ungulates—giraffes and rhinos, for instance. The smallest ungulates are those such as the lesser mouse deer of tropical Asia and the royal antelope of Africa. Ungulates are among the most obvious of mammals since most are diurnal and many roam open grasslands in large herds; zebra of Africa, for example, gather in groups of 25,000 and more during migratory periods. Ungulates typically are grazing or browsing herbivores of grassland or open woodland (Figure 17.14). Many have been domesticated for use as food (cows, pigs, sheep), transportation (horses, camels), or work animals (mules, oxen, water buffalo).

The elephant is a familiar mammal to everyone. There are two species, the African and Asian elephants, and they are the largest of

Figure 17.14 Gemsbok gazelle. Along with many other species of gazelles and related ungulates, the Gemsbok gazelle is found in East Africa.

any living land animals. Early elephant ancestors had elongated jaws, which evolved into long trunks capable of reaching into trees for leaves. Their incisors have developed into elongated tusks for defense and perhaps for knocking over trees to feed on the leaves. Large molars serve to grind the fibrous food. Due to poaching and habitat destruction, elephant populations have declined severely and the elephant may become an endangered species in the near future.

There are a number of very unusual mammals such as the aardvark, manatee, and pangolin. The African aardvark, or bush pig, resembles a 60-kilogram pig with an arched back, pointed snout, thick tail, and long ears. It has very small nonfunctional teeth, and feeds by extending its long sticky tongue into termites' nests for the insects. The two groups of mammals that are totally aquatic are the **cetaceans**—the whales, dolphins, and porpoises—and the **sireneans**—the manatees and dugongs. The four species of sireneans living today are found in Africa, South America, the Indo-Pacific area, and the Caribbean. They eat shallow water plants and are thus restricted to coastal, estuary, or freshwater habitats. In Florida they eat the water hyacinths, which tend to clog the waterways. Looking much like a walrus without tusks, they remain underwater for 10 to 15 minutes at a stretch. Their sluggish movements and proximity to human habitats have led to their decrease. Once eaten and used for oil, they are now threatened by fish nets, water pollution, and speedboats. In Florida, a manatee without scars from a propeller is rare. Pangolins, like the aardvarks, have no close relatives. Native to tropical forests of Asia and Africa, they are as large as a fox, covered with large overlapping scales, and have no teeth. Like the nocturnal aardvark, they are also insect eaters, have a long sticky tongue, and live in burrows during the day. They may eat 200,000 insects a day! The word "pangolin" comes from a Malaysian word meaning to curl up, which is what the pangolin does when threatened. Its soft belly is protected and the sharp scales prevent a predator from trying to unroll the pangolin. Unfortunately, like many exotic animals, the pangolin is threatened by humans. Its hide is used for cowboy boots; a total of 60,000 skins were imported into the United States in 1981 and 1982. In Asia, it is eaten and its scales are used as good luck charms and home remedies for snakebites.

The primates are the most advanced group of mammals, composed of apes, monkeys, lorises, lemurs, and humans. Like many other animal groups, many species are threatened by pollution, destruction of habitat, the pet trade, and illegal hunting for a variety of other reasons. Primates are native to Africa, Asia, and South America. Their behavior, social structure, anatomy, and physiology have been extensively studied, not only to learn about primates but also because they are our closest living relatives, and much of what we learn about our wild relatives applies to us in some way. Humans are the most modern primates.

Humans and Mammals

Cows, sheep, and pigs provide us with a domesticated source of meat, but all human cultures eat some wild mammals for subsistence. Horses, llamas, mules, camels, and elephants are used for transportation, as riding or pack animals. Many kinds of big and small "game" are hunted for sport. Scientific institutions use rats, mice, rabbits, dogs, monkeys, and apes to learn more about the way mammalian bodies work; to test the effects of drugs; and to determine the effects of exposure to pollution, smoking, and so on. Humans keep mammalian pets of all sorts, most notably dogs and cats. Mammals are undoubtedly the most obvious of all animals—especially when you consider that we, too, are mammals.

CONCLUSION

We have only cracked open the door to take a peak at the kingdoms of organisms. A comprehensive look at any one kingdom, or even a small group such as the mammals, would require volumes. We need to realize that there are innumerable species of organisms filling virtually an infinite number of niches; simply comprehending nature's diversity is an intellectual achievement worthy of praise.

SUMMARY

The backboned vertebrates are the most familiar of all animals. Although not always of great significance in every ecosystem, their size, colors, and sounds make them conspicuous. They are economically very important. The five groups of vertebrates are the fishes, amphibians, reptiles, birds, and mammals. The birds and mammals are homeotherms and the others poikilotherms.

STUDY QUESTIONS

1. Define spawn, oviparous, fry, anadromous, parthenogenesis, altricial, marsupium.
2. Differentiate between the terms ovoviviparous, viviparous, and oviparous.

3.   What characteristics do the monotremes, marsupials, and placentals have in common? How do they differ?

4.   Differentiate among the three major groups of fishes.

5.   List the major characteristics of each vertebrate group.

6.   How can you differentiate among the three major groups of amphibians?

7.   What adaptations to terrestrial living do reptiles have that amphibians do not?

8.   What characteristics do mammals and birds share and how are they different?

9.   In what ways is it advantageous to go through a larval stage before metamorphizing into an adult, and how is this type of life cycle disadvantageous?

10.   Which group of vertebrates has evolved into the greatest number of niches? Justify your anwer.

Glossary

abiotic Nonliving or derived from nonliving processes.

abscissin A plant hormone that is associated with the phenomenon of leaf fall.

absolute humidity The amount of water vapor in the air.

acclimation The process by which changes occur in response to the environment over an organism's lifetime; not genetic.

adaptation The evolution of characteristics that make organisms more successful in their environment; genetic.

aerobic bacteria Bacteria that need atmospheric oxygen to live.

aggregation A group of individuals which has come together for reasons other than social interaction.

albinism The absence of pigmentation.

allele Alternative form of a gene.

allelopathy The inhibition of one organism by another via the release of chemicals.

Allen's Rule Homeothermic animals of cooler regions have shorter appendages than those in warmer areas.

allopatric In separate areas or in isolation, as *allopatric populations.*

allopatric speciation The origination of different new species from two geographically isolated populations of the same parent species.

alluvium Sediment deposited by flowing water.

alternation of generations The alternation of sporophyte and gametophyte phases in a plant's life cycle.

altricial Helpless and featherless when hatched; pertains to birds.

anadromous Migrating up freshwater rivers from the sea to spawn.

anaerobic Not requiring oxygen; an environment without oxygen.

anaerobic bacteria Bacteria that survive in the absence of free oxygen.

analogous Similar in function but of different evolutionary origins.

angiosperms The flowering plants.

anthocyanin A pigment capable of producing red and blue colors.

anting A behavior exhibited by birds in which they rub ants through their feathers.

aphotic zone The area in an aquatic system devoid of light.

apical dominance The bud at the tip, or apex, of a plant produces hormones that inhibit the growth of lateral branches.

arboreal Tree-dwelling.

Aristotle's lantern A chewing device in the mouth cavity of sea urchins.

artificial selection Selective breeding for hereditary traits in animals and plants by humans.

atmosphere The mass of gases surrounding the earth (or other celestial body).

autotroph An organism capable of producing its own food.

auxin A plant hormone that affects growth.

bacillus Rod-shaped bacterium (plural: bacilli).

bacteroid A bacterial organism, or an organism similar to a bacterium.

Beltian bodies Packets of protein material produced at the tips of some species of acacia trees.

benthos Organisms inhabiting the bottom of an aquatic system.

Bergmann's Rule Homeothermic animals in cooler regions have larger body sizes than animals in warmer areas.

biodegradability The ability to be degraded, or decomposed, by biological processes.

biogeochemical Cycles of matter that involve biological, geological, and chemical processes.

biogeography The study of the distribution of flora and fauna.

biological control Control of pests by other biological organisms.

biomass The weight of biological organisms or their parts.

biome One or more large ecosystems that are similar in form and function.

biosphere The parts of the earth in which life is found.

biotic Living or derived from living organisms.

bog Acid, marshy areas of northern regions, usually formed by movement of glaciers.

brackish Slightly salty.

bryophyte A group of nonflowering plants that includes the mosses, liverworts, and hornworts.

bunchgrass Grass that tends to grow in separate clumps.

carnivore A meat-eating organism.

carotene An orange pigment.

carrying capacity The maximum number of individuals that can be maintained by a habitat; occurs at the equilibrium level of the growth phase.

catadromous Living in fresh water and moving to the ocean to breed.

cellular respiration The process whereby electrons are transferred from nutrient molecules to molecular oxygen, releasing large amounts of energy and converting O_2 to H_2O.

cercaria The last, free-swimming tailed larval stage of a parasitic fluke.

cetacean A member of the whale and porpoise group of mammals.

chaparral A dense thicket of shrubs and small trees.

chemosynthetic Capable of deriving energy for biological processes by breaking chemical bonds, as *chemosynthetic bacteria*.

chitin A sugar–protein compound that is found in the exoskeleton of arthropods and the cell walls of fungi.

chloroplast A chlorophyll-containing organelle in plant cells.

circumpolar Around the world in the polar regions; refers to the distribution of organisms.

climax Describes the final, community stage (sere) in the process of succession.

cline A geographic continuum exhibiting physiological and/or morphological differences.

clones Genetically identical individuals.

closed system A system in which materials recycle and never are lost.

cnidocyctes The cells of coelenterates that produce nematocysts.

coccus Spherical bacterium (plural: cocci).

coevolution The development of interdependent features by two or more species; e. g., hummingbirds and hummingbird flowers.

commensalism A symbiotic relationship in which one organism benefits and the other is neither benefitted nor harmed.

community All the living organisms of an area.

competition The striving for resources that are in short supply.

competitive exclusion principle No two species can coexist indefinitely on the same resource; no two species can occupy the same niche.

conifers Cone-bearing, mostly evergreen, trees and shrubs.

conspecific Of the same species.

consumer Any organism that ingests food; any organism that is not a producer. All animals and nongreen plants are consumers.

continental drift The theory that one large land mass split apart into continents which then moved apart into their present positions, and are still moving.

convergent evolution Unrelated organisms evolving similar adaptations.

corridors Land connections that allow free movement of organisms from one land mass to another.

crossing over The exchanging of pieces by two homologous chromosomes during meiosis.

crustose Crustlike; often refers to a growth form of lichen.

cyst A resting stage of certain organisms, such as bacteria and protozoa, in which they are resistant to environmental extremes.

deciduous Falling off or shedding at a particular season, as the leaves on a tree.

decomposer Any organism that serves to break down organic dead or waste matter into simpler components.

definitive host The host organism in which a parasite reproduces sexually.

dendrochronology The science of dating historical events by the use of tree rings.

denitrifying Breaking down nitrogen-containing compounds, releasing gaseous nitrogen.

density The number of organisms (or objects) per unit area.

density-dependent factors Those environmental factors whose effect on a population is proportional to the density of that population.

detritus Loose fragments or particles of rock, sand, or organic matter produced by their decomposition; debris.

diapause A period during which growth or development is suspended.

diploid Having two of each chromosome.

dispersal Movement away from an area.

dispersal barrier Physical, biotic, or climatic obstructions to the movement or establishment of organisms.

diurnal Active during the day.

dormin A plant hormone associated with initiating winter dormancy.

down The soft undercoating of feathers. The soft, fluffy feather.

dwarfism A condition of arrested growth.

ebb A period of fading or diminishing; as *ebb tide*.

ecesis The germination, growth, and reproduction of migrules; the successful establishment of an organism.

ecological density The number of organisms per unit area of appropriate habitat.

ecological equivalents Unrelated organisms that perform similar roles in different ecosystems.

ecological succession *See* succession.

ecology The study of the interrelationships of organisms and the environment.

ecosystem All organisms, the surrounding environment, and their interactions in a stable situation.

ecotone A transition zone between biomes or different habitats.

ectoparasite A parasite that lives on the outside of a host's body.

elephantiasis A disease characterized by the enlargement of appendages due to an accumulation of lymph fluid, caused by lymph ducts blocked by the nematode worm, *Wuchereria bancrofti*.

endoparasite A parasite that lives on the inside of a host's body.

endosperm Tissue surrounding the ovule and providing food for it.

entropy The tendency of all systems to move toward randomness and disorganization.

environment All the biotic and abiotic factors that surround an organism.

environmental plasticity The ability of an organism to adapt to the environment.

epilimnion The upper, warm, circulating layer of water in a lake.

epiphyte A plant that grows on another plant for support.

equilibrium phase That part of the growth phase when birth rates decrease and death rates increase, and the population size stabilizes.

ergotism Poisoning produced by an ergot fungus infection of grains.

estivation A period of dormancy during the summer or hot periods.

estuary A body of water, typically marshlike, that receives river flow and usually empties into an ocean.

eukaryote Those organisms with eukaryotic cells.

eukaryotic cells Those cells that have a nuclear membrane and other organelles; present in all organisms except bacteria and blue-green algae.

eutrophic Having accumulated a substantial amount of nutrients, as *eutrophic lake*.

eutrophication The process of accumulating nutrients.

exponential growth phase That part of the growth curve where the population nearly doubles every generation.

extant Living today.

extinct No longer existing; refers to a species or other group.

extinction level The population size below which a species cannot maintain its numbers and could become extinct.

fall overturn The mixing of nutrients through all water layers of a lake in the autumn.

fertilization The fusion of gametes.

filter bridges Land connections over which only some organisms are able to cross.

flagging The formation of a tree into a form that resembles a flag—the branches are missing on one side due to constant drying or cold winds.

fledging The leaving of the nest by a young bird (a fledgling) that has recently acquired flight feathers.

flood Inundation by water.

floodplain The area on either side of a river that is regularly inundated by the river.

foliose Leaflike form, often used to describe a form of lichen.

food chain A line of organisms, each one feeding on the one after it and being fed on by the one before it.

food web A representation of the predator–prey interactions of many organisms in an ecosystem.

fossil A piece or impression of an organism from the geologic past.

frond The leaflike structure of a fern or palm.

fry A recently hatched fish.

fruticose Hanging or erect growth forms; often used to describe some lichen.

fundamental niche The niche that can be fully occupied by an organism in the absence of competition; its potential niche. *See* niche.

gamete A sexual reproductive cell such as a sperm or egg.

gametophyte The gamete-producing generation of a plant's life cycle.

gaseous In the state of being a gas.

gene pool All the genes of a population.

genetic death Inability to produce offspring.

genetic recombination The shuffling of genes and chromosomes to produce different genetic combinations.

genotype Genetic makeup.

geographical isolation Separation of populations by geographic features.

geothermal vent Openings in the earth's surface where heat from the earth's core is released.

geotropism The reaction of plants to gravity.

gestation period The period of carrying developing young (in the uterus of mammals).

glochidium A larval form of clams.

Gloger's Rule Homeothermic animals in warm regions have more dark pigmentation than do those in cooler areas.

Gondwana The ancient southernmost land mass that split into our present southern continents.

gross primary production (GPP) The total amount of energy fixed by a plant.

growth curve A plot of a population increasing and decreasing over time.

growth form An organism's shape as determined by the environment.

guano Bird droppings.

guild A group of organisms with similar niches.

gymnosperm Seed-producing plant without flowers or fruits.

habitat A set of conditions suitable for an organism's survival; the physical features of an organism's immediate environment. *See* micro- and macrohabitat.

haploid Having only one of each kind of chromosome.

hemotoxin A poisonous substance secreted by organisms that affects the circulatory system of the enemy or prey.

herbaceous Having a fleshy stem.

herbivore A consumer that eats plants.

heterogeneous Having dissimilar parts; of mixed composition, as *heterogeneous habitat.*

heterotroph An organism incapable of making its own food, which thus must obtain food from other organisms.

heterozygous Having two different alleles for the same gene.

hibernation The dormant state during which many animals spend the winter in a condition of decreased metabolism.

hierarchical An arrangement according to rank.

holdfast The rootlike structures that attach large algae to rocks.

homeothermic Capable of maintaining a constant body temperature.

home range The entire area over which an animal moves.

homogeneous Uniform in composition throughout; as *homogeneous habitat.*

homologous Similar in structure, function, and evolutionary origins.

homozygous Having two identical alleles on a pair of chromosomes.

humus Partially decayed organic matter, rich in nutrients, that forms the uppermost layer of soil.

hydrocarbon A chemical compound that contains only carbon and hydrogen.

hydrologic Referring to water and the study of water flow.

hydrosphere The aqueous environment of the earth.

hypha Fungal filament (plural: hyphae).

hypolimnion The colder, noncirculating water at the bottom of a lake.

hypothesis An untested idea about the cause of observed events; a foundation explanation for a set of facts that can be used as a base for experimentation.

imprinting A learning process occurring very early in the life of an animal; the process happens very quickly, and what is learned is retained by the organism for its lifetime.

incomplete metamorphosis A partial change in body form.

individual distance The space maintained by an animal between itself and other individuals.

insectivorous Insect-eating.

interspecific competition Competition between individuals of different species.

interspecific territory Territory that is defended against individuals of different species.

intraspecific competition Competition between individuals of the same species.

intraspecific territory Territory that is defended against other individuals of the same species.

Krummholz The gnarled, twisted forms of trees caused by wind, cold, or salt spray.

lag plase The initial part of the growth curve where birth and death rates are equal and close to zero.

lake A large inland body of water.

Laurasia The ancient northernmost land mass that split into our present northern continents.

Laws of Thermodynamics Principles governing the conservation of matter and energy.

lentic Standing water, as in bogs, ponds, and marshes.

life table A table of data pertaining to birth rates, death rates, and the longevity of a population.

limnetic zone The open water zone of a lake partially penetrated by sunlight.

lithosphere The solid crust of the earth.

litter Dead organisms, their parts, and waste products that lie on top of the soil.

littoral zone The shallow water area of a lake where light penetrates to the bottom.

long-day plants Those plants that require a day length of at least 12 hours before blooming.

lotic Running water, as in streams and rivers.

macroevolution Evolution in large increments.

macrohabitat The general set of environmental factors suitable for an organism's survival.

macronutrients Chemicals needed in relatively large amounts by an organism.

magma Molten matter under the earth's crust.

mammary Of or relating to milk-producing glands of female mammals.

mark-recapture method A technique by which animals are trapped, marked, released, and retrapped to determine their population size.

marsh A small, shallow body of water high in nutrient content and containing emergent vegetation.

marsupial A mammal whose young are born at a very early stage and suckled outside the mother's body in a pouch.

marsupium The abdominal pouch of marsupial mammals.

medusa The free-floating or swimming umbrella-shaped form of a coelenterate.

mesic Wet or moist.

metacercaria A fluke larva that has lost its tail as it burrowed into its host. *See* cercaria.

metamorphosis A change in body form as larvae metamorphose into adults.

microevolution Evolution by small increments.

microhabitat A specific set of environmental factors in an organism's immediate surroundings.

micronutrients Chemicals needed in very small amounts by an organism.

migration the movement of organisms into or out of an area.

migratory restlessness A twice-yearly period of restlessness seen in caged migratory birds that indicates their readiness to migrate.

migrule An organism or its offspring capable of dispersal.

miracidium The first larval stage of parasitic flukes.

mitosis The process of cell division resulting in two identical cells.

mobile Capable of being moved.

monoculture A culture or agricultural system of only one species.

monotreme A member of a primitive group of mammals that lays eggs.

mortality rate Death rate.

motile Capable of moving.

mutagen Any agent that causes mutations to occur.

mutation A change in genetic material that is heritable.

mutualism A symbiotic relationship in which both organisms benefit.

mycelium A mass of hyphae that forms the body of a fungus.

natality rate Birth rate.

natural history A descriptive body of facts about the life histories of organisms.

natural selection The differential reproduction of genotypes.

n-dimensional hypervolume Of or relating to the niche and the infinite set of variable factors that affect it.

nectary The nectar-producing organ of a flower.

negative geotropism A movement or response opposite to the pull of gravity.

nekton Swimming organisms not at the mercy of the current.

nematocyst The hairlike stinging structure of coelenterates.

net primary production (NPP) The amount of energy fixed by a plant minus that plant's metabolic needs; the amount available to consumers.

neurotoxin A poisonous substance secreted by organisms that affects the nervous system of the enemy or prey.

neuston Organisms inhabiting the surface of the water.

niche The role an organism plays in a community; its job.

niche breadth The range of environmental conditions utilized by a species.

niche segregation The division of niches, or roles, in a community by different organisms or species.

niche width The range of resources utilized by a species.

nitrification The process by which nitrogen is fixed (oxidation of ammonium salts → nitrites → nitrates).

nitrifying bacteria Bacteria that fix atmospheric nitrogen.

nitrogen fixation The incorporation of gaseous nitrogen in the air into compounds usable by plants.

nitrogen fixing Capable of incorporating gaseous nitrogen into nitrogen compounds.

nocturnal Active during the night.

nudation The disturbance of an environment in such a way as to allow the invasion of organisms not previously present.

nutrient Any chemical element or compound needed for an organism to survive and grow.

ocean hot springs Openings in the ocean floor that release heat from the earth's core.

oceanography The study of ocean systems.

oligotrophic Lacking nutrients, relatively sterile.

omnivore An organism that eats both plants and animals.

open system A system to which materials may be added and from which they may be lost.

ophidophobia The fear of snakes.

orb A circular spider web with strands of silk radiating from the center.

organic Containing carbon.

oviparous Producing eggs that hatch outside of the body.

ovoviviparous Producing eggs that hatch within the female's body.

ovule The plant structure that develops into a seed after fertilization.

Pangaea The original one-piece mass of land that gave rise to all continents.

panmixis Random mating among individuals of a population.

parapatric speciation The origination of new species at one end of a cline.

parasite An organism that lives on or in another organism (the host), usually harming it.

parasitism A symbiotic relationship in which one organism (the parasite) benefits and the other (the host) is harmed by the feeding of the parasite.

parthenogenesis Development of eggs without fertilization by sperm.

pathogen Disease-causing organism.

peat Partially decomposed plant matter, especially mosses, found in bogs.

pelagic Living in open waters.

perennial Having a life span of 2 or more years.

permafrost Permanently frozen soil found in the tundra.

phenology The study of periodic biological phenomena, such as flowering.

phenotype The physical makeup of an organism.

pheromone A chemical, often a hormone, that is released to affect the behavior of other animals, usually of the same species.

phloem The food-conducting tissue of vascular plants.

photic zone The zone in an aquatic system through which light penetrates (also *euphotic*).

photoperiod The relative length of periods of light within a day.

phycobilin A red pigment found in algae.

phylogeny The arrangement of organisms by their presumed evolutionary history.

-phyta Plant (suffix).

phytochrome A blue pigment controlling the time of flowering and leaf drop in plants.

phytoplankton Plant plankton.

pistil The female part of a flower.

placental A member of a group of mammals whose young are nourished inside the mother's body by a series of membranes called the *placenta*.

plankton Small organisms that move at the mercy of the water currents.

plasmodium A mass of multinucleate protoplasm in a stage of slime molds.

plate tectonics The study of the movement of the earth's crust—movement that causes the spreading of the sea floor and movement of continents.

plot A designated area of land, usually of rectangular shape.

poikilothermic Unable to regulate body temperature.

pollen The powderlike substance produced by the male part of a seed plant that functions in the process of fertilization.

pollution The contamination or overwhelming of natural systems by toxic products or large amounts of natural materials.

polyploidy The presence of duplicate sets of chromosomes.

population A group of organisms of the same species; a subset of a species in a particular geographic area.

positive geotropism A movement or response in the direction of gravitational pull.

prairie A habitat of flat or rolling grassland.

precocial Capable of walking around immediately after birth; as *precocial birds.*

predator Any organism that kills and eats another organism.

prey Any organism that is killed and eaten by a predator.

primary consumers Those consumers that eat the producers; the herbivores.

primary production The fixation of energy by plants.

primary succession Successional processes that occur on land that previously supported no life.

proboscis A long, tubular tongue for sucking or piercing, as occurs in insects or flatworms.

producer A green plant that produces organic matter via photosynthesis.

profundal zone The bottom or deep water area not reached by light.

prokaryotic cells Cells that lack a nuclear membrane and other organelles; only found in bacteria and blue-green algae.

propagule An organism or group of organisms capable of breeding and causing a population increase.

pseudopodium "False foot"—a temporary extension of the cytoplasm used in locomotion and feeding.

putrefaction The decomposition of organic matter by microorganisms.

quadrat Rectangular plot used in census techniques.

radula A rasplike tongue of some molluscs.

reaction The change in a habitat due to the effects of invading organisms.

realized niche The actual or real niche that an organism occupies; always smaller than the fundamental niche.

redwater fever A disease of cattle caused by protozoa.

relative humidity The amount of water vapor in the air as a percentage of the maximum amount of water vapor that could be held in the air.

respiration The chemical breakdown of organic matter to produce energy.

rete mirable A network of small arteries and veins at the junction of extremities and the trunk of vertebrates.

rheotaxis A directed movement against water current, as fish swimming upstream.

rhizoids Rootlike attachments of mosses, liverworts, and ferns.

rhizome A below-ground, horizontal stem.

riparian Relating to or living in the habitat bordering a lake, river, or stream.

rosette Rose-shaped, usually referring to a circular cluster of leaves or of other plant parts.

rumen A portion of the stomach of an animal such as a cow in which partial digestion occurs.

saprophytic Feeding on dead or decaying plant matter.

saprozoic Feeding on dead or decomposing animal matter.

scarification The breaking or scarring of the outer coat of a seed.

schistosomiasis A human disease caused by an infection of blood flukes of the genus *Schistosoma*.

sclerophyll Thick-leaved; usually refers to scrubby plants of the chaparral.

secondary consumers Those consumers that eat plants and/or primary consumers.

secondary succession Successional processes that occur on land that once supported life; some soil is present.

sedimentary Pertaining to rocks, the earth's crust, or sediment deposited by geological processes involving movement of minerals by water.

seed A structure that is composed of an embryo and stored food and is capable of developing into a mature plant.

sere A stage in the process of succession; referring to a community coming before the climax stage.

sessile Incapable of moving from place to place.

short-day plants Those plants that bloom before the day length reaches 12 hours.

silt Sediment deposited by a river or stream; consists of fine organic and inorganic particles.

sirenean A member of the manatee and dugong group of mammals.

sludge Fine-particled sediment left over after sewage treatment that is composed of bacteria and organic matter.

sod-forming grass Grass that tends to grow in a uniform distribution, forming a layer over the soil.

spawn The depositing of eggs by an aquatic animal; the eggs of those animals.

speciation The origination of new species.

species A group of individuals who are all capable of interbreeding.

spicules Small needlelike or crystallike structures that form the skeleton of some sponges.

spirillus Spiral-shaped bacteria (plural: spirilla).

spontaneous generation The idea that life arises from nonliving matter.

sporangia Spore-bearing structures on the underside of fern fronds.

sporocyst A saclike larval stage of many parasitic flukes which develops from a miracidium.

sporophyte The spore-producing generation of a plant's life cycle.

spring overturn The mixing of nutrients through all water layers of a lake in the spring.

stabilization The point at which succession produces a relatively stable and unchanging climax community.

stamen The male part of a flower which produces pollen.

statocysts Organs of equilibria (found in several animal groups).

stolon An above-ground horizontal stem.

stomate An opening in a leaf that controls the rate of transpiration (plural: stomates or stomata).

stratum A vertical layer of, for instance, sediment in a geological formation, or a layer of forest vegetation (plural: strata).

stratification Vertical layering, as in vegetation in a forest.

stream A fast-moving, narrow body of water, usually with a rocky bottom.

subclimax A long-lasting stage that immediately precedes the climax stage.

subspecies A population of a species showing some differences from other populations.

succession The gradual and predictable changes in communities, leading to a climax community.

succulent Having thick or fleshy leaves that retain moisture.

survivorship curve A plot of the death rate of a population over a period of time.

swamp A shallow body of water, high in nutrients, with floating vegetation and trees.

sweepstakes route A chance dispersal to isolated areas.

symbiosis A close, long-term, dependent relationship between two or more organisms. *See* commensalism, mutualism, and parasitism.

sympatric In the same area, as *sympatric populations.*

taiga The northernmost part of the coniferous forest biome.

tannin A chemical common to many plants in their bark, leaves, or fruits. Used in tanning. Also, *tannic acid.*

taxonomy The science of naming and classification.

territory Any area defended by an organism against other organisms.

tertiary consumer A consumer that eats secondary and sometimes primary consumers.

test The external shell of sea urchins.

thallus The body of a plant that lacks true stems, roots, or leaves, such as occurs in mosses and liverworts.

theory A tentative explanation of observed phenomena, subject to change with the addition of further information.

theory of acquired characteristics The theory that an organism will inherit traits acquired by its parents during their lifetime.

thermal stratification Layering of water or air by temperature.

torpor A condition of physical and mental inactivity.

tracheophyte Any plant with a vascular system.

transect A line or narrow path through a designated area; used in census techniques.

transpiration The loss of water through leaves.

treeline The altitudinal or latitudinal level at which climatic conditions preclude trees from establishing themselves and growing.

trichinosis A disease caused by the infection of mammals by *Trichinella spiralis* nematodes.

trophic Of or relating to food or feeding.

troposphere The lowest region of the earth's atmosphere.

true census A count of all the organisms in a given area.

turbidity Having sediment or particles suspended in the water, reducing clarity.

ungulate A hoofed mammal.

upwelling The movement of nutrients from the bottom to higher levels of aquatic systems, especially in the ocean.

vascular Supplied with vessels, as occurs in circulatory systems of organisms.

vascular bundle A cylinder of cells composed of xylem and phloem.

vascular cambium A cylinder of cells that produces the xylem and phloem.

viviparous Producing live young.

water vascular system Network of canals used for locomotion in the echinoderms.

xanthophyll A yellow pigment.

xeric Dry.

xylem The water-conducting tissue of vascular plants.

zooplankton Animal plankton.

zygote A fertilized egg.

Further Reading

Audubon. The magazine of the National Audubon Society, New York. *Monthly magazine with articles on various natural history topics.*

Ayala, Francisco J., and James W. Valentine. 1979. *Evolving.* Benjamin/Cummings, Menlo Park, Calif. *An intermediate level text covering the fundamentals of evolution.*

Barbour, Michael G., Jack H. Burk, and Wanna D. Pitts. 1980. *Terrestrial Plant Ecology.* Benjamin/Cummings, Menlo Park, Calif. *Excellent coverage of plant ecology, plant adaptations, and habitats.*

Barnes, Robert D. 1966. *Invertebrate Zoology.* W. B. Saunders, Philadelphia. *A detailed overview of invertebrate phyla.*

Benton, A. H., and W. E. Werner, Jr. 1974. *Field Biology and Ecology.* McGraw-Hill, New York. *A simple and basic ecology text.*

Benton, A. H., and W. E. Werner, Jr. 1972. *Manual of Field Biology and Ecology.* Burgess, Minneapolis, Minn. *A manual of field and lab exercises.*

The Biosphere. A Scientific American Book. W. H. Freeman, San Francisco, Calif. 1970. *Readings from* Scientific American *with emphasis on bio-geochemical cycles and human food production.*

Brewer, Richard. 1979. *Principles of Ecology.* W. B. Saunders, Philadelphia. *A lower level ecology text.*

Brower, J. E., and J. H. Zar. 1977. *Field and Laboratory Methods for General Ecology.* Wm. C. Brown, Dubuque, Iowa. *A rather sophisticated set of field and lab exercises.*

Byrne, Frank. 1974. *Earth and Man.* Wm. C. Brown, Dubuque, Iowa. *A good general geology text.*

Clapham, W. B., Jr. 1973. *Natural Ecosystems.* Macmillan, New York. *A basic ecology text with emphasis on habitats.*

Cody, Martin, and Jared Diamond, eds. 1975. *Ecology and Evolution of Communities.* Belknap Press, Cambridge, Mass. *A collection of advanced readings in ecology.*

Colinvaux, Paul. 1973. *Introduction to Ecology.* John Wiley and Sons, New York. *A beginning/intermediate level ecology text.*

Collier, B. D., G. W. Cox, A. W. Johnson, and P. C. Miller. 1973. *Dynamic Ecology.* Prentice-Hall, Englewood Cliffs, N. J. *A beginning/intermediate level ecology text.*

Continents Adrift. Readings from *Scientific American.* W. H. Freeman, San Francisco, Calif. 1970. *Various readings covering the concept of continental drift.*

Cox, G. W. 1980. *Laboratory Manual of General Ecology.* Wm. C. Brown, Dubuque, Iowa. *An intermediate level collection of exercises.*

Curtis, Helena, and N. Sue Barnes. 1981. *Invitation to Biology,* 3d ed. Worth, New York. *Covers the fundamentals of biological science, including a review of plant and animal groups.*

Darnell, R. M. 1971. *Organisms and Environment.* W. H. Freeman, San Francisco, Calif. *A beginning/intermediate manual of field exercises in ecology.*

Dasmann, Raymond F. 1981. *Wildlife Biology,* 2d ed. John Wiley and Sons, New York. *Ecological principles as related to wildlife populations.*

Discover. The Newsmagazine of Science, New York. *A monthly magazine containing articles on all areas of science.*

Ecology, Evolution, and Population Biology. Readings from *Scientific American.* W. H. Freeman, San Francisco, Calif. 1973. *Various readings from past issues.*

Emlen, J. Merritt. 1973. *Ecology: An Evolutionary Approach.* Addison-Wesley, Reading, Mass. *An advanced text in ecology, emphasizing the evolutionary process.*

Emmel, Thomas C. 1973. *Ecology and Population Biology.* W. W. Norton, New York. *A simple, short, ecology text for high school or lower level college students.*

Evolution. A Scientific American Book. W. H. Freeman, San Francisco, Calif. 1978. *Readings from* Scientific American *dealing with evolution.*

Fell, Barry. 1974. *Life, Space, and Time.* Harper & Row, New York. *An emphasis on descriptions of environments of the world.*

Giesel, James T. 1974. *The Biology and Adaptability of Natural Populations.* C. V. Mosby, St. Louis, Mo. *An introduction to population biology.*

Good, Ronald. 1974. *The Geography of the Flowering Plants.* Longman Group Ltd., London. *A general text on the distribution and ecology of flowering plants.*

Gould, Stephen J. 1977. *Ever Since Darwin.* W. W. Norton, New York. *A compilation of natural readings; extremely interesting.*

Gould, Stephen J. 1980. *The Panda's Thumb.* W. W. Norton, New York. *A compilation of natural history readings; extremely interesting.*

Grant, Verne. 1963. *The Origin of Adaptations.* Columbia University Press, New York. *Fundamentals of evolution and the origin of organisms' adaptations.*

Gunderson, Harvey L. 1976. *Mammalogy.* McGraw-Hill, New York. *Natural history, anatomy, physiology, and distribution of mammals.*

Hickman, C. P., Sr., C. P. Hickman, Jr., F. M. Hickman, and L. S. Roberts. 1979. *Zoology.* C. V. Mosby, St. Louis, Mo. *Fundamentals of zoology, including a review of the animal kingdom.*

Keeton, William T., and Carol H. McFadden. 1983. *Elements of Biological Science.* W. W. Norton, New York. *Fundamentals of biology.*

Kellman, Martin C. 1975. *Plant Geography.* St. Martin's Press, New York. *The distribution of plant groups.*

Kendeigh, S. Charles. 1974. *Ecology.* Prentice-Hall, Englewood Cliffs, N.J. *A biome-oriented approach to ecology.*

Kormondy, Edward J. 1976. *Concepts of Ecology,* 2d ed. Prentice-Hall, Englewood Cliffs, N.J. *Fundamentals of ecological principles for beginning level students.*

Kucera, Clair L. 1978. *The Challenge of Ecology,* 2d ed. C. V. Mosby, St. Louis, Mo. *Fundamentals of ecology with an emphasis on human activities.*

Life: Origin and Evolution. Readings from *Scientific American.* W. H. Freeman, San Francisco, Calif. 1978. *Various readings from past issues.*

Luria, S. A., S. J. Gould, and S. Singer. 1981. *A View of Life.* Benjamin/Cummings, Menlo Park, Calif. *Basic concepts of biology covered via an evolutionary approach.*

Man and the Ecosphere. Readings from *Scientific American.* W. H. Freeman, San Francisco, Calif. 1971. *Emphasizes the human influence on ecosystems.*

Margulis, Lynn. 1982. *Early Life.* Science Books International, Boston. *The evolution of life and the earliest forms of life are discussed.*

Mayr, Ernst. 1966. *Animal Species and Evolution.*

Margulis, Lynn. 1982. *Early Life.* Science Books International, Boston. *The evolution of life and the earliest forms of life are discussed.*

Mayr, Ernst. 1966. *Animal Species and Evolution.* Belknap Press, Cambridge, Mass. *A classic study of evolution and speciation.*

McNaughton, S. J., and L. L. Wolf. 1979. *General Ecology.* Holt, Rinehart and Winston, New York. *An upper level text in ecology.*

Morse, Douglass H. 1980. *Behavioral Mechanisms in Ecology.* Harvard University Press, Cambridge, Mass. *The behavior of organisms in an ecological context.*

Natural History. American Museum of Natural History, New York. *A monthly magazine with articles on all varieties of natural history subjects.*

Odum, Eugene P. 1971. *Fundamentals of Ecology.* W. B. Saunders, Philadelphia. *A thorough coverage of ecological principles for intermediate to advanced students.*

Palmer, E. Laurence, and H. Seymour Fowler. 1975. *Fieldbook of Natural History.* McGraw-Hill, New York. *The natural history of thousands of organisms is described in encyclopedic form.*

Pianka, Eric R. 1978. *Evolutionary Ecology,* 2d ed. Harper & Row, New York. *An advanced text that discusses in depth some principles of ecology and evolution.*

ice, Peter. 1975. *Insect Ecology.* Wiley-Interscience, New York. *Excellent discussion of ecological dynamics, using insects as examples.*

Ri ardson, Jonathan L. 1977. *Dimensions of ology.* Williams and Wilkins, Baltimore, Md. *logical theory with an emphasis on com- ities and ecosystems.*

Ric Robert E. 1973. *Ecology.* Chiron Press, Ore. *A detailed discussion of major principles for intermediate and ad- nts.*

Rickl 1976. *The Economy of Nature.* Ch nd, Ore. *Principles of ecol- ogy the dynamics of popu- lation, osystems.*

Rolan, R. G. Field Investi- gation in Gen illan, New York. *Lab and fie of vari- ous techniques and ins tatisti- cal analyses are discussed.*

Savage, Jay M. 1977. *Evolution,* 3d ed. Holt, Rinehart & Winston, New York. *A succinct discussion of evolutionary principles and processes.*

Science 83. American Association for the Advancement of Science, Washington, D.C. *A monthly science magazine for the interested public.*

Smith, Robert Leo. 1974. *Ecology and Field Biology.* Harper & Row, New York. *A thorough coverage of ecological principles; well written.*

Smith, Robert Leo. 1977. *Elements of Ecology and Field Biology.* Harper & Row, New York. *A condensed version of Smith's* Ecology and Field Biology.

Stebbins, G. Ledyard. 1966. *Processes of Organic Evolution.* Prentice-Hall, Englewood Cliffs, N.J. *Discusses the major principles and processes of evolution.*

Stern, Kingsley R. 1982. *Introductory Plant Biology.* Wm. C. Brown, Dubuque, Iowa. *An excellent review of the plant kingdom and principles of botany.*

Stonehouse, Bernard, and Christopher Perrins, eds. 1977. *Evolutionary Ecology.* University Park Press, Baltimore, Md. *A series of readings for the advanced student of ecology.*

Storer, Tracy I., and Robert L. Usinger. 1963. *Sierra Nevada Natural History.* University of California Press, Berkeley. *A natural history guide to the organisms and geology of the Sierra Nevada.*

Townsend, C. R., and P. Calow, eds. 1981. *Physiological Ecology.* Sinauer Associates, Sunderland, Mass. *Physiological aspects of ecology for the advanced student.*

Vernberg, F. John, and Winona B. Vernberg. 1970. *The Animal and the Environment.* Holt, Rinehart and Winston, New York. *Emphasizes animal adaptations to the environment.*

Williams, George C. 1966. *Adaptation and Natural Selection.* Princeton University Press, Princeton, N.J. *An evolutionary approach to ecology for the advanced student; contains stimulating and novel ideas.*

Wilson, David L. 1980. *The Natural Selection of Populations and Communities.* Benjamin/Cummings, Menlo Park, Calif. *For the advanced student in ecology.*

Index of Scientific Names

To avoid confusion, organisms in the text have generally been referred to by common names. But since common names vary, this index gives the scientific names of these organisms. If a particular species is meant, the full scientific name (genus and species) is given. If the common name refers to several closely related organisms, the genus alone is listed. For example, the raccoon is listed as *Procyon lotor*, but the zebra is listed as *Equus sp.* since there is more than one species of zebra.

When the text refers to broader groups of organisms, such as spiders or mushrooms, no name is listed as there are hundreds of species and dozens of genera of each.

Common Name/Scientific Name

Aardvark, *Orycteropus afer*

Acacia ants, *Pseudomyrmex* sp.

African egg-eating snake, *Dasypeltis scabra*

African elephant, *Loxodonta africana*

African ostrich, *Struthio* sp.

Alder, *Alnus* sp.

Alfalfa, *Medicago sativa*

Aloe, *Aloe* sp.

Alpaca, *Lama pacos*

American alligator, *Alligator mississipiensis*

American coot, *Fulica americana*

American robin, *Turdus migratorius*

Anglerfish, *Photocorynus spiniceps*

Anole lizard, *Anolis carolinensis*

Antelope (U.S.), *Antilocapra americanus*

Archeopteryx, *Archeopteryx lithographica*

Archerfish, *Toxotes jaculator*

Arctic fox, *Alopex lagopus*

Arctic tern, *Stérna paradisaêa*

Common Name/Scientific Name

Meadow frog, *Rana pipiens*

Mealworm (larvae of darkling beetle), *Tenebrio* sp.

Medicinal leech, *Hirudo medicinalis*

Mediterranean fruit fly, *Ceratitis capitata*

Mexican beaded lizard, *Heloderma horridum*

Mexican horncone, *Ceratozamia* sp.

Midwife toad, *Alytes obstetricans*

Mistletoe, *Phoradendron* sp.

Mockingbird, *Mimus polyglottos*

Monarch butterfly, *Dánaus plexíppus*

Mongoose—Golden brown, *Herpestes javanicus*

Moose, *Alces palces*

Mudpuppy, *Necturus maculosus*

Musk ox, *Ovibos moschatus*

Narrow-mouthed toad, *Gastrophryne olivacea*

Nautilis, *Nautilis* sp.

Nettle, *Urtica* sp.

Nightcrawler, *Lumbricus* sp.

Northern oriole, *Icterus galbula*

Ocotillo, *Fouquieria splendens*

Oilbird, *Steatornis guáchara*

Oil palm, *Elaeis quineensis*

Onion, *Allium cepa*

Opossum, *Didelphis marsupialis*

Osprey, *Pandion halietus*

Paddlefish (American), *Polyodon spathula*

Painted turtle, *Chrystemys picta*

Palolo worm, *Eunice viridis*

Palo verde, *Cercidium floridum*

Panda, *Ailuropoda melanoleuca*

Pangolin, *Manis* sp.

Paramecium, *Paramecium caudatum* and *Paramecium aurelia*

Passenger pigeon, *Ectopistes migratorius*

Passion-flower vine, *Passiflora Coerulea*

Pearlfish, *Carapus* sp.

Penicillium fungus—cheese, *P. roqueforti* and *P. camemberti*

Penicillium fungus—penicillin drug, *P. notatum* and *P. chrysogenum*

Peppered moth, *Biston betularia*

Peregrine falcon, *Falco peregrinus*

Philippine goby, *Pandaka pygmea*

Philodendron, *Philodendron* sp.

Pine drops, *Pterospora andromeda*

Pine sawfly, *Neodiprion sertifer*

Pintail duck, *Anas acuta*

Pitcher plant, *Sarracenia purpurea*

Pitcher-plant mosquito, *Wyeomyia* sp.

Pocket-book mussel, *Lampsilia ventricosa*

Poinsettia, *Euphorbia pulcherrima*

Polar bear, *Thalarctos maritimus*

Pomarine jaeger, *Stercorarius pomarinus*

Ponderosa pine, *Pinus ponderosa*

Pond slider, *Pseudemys scripta*

Puffball, *Geastrum* sp., *Lycoperdon* sp., *Calvatia* sp.

Puss moth, *Cerura vinula*

Pygmy shrew, *Suncus etruscus*

Quillwort *Isoetes engelmanni*

Raccoon, *Procyon lotor*

Rainbow trout, *Salmo gairdneri*

Rattlesnake, *Crotalus* sp.

Rattleweed, *Daucus pusillus*

Red-backed salamander, *Plethodon cinereus*

Red-breasted nuthatch, *Sitta canadénsis*

Red mangrove, *Rhizophora mangle*

Red-necked wallaby, *Macropus* sp.

Red-winged blackbird, *Agelaius phoeniceus*

Redwood, *Sequoia sempervirens*

Rhinoceros, *Dicerus bicornis*

Rose, *Rosa* sp.

Royal antelope, *Neotragus pygmaeus*

Rubber tree, *Ficus elastica*

Sagebrush lizard, *Sceloporous graciosus*

Salmon, *Oncorhynchus* sp.

Sawfish, *Pristophorus* sp.

Scarab beetle, *Melontha* sp.

Scarlet macaw, *Ara macao*

Scotch pine, *Pinus sylvestris*

Scouring rushes, *Equisteum hiemale*

Seahorse, *Hippocampus obtusus*

Sea snake, *Pelamis* sp., *Laticauda* sp.

Sea turtle, *Chelonia* sp.

Senioritas (fish), *Oxyjulis californica*

Shaggy mane, *Coprinus comatus*

Siamese fighting fish, *Betta splendens*

Sidewinder, *Crotalus cerastes*

Silverleaf oak, *Quercus hypoleucoidea*

Skunk cabbage, *Symplocarpus foetidus*

Smoke tree, *Rhus coggygria*

Snapping turtle, *Chelydra serpentina*

Snow plants, *Sarcodes sanguinea*

Snowshoe hare, *Lepus americanus*

Snowy owl, *Nyctea scandiaca*

Common Name/Scientific Name

Soldier beetle, *Chauliognathus* sp.

South American marsupial toad, *Gastrotheea ovifera*

Spanish moss, *Tillandsia usneoides*

Sphagnum moss, *Sphagnum* sp.

Spider mite, *Sarcoptes scabiei*

Spiny echidna, *Tachyglossus* sp.

Spitting spider, *Scytodes* sp.

Spoonbill—Roseate, *Ajáia ajája*

Spotted hyena, *Hyena crocuta*

Starfish, *Pisaster ocharaceus* and *Lepasterias hexactis*

Starling, *Sturnus vulgaris*

Steelhead, *Salmo* sp.

Stentor, *Stentor* sp.

Stickleback—Brook, *Euclaia inconstans*

St. John's wort, *Hypericum perforatum*

Sundew, *Drosera* sp.

Swainson's thrush, *Catharus ustulata*

Swamp rabbit, *Sylvilagus aquaticus*

Tarantula (giant), *Dugesiella hentzi*

Tarpon, *Tarpon atlanticus*

Teredo clam (shipworm), *Teredo* sp.

Tiger snake, *Notechis scutatus*

Toucan, *Ramphastos* sp.

Touch-me-not, *Impatiens pallida*

Townsend's solitaire, *Myadestes townsendi*

Trichinella worm, *Trichinella spiralis*

Triticale, *Triticum* sp. and *Secale* sp.

Tropical palm, *Euterpe globora*

Truffle, *Tuber aestivum*

Tuatara, *Sphenodon*

Venus flytrap, *Dionaea muscipula*

Viceroy butterfly, *Limenìtis archippus*

Vicuña, *Lama vicugna*

Vinegar eel, *Turbatrix aceti*

Volvox, *Volvox* sp.

Walking catfish, *Clarius batrachus*

Wallaby (see Red-necked wallaby), *Macropus* sp.

Wandering albatross, *Diomedea exulans*

Wandering Jew, *Tradescantia* sp.

Water buffalo, *Bubalus bubalis*

Water flea, *Daphnia longispina*

Water hyacinth, *Eichhornia crassipes*

Welwitschia, *Welwitschia* sp.

Western budworm, *Choristoneura fumiferana*

Western meadowlark, *Sturnella neglecta*

Whale shark, *Rhineodon typus*

White fir, *Abies concolor*

White pelican, *Pelecanus erythrorhynchos*

Wildebeest, *Connochaetes* sp.

Willow, *Salix* sp.

Willow ptarmigan, *Lagopus lagopus*

Witch hazel, *Hamamelis virginiana*

Wolf spider, *Lycosa* sp.

Wombat, *Phascolomis* sp.

Woodchuck, *Marmota monax*

Wood stork, *Mycteria americana*

Yarrows spring lizard, *Sceloporus jarrovii*

Yeast, *Saccharomyces cerevisiae,*

Yellow-bellied marmot, *Marmota flaviventris*

Yellow-headed blackbird, *Xanthocephalus xanthocephalus*

Zebra, *Equus* sp.

Zebra butterfly, *Heliconius charitonius*

Index

Boldface numbers indicate pages on which terms are defined.

Photo and Illustration Credits

Fig. 4.3   courtesy of Carolina Biological Supply

Ex. 4.2   micrograph by S. E. Frederick, courtesy of E. H. Newcomb, University of Wisconsin/BPS

Fig. 5.2   Field Museum of Natural History, Chicago

Fig. 5.3   Grant Heilman, Inc.

Fig. 5.18   Sheldon N. Grebstein (ed.): Monkey Trial, p. 3 Houghton Mifflin Research Series. Copyright © 1960 by Houghton Mifflin Company. Used by permission.

Fig. 6.6   from *Insect Ecology* by Peter Price, p. 422. Copyright © 1975 by John Wiley & Sons, Inc.

Fig. 6.10   from *The Origin of Adaptation* by V.E. Grant, 1963. By permission of the Columbia University Press.

Fig. 7.1a   R. J. Lederer

Fig. 7.6   Copyright 1981 by the National Wildlife Federation. Reprinted from the November-December issue of International Wildlife Magazine.

Fig. 8.1   R. J. Lederer

Fig. 8.2   courtesy of Jerome Wyckoff

Fig. 8.3   courtesy of Jerome Wyckoff

Fig. 8.5   courtesy of Steve Radosevich, University of California, Davis

Fig. 8.7a   J. Cunningham

Fig. 8.7b   courtesy of Joe McConnell

Fig. 8.8   J. Cunningham

Fig. 8.9   courtesy of USDA, Forest Service

Fig. 8.10   J. Cunningham

Fig. 8.12a   J. Cunningham

Fig. 8.12b   R. J. Lederer

Fig. 8.13a   J. Cunningham

Fig. 8.13b   R. J. Lederer

Fig. 8.14a   J. Cunningham

Fig. 8.14b   Field Museum of Natural History, Chicago

Fig. 8.15   R. J. Lederer

Fig. 8.16   R. J. Lederer

Fig. 8.17   R. J. Lederer

Fig. 8.18   R. J. Lederer

Fig. 8.19   R. J. Lederer

Fig. 8.20   R. J. Lederer

Fig. 8.23   R. J. Lederer

Fig. 8.24   R. J. Lederer

Fig. 8.26   R. J. Lederer

Fig. 8.27   courtesy of Megan Dethier, Friday Harbor Labs, University of Washington

Fig. 10.1   Gary R. Williams, University of Michigan

Fig. 10.2   J. Cunningham

Fig. 10.5   Norman Owen Tomalin/Bruce Coleman, Inc.

Fig. 10.6   courtesy of Stephen Wilson

Fig. 10.7   Reprinted from the Laboratory Manual for *General Biology* by Oswald et al. Copyright © 1980 by Kendall/Hunt Publishing Company. Used by permission.

Fig. 10.8   courtesy of Fannie Toldi

Fig. 10.9   Dr. Vernon Oswald, CSU Chico

Fig. 10.10   R. J. Lederer

Fig. 10.11   courtesy U.S. Fish and Wildlife Service. Peter J. Nam Huiaen.

Fig. 11.2   modified from American Journal of Physical Anthropology 11:533–588, 1953

Fig. 11.3   R. J. Lederer

Fig. 11.5   R. J. Lederer

Fig. 11.8a   R. J. Lederer

Fig. 11.8b   courtesy of T. S. Shaffer

Fig. 11.10   modified from "Some Aspects of the Ecology of Migrant Shorebirds" by H.F. Recher, 1966. *Ecol.* 47:393–407.

Fig. 11.11   "Coexistence, Coevolution, and Convergent Evolution in Seabird Communities" from *Ecol.* 54 No. 1: 31-44. Copyright © 1973, Ecological Society of America. By permission of Duke University Press.

Fig. 11.12   Copyright 1982 by the National Wildlife Federation. Reprinted from the March-April issue of International Wildlife Magazine.

Fig. 12.1ab   Courtesy of Z. Kobe, Forsyth Dental Center, Boston, MA

Fig. 12.1c   Courtesy of J.B. Baseman, University of North Carolina School of Medicine. Appeared in *Infect. Immun.* 17:74–186, 1977.

Fig. 12.2   United States Dept. of Agriculture, Office of Information

Fig. 12.3   courtesy of Jim Ueda and Leroy Baker

Fig. 12.4   © Paul W. Johnson, University of Rhode Island/BPS

Fig. 13.1   R. J. Lederer

Fig. 13.2   Field Museum of Natural History, Chicago

Fig. 13.3   Carolina Biological Supply Co.

Fig. 13.4a   Carl May

Fig. 13.4b   Carolina Biological Supply Co.

Fig. 13.4c   Dr. Vernon Oswald, CSU Chico

Fig. 13.5a   Don Kowalski, CSU Chico

Fig. 13.5b   Don Kowalski, CSU Chico

Fig. 14.1   courtesy of Stephen Wilson

Fig. 14.2   Courtesy of D. Pramer. Rutgers University. Appeared in "Nematode-Trapping Fungi" by D. Pramer. *Science* 144:382–388, April 24, 1964. Copyright 1964 by the *American Association for the Advancement of Science.*

Fig. 14.3   R. J. Lederer

Fig. 14.4a   R. J. Lederer

Fig. 14.4b   R. J. Lederer

Fig. 14.5   Centers for Disease Control, Atlanta, Ga. 30333

Fig. 14.6   Field Museum of Natural History, Chicago

Fig. 14.7   R. J. Lederer

Fig. 14.8   R. J. Lederer

Fig. 15.1   R. J. Lederer

Fig. 15.3   National Parks Service, Kings Canyon National Park

Fig. 15.4a   © Paul W. Johnson, University of Rhode Island, 1981/BPS

Fig. 15.4b   Carolina Biological Supply Co.

Fig. 15.4c   R. J. Lederer

Fig. 15.5   courtesy of Jerome Wyckoff

Fig. 15.6   Field Museum of Natural History, Chicago

Fig. 15.8   Field Museum of Natural History, Chicago

Fig. 15.12   R. J. Lederer

Fig. 15.13   Field Museum of Natural History, Chicago

Fig. 15.14   Field Museum of Natural History, Chicago

Fig. 15.15   art courtesy of St. Regis Paper Co.

Fig. 15.16   Field Museum of Natural History

Fig. 15.17   R. J. Lederer

Fig. 15.20   from *The Economy of Nature* by R.E. Ricklef. By permission of Chiron Press.

Fig. 15.21   R. J. Lederer

Fig. 15.22   R. J. Lederer

Fig. 15.23   Field Museum of Natural History

Fig. 15.24   courtesy of U.S. Dept. of the Interior, Yosemite National Park

Fig. 15.25   courtesy of J. C. Allen & Son

Fig. 16.2   J. Cunningham

Fig. 16.3   courtesy of Carolina Biological Supply Co.

Fig. 16.4   courtesy of P. V. Rich

Fig. 16.8   R. J. Lederer

Fig. 16.11   R. J. Lederer

Fig. 16.12   R. J. Lederer

Fig. 16.13   From *Animals, Animals.* © Breck P. Kent.

Fig. 16.15   Dept. of Fish, Game & Wildlife

Fig. 16.16   Anne Wertheim

Fig. 17.1   From "The Compleat Angler: Aggressive Mimicry in an Antennariid Anglerfish," by Pietsch, T. W. and Grobecker, D. B., Science 201: 369–371 28 July 1978. Copyright 1978 by the American Association for the Advancement of Science

Fig. 17.2   courtesy of Fred Conte

Fig. 17.3   R. J. Lederer

Fig. 17.4   R. J. Lederer

Fig. 17.5   Florida Dept. of Commerce, Div. of Tourism

Fig. 17.6   Field Museum of Natural History

Fig. 17.7   R. J. Lederer

Fig. 17.8   Gary R. Williams, University of Michigan

Fig. 17.9   R. J. Lederer

Fig. 17.10   Field Museum of Natural History, Chicago

Fig. 17.11   National Park and Wildlife Service, Sydney, Australia

Fig. 17.12   National Park and Wildlife Service, Sydney, Australia

Fig. 17.13   Kenya Tourist Service

Fig. 17.14   J. Cunningham

p. 254   quote from ON BEYOND ZEBRA, Dr. Seuss, Random House, Inc.